Semiconductors and the Information Revolution: Magic Crystals that made IT happen

Semiconductors and the Information Revolution: Magic Crystals that made IT happen

JOHN ORTON
Emeritus Professor
School of Electrical and Electronic Engineering
University of Nottingham

ELSEVIER

AMSTERDAM • BOSTON • HEIDELBERG • LONDON • NEW YORK • OXFORD
PARIS • SAN DIEGO • SAN FRANCISCO • SINGAPORE • SYDNEY • TOKYO
Academic Press is an imprint of Elsevier

ACADEMIC
PRESS

Academic Press is an imprint of Elsevier
Radarweg 29, PO Box 211, 1000 AE Amsterdam, The Netherlands
Linacre House, Jordan Hill, Oxford OX2 8DP, UK
32 Jamestown Road, London NW1 7BY, UK
30 Corporate Drive, Suite 400, Burlington, MA 01803, USA
525 B Street, Suite 1900, San Diego, CA 92101-4495, USA

First edition 2009
Copyright © 2009 Elsevier B.V. All rights reserved

British Library Cataloguing in Publication Data
A catalogue record for this book is available from the British Library

Library of Congress Cataloging-in-Publication Data
A catalog record for this book is available from the Library of Congress

ISBN: 978-0-444-53240-4

For information on all Academic Press publications
visit our website at elsevierdirect.com

Printed and Bound in the United States of America

Transferred to Digital Printing 2011

Working together to grow
libraries in developing countries

www.elsevier.com | www.bookaid.org | www.sabre.org

ELSEVIER   BOOK AID
            International   Sabre Foundation

In 2004 Oxford University Press were generous enough to publish my book "The Story of Semiconductors" which was written for final year undergraduates, research students and research workers in the many aspects of semiconductor physics and applications. For my part it was a labour of love following my retirement from a working life spent almost entirely in the field of semiconductor research, firstly with the Philips company, later at the University of Nottingham. It was motivated by an interest in the fascinating question as to how things happened in a field of such tremendous importance, not only for the development of a vital branch of solid state physics but also for the general betterment of mankind. Indeed, there can be little doubt that the invention of the transistor and its many counterparts turned western civilisation on its head during the second half of the twentieth century and it seemed to me that students of physics, electrical engineering and materials science should have an opportunity to understand, not only the technical aspects of the subject but at least something of its history. There are, of course, many technical books on semiconductor physics and semiconductor devices and a small number that provide a historical look at certain specific happenings, such as the inventions of the transistor and the integrated circuit but mine was, I believe, the first to cover the whole gamut of semiconductor developments from their early faltering steps in the nineteenth century right through to the millennium. I firmly believe that scientists and engineers should be able to communicate with their peers in the non-scientific world and an appreciation of the context and history of their specialist subjects must surely constitute a vital part of such interaction. It represents a natural point of contact between the two 'sides'.

The book was quite favourably received and has, I hope, helped a modest number of scientists to give thought to the less technical side of their specialities, even, perhaps, to stimulate discussion between them and their non-scientific friends. However, there can be no question that it was a text addressed specifically and exclusively to scientists. There was far too much mathematics and physics in the presentation to permit the general reader to approach it with any hope of serious enlightenment. I might have addressed one side of the divide with modest success but there was still an important, and probably much larger, readership needing help to jump the gap from the opposite direction. Why not attempt to provide that help? Why not, indeed? I now had more leisure, with plenty of time to think about the nature of the problem of communication between the 'Two Cultures' and the idea gradually took shape for a second attempt to describe the history of my subject with all the mathematics and much of the physics removed. Was it possible, I wondered? The answer was a qualified 'yes' – I was, at least, in a good position to make the attempt, having given much thought to the historical and political aspects of the subject already. It might be far from easy to maintain a scientifically accurate presentation but I decided to try and, with the welcome encouragement of Elsevier, I have taken the plunge. The present book therefore celebrates the exciting progress made by semiconductor electronics during the second half of the twentieth century and sets it in the wider context of the semiconductor physics which made it all possible. It attempts to explain the complexities of highly technical developments in such a manner as to be intelligible to the general reader who is prepared to make a serious effort but above all it tells a fascinating story. I sincerely hope that it will help him or her to gain a deeper understanding and, in so doing, find the same kind of pleasure and satisfaction experienced by those scientists who have been more directly involved in the chase. As an example of human achievement it must rank alongside the Beethoven symphonies, Concord, Impressionism, mediaeval cathedrals and Burgundy wines and we should be equally proud of it. I only hope that my attempt to explain something of its appeal will help the layman to obtain the same kind of enjoyment from an understanding of semiconductor electronics that he or she might experience in contemplation of any of these.

My own experience in both Industry and University leant ever so slightly towards material science and I have therefore adopted an approach which highlights this aspect. It is part of my thesis that materials are vital to any form of engineering and a proper appreciation of the part they play is essential to any real understanding of such developments. I have tried, in a very brief introduction, to highlight the role of materials in a range of other human activities and have used this to point up their very similar role in the field of semiconductor devices and systems. I hope this will help the reader's appreciation of the many parallels to be found between the present subject and others with which he or she may be more familiar. It is also a central tenet of my presentation that materials form a link between the various different aspects of the subject, namely those of basic physics, technology and practical engineering. There is a vital interplay between each of these which enhances all three together and many examples will be found in the following account. Again, I hope that I have made clear just how things happen in the real world of science and technology and how scientists themselves influence this. It should never be forgotten that scientists are only human and that science makes progress in much the same chaotic fashion as do most other human activities.

The idea that science is somehow 'different' should be firmly dismissed and I hope that this book may help, if only slightly, in dismissing it.

While I take full responsibility for all errors, inconsistencies, obscurities, misrepresentations and clumsiness of expression that may have crept into the text, I should none-the-less like to record my grateful thanks to various people who have helped me in putting this book together. Brian Fernley, Mike Seymour and Maurice Tallantyre read sections of the manuscript and offered helpful comments. My wife, Joyce, not only put up graciously with long hours of effective separation but also helped with the writing of this preface. My thanks are also due to all those people with whom I have enjoyed stimulating working contact over the years, particularly those Philips colleagues in Redhill and in Eindhoven and Limeil. Without their help I could scarcely have acquired an understanding of semiconductor physics adequate to the task of explaining it to others.

Orchard Cottage
October 2008

CONTENTS

CHAPTER 1

What Exactly is a Semiconductor: and what can it do?

For many people, the run-up to Christmas 1947 probably represented a welcome return to some semblance of peaceful normality following the end of the Second World War hostilities. But at the Bell Telephone Laboratories in Murray Hill, NJ something altogether more significant was in the air. On 16 December Walter Brattain and John Bardeen, senior members of William Shockley's Solid State Physics Group, observed for the very first time the phenomenon of electronic power gain from their Heath Robinson arrangement of springs and wires, connected to a small piece of germanium crystal. The culmination of two years of concentrated effort, this was the world's first solid state amplifier and the world was never to be quite the same again. In terms of its long-term impact on human life, the transistor (as it soon came to be known) was probably of far greater significance even than the war which had so recently ended – and which had, incidentally, contributed considerably to its development.

These early Bell devices depended on the less than wholly reliable behaviour of metal point-contacts on carefully selected germanium samples and left a great deal to be desired from the viewpoint of the production engineers who were entrusted with the task of turning them into a commercially viable product. However, an essential 'existence theorem' had been demonstrated. It would now be possible to replace the existing vacuum tube amplifiers with very much smaller and (eventually) very much more reliable solid state equivalents. Christmas 1954 was to be enlivened for many Americans by the availability of the first transistor radios – small enough indeed to be included in an averagely generous Christmas stocking – and by 1971 the first pocket calculators were being marketed by a rapidly emerging small company known as Texas Instruments, an achievement specifically designed to utilise the

Semiconductors and the Information Revolution: Magic Crystals that made IT happen   © 2009 Elsevier B.V.
DOI: 10.1016/B978-0-444-53240-4.00001-5

newly developed integrated circuits which were about to revolutionise so many aspects of our everyday lives. The first of these ICs was also a product of TI research, being the brainchild of Jack Kilby, a relatively new employee who, in 1958, needed something to occupy his busy mind and fingers while his colleagues were away on vacation! However, it was a 1959 patent application, describing the planar process, by Robert Noyce of TI's rival, Fairchild, which set the integrated circuit squarely on the road to success. The first microprocessor, the invention of Ted Hoff at Intel, followed in 1971 and the first personal computer in 1975. The information age was well and truly launched and the next 25 years were to see quite unprecedented changes in mankind's handling of data storage, arithmetical calculation, telecommunications, sound and vision reproduction, automobile engine control, electrical machine control and, of course, a multiplicity of military requirements.

Nor should we overlook the small matter of the transistor-based NASA programme which put a man on the moon in July 1969! Available rocket power was such that vacuum-tube-based electronic control systems were ruled out on the grounds of excessive weight (never mind high power demands and poor reliability). A solid state electronic solution was essential to overall success and it was probably also true that NASA funding was essential to the success of solid state electronics. There can be few better examples of two fledgling technologies providing mutually beneficial stimuli but when the world goggled at the sight of Neil Armstrong and 'Buzz' Aldrin hopping drunkenly about on the lunar surface there was, perhaps, an understandable lack of appreciation of the essential contribution made by the new electronics. Unsurprisingly, the more tangible aspects of NASA's success came naturally to the fore but, make no mistake about it, without the transistor the moon landing would certainly have been 'mission impossible'.

Moving forward to today, it is still difficult for us to accept just how many of our everyday activities depend wholly or in part on modern electronic wizardry. At home we take for granted that our worldly goods will be protected by an electronic burglar alarm, based on some form of infra-red detector and activated by a simple, but highly reliable, electronic control mechanism. We sit down to meals cooked (all too often!) in an electronically controlled microwave cooker which decides for itself just how the chicken, beef or lamb should be processed and we consume them (alas!) while watching our favourite soap opera on the latest flat screen television receiver. We spend the rest of the evening listening to music provided by a laser-based compact disc player (and marvelling, perhaps, at the amazing sound quality provided by our background-noise-cancelling earphones), or watching a favourite video recording or, even better, a film on DVD (also laser-based). When one of our friends rings from the other side of the world right in the middle of an exciting sequence, we merely 'pause' the programme and listen to his (or her) far-away news over a crystal clear telephone line, no matter how many thousand miles (or kilometres) long that line may happen to be. At bedtime we set our satellite-controlled digital alarm clock to rouse ourselves more or less gently the next morning. Before rising, we listen to the 'Today' programme on a superbly interference-free digital radio (which also tells us the time with satellite controlled precision). When we do, finally, get out of bed, we take it for granted again that the house will be warm and comfortable from the ministrations of an electronically controlled central heating system.

Time to leave for work and we blearily open the garage door by means of a handy remote control which saves any unnecessary fiddling with freezingly cold door handles or locks. We gain entry into the engine-immobilised family car by gently pressing the appropriate spot on the ignition key (which works at an amazing distance from the vehicle and even through closed garage doors!), take for granted (again!) the fact that the electronic ignition system ensures the engine starts immediately and that its electronic management system will ensure it continues to run smoothly and powerfully. The instrument panel tells us that the outside temperature is close to zero (Celsius) so we should be wary of possibly icy roads, the level of oil in the engine, the number of miles to the next service, the air pressures in all four tyres, the fact that we have forgotten to let off the handbrake and that one of the car doors is not properly shut. Once moving, we learn from the in-car computer just how long it will take us to cover the known number of miles to our destination and how many miles per gallon (or litres per kilometre) we can hope to achieve in the process of getting there. We help to alleviate the tedium of the inevitable traffic jam by switching on one of the six pre-loaded CDs located somewhere in the depths of the boot (or trunk?). When the rain begins (thus dampening further our early morning gloom?) we simply switch on the automatic wiping system which knows exactly when the windscreen is in need of attention. And all this without the additional wonders provided by the in-car satellite navigation system which settles for ever those apparently inevitable (and usually heated) arguments with our loving partners concerning the wisdom of making a left (or should it have been a right?) turn at the next intersection. And, whilst waiting in frustration in front of yet another red traffic light, we might just notice the smart new colours provided by the light-emitting-diodes which have now replaced those old fashioned light bulbs. It may not make the wait any less tedious but we may take consolation from the knowledge that the city council is saving thousands of pounds (or dollars or euros or —!) of our tax money as a result of their greater efficiency and reliability. Nor let us forget that similar bright red LEDs have replaced the often fallible light bulbs which originally served as car brake lights (see Fig. 1.1).

FIG. 1.1 Light emitting diodes have largely taken over the role of brake lights on modern cars and, in many cases, that of parking lights and direction indicators. Headlights, too, are soon likely to take the form of white LEDs. Courtesy of the author.

Having reached the workplace and having let ourselves into the office by remembering the necessary electronic key code, it is time to switch on the computer (never left on overnight, in the interest of global warming) and check the status of the e-mail system. A message from a client requires immediate attention by phone but there is no need to look up and dial a lengthy number – the desk phone remembers it well and allows easy connection. A number of letters and memos can then be recorded for later attention by one's ever reliable secretary before a hasty preparation for the day's main business, a sales presentation by the marketing group. They give it, of course, using Power Point and, whether or not the proposed new initiatives are altogether convincing, there can be no doubt that the presentation looks highly professional. Lunch is followed by a dash to the station to catch the afternoon train to London, a journey just long enough for the latest sales figures to be checked and analysed on the faithful lap-top computer, leaving time for a quick mobile phone call to one's nearest and dearest to confirm one's time of return the following day. The London meeting having passed without major catastrophe, it is nice to relax in the comfort of the well appointed hotel booked some days ago via the internet. With luck, it may be possible to watch the highlights of today's test match on Terrestrial TV, admiring once again the remarkable improvement in cricket-watcher comfort afforded by their use of some highly appropriate technology. At long last the LBW rule becomes very nearly clear and the mystery of 'reverse swing' is explained, at least to most people's satisfaction. One really has to admire the video shots of the ball in rapid flight, taken from maybe fifty yards distance, which show quite clearly the orientation of the seam and the distinction between the rough and shiny sides. (If only the batsman had access to similar information, how much easier would be *his* shot selection!) Though no-one who has seen satellite pictures of his or her local environment could be seriously surprised by this amazing clarity.

On the return train journey, next day one might well speculate on the power of electric traction and marvel at its smooth control – all achieved with the connivance of surprisingly large-scale transistor-like devices. While the active elements in the integrated circuit have been shrinking steadily in size over the years and are now measured in microns (1 micron $= 10^{-6}$ m), their power device relatives have been growing in the opposite direction, the better to handle the megawatt power levels associated with typical electric locomotives. (Their somewhat smaller brethren, by the way, cope effectively with switching on and off the car headlights in many of today's more sophisticated vehicles.) On finally reaching the sanctuary of home, there is barely time to down a bowl of home-made soup before rushing off to the evening's Parents and Teachers meeting – soup made possible, of course, by the ministrations of an electronically controlled liquidiser! (this being only one of many domestic items dependent on semiconductor electronics for their functional control – see Fig. 1.2). All the more interesting, therefore, when the school's technology master later demonstrates the exciting range of power tools available for little Johnny's greater learning experience in the school workshop. You may not have realised the importance of speed control in the proper functioning of the router with which your aspiring offspring has recently cut that beautifully smooth and precisely positioned wooden moulding but he, having initially set the speed far too high, now does. Need I say it? It's all done with transistors. Next stop is the Art Room where the Fourth Form have mounted a truly impressive exhibition of photographs relating to everyday life in school. They were taken, of course, with digital cameras (see Fig. 1.3) and, no doubt, electronically touched-up on one of the many computers now accepted as standard equipment in any self-respecting educational establishment.

Have I forgotten anything? Yes, of course, there are such things as i-Pods and MP3 players with which our young entertain themselves when not frantically texting one another on their up-to-the-minute mobile phones. Meanwhile, we ourselves are amused by the digital weather stations which allow us to record both inside and outside temperatures by radio connection and which provide a rather basic weather forecast by the courtesy of some friendly satellite in the far flung purlieus of our earthly atmosphere. The crossword fanatic may readily purchase a splendid electronic thesaurus, giving him instant access to any number of invaluable synonyms; the bridge enthusiast, likewise, is catered for by a

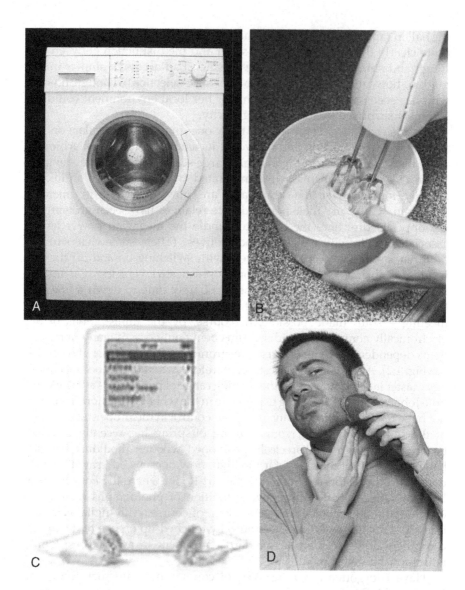

FIG. 1.2 Modern semiconductor electronics permeates a surprisingly large range of domestic and personal artifacts. Here we demonstrate no more than a very few such applications. (a) Courtesy of iStockphoto, © Adisa, image 6504202. (b) Courtesy of iStockphoto, © Don Bayley, image 4857221. (c) Courtesy of Google. (d) Courtesy of iStockphoto, © Franck Camhi, image 7748200.

FIG. 1.3 Digital photography has revolutionised our approach to both holiday snaps and serious imagery. The ability of even the amateur photographer to computer-edit his output adds an intriguing spice to the subject – though totally disqualifying the legitimacy of photographic evidence. Courtesy of iStockphoto, © Jess Wiberg, image 7808545.

hand-held hand-player with which he or she may while away the lonely hours between matches; the aspiring linguist can rely on a convenient electronic dictionary to help him or her with his or her translations; while the range of computer games available to those with time on their hands seems to grow at an unbelievable rate. And if one of us is unfortunate enough to be a hospital patient, we come into contact with a veritable phalanx of medical electronics designed to monitor or control our every movement (see Fig. 1.4). No doubt the reader can bring to mind yet further examples of the electronic arts which impinge in one way or another on our every day and not so every day lives. I suggest, though, that the point is already well made – electronics is central to the very existence of anyone fortunate enough to be born into a western democracy and is, without doubt, spreading rapidly throughout the rest of the civilised world. As for the emerging world, it may be worth contemplating the importance of solar electricity in providing much needed energy in far away places which power lines fail to reach – it's all done by 'semiconductors'.

All this, and I have not even mentioned the almost infinite range of military applications – bomb sights, night vision systems, missile guidance, navigational aids, field communications, gun controls, satellite surveillance, radar and infra-red detection systems are just a few of these. Military needs were, in fact, some of the first to be satisfied by the fledgling solid state electronics industry. For example, the notion of mounting an electronic guidance system on a missile surely demanded a reliable solid state device rather than the fragile vacuum tube equivalent

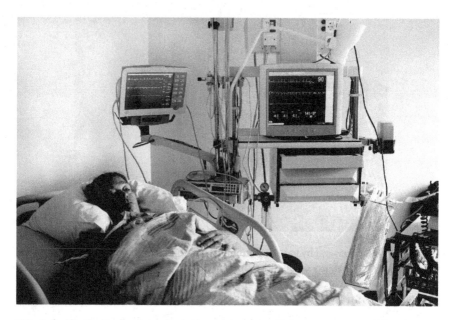

FIG. 1.4 Applications of medical electronics have burgeoned, particularly those concerned with patient monitoring in the modern hospital. Courtesy of iStockphoto, © Ariadna de Raadt, image 5850146.

and military money certainly made a huge contribution to the industry's early development. The relative importance of military investment today may be significantly less than it was during the 1960s, but there can be little doubt that it still represents a major factor in stimulating new developments. And, given the confrontational state of much of the world, it looks like remaining so for a long time yet.

So much for the obvious importance of modern day electronics – in the remainder of the book, I hope to address the question of how it all came to pass. What was its basis? What were the specific new developments which made it possible? And, in particular, what new materials were essential to its success. Note that this last question is crucial. Throughout man's long history, many new technological steps have required him to discover, and learn how to control new materials. The first wheel was probably made in Mesopotamia in the 5th millennium BC and was almost certainly made from wood because this was widely available and, at the same time, the easiest material to process. Some early weapons, in the form of clubs and spears, may also have been made from wood but all these artefacts depended crucially on the still earlier development of flint working, which represented a major step forward in man's ability to make effective tools. Flint possessed the appropriate combination of hardness and 'chipability' which allowed the production of sharp edges, thus enhancing his capabilities in the fields of both hunting and warfare. Indeed, the acquisition of the skills necessary to apply each newly discovered material increased man's ability, not only to survive, but to improve his standard of living. It remains as true today as it was in 5000 BC that both our artistic and utilitarian fulfilments depend on the way in which we develop control over new materials. The quite remarkable contrast between then and now concerns not the need for such new skills but the speed with which we now master them.

Before plunging into the complexities of modern semiconductor materials (on which the wonders of the new electronics depend), it is worth examining a little more carefully the development of one or two other important materials which have proved vital to mankind's worldly success and which, incidentally, are more familiar to the general reader. We shall look, very briefly, at the story of copper, of bronze and of iron and steel. In doing so, we shall notice a number of similarities (and some stark contrasts!) with the more recent story of semiconductor development, comparisons which both add interest in themselves and also help us to comprehend the subtleties of these newer materials which form the subject of our present study.

Copper was known to the ancient world as long ago as 8000 BC in the form of small pieces of metal mixed with copper ores and may have acquired a certain prominence as a result of early attempts to fashion cheap jewelry. Lapis lazuli, a beautiful bluish-green mineral, was much prized by those who could afford it but, in the interest of providing an acceptable substitute for the less wealthy, soapstone was provided with a somewhat similar turquoise glaze by heating it with copper ores. Later, the soapstone was replaced by glass (probably the first example of man's use of a purely synthetic material) to yield what is still well known as Egyptian faience (even though it was first produced in Mesopotamia!) but, from our point of view, the important discovery was that copper could be produced in metallic form by heating the ore. Copper smelting is known to have existed from about 5000 BC and was widely used in the Near East. In pursuit, perhaps(?), of the essential principle that the dead Pharaohs should be provided with everything necessary for their journey into the after-life, one Egyptian pyramid, for example, built round about 3000 BC, was provided with a complete copper plumbing system.

Copper was initially shaped simply by hammering it, there being no method of joining pieces together, and it soon became clear that this working of the metal caused it to become hard and brittle. Softening the work-hardened material required heat and copper was heated, at first, in simple domestic hearths, though, once the demand for larger products was realised, special forges were developed in which workers raised the temperature by blowing air through blow-pipes. Pure copper melts at 1083°C which is a much higher temperature than can be achieved with any simple furnace; hence, for centuries it was impossible to melt the metal and therefore impossible to use the much more flexible fabrication process of casting. Success was finally achieved around 3000 BC, using furnaces based on those developed for firing pottery and from then on copper artefacts of considerably greater sophistication became possible, there no longer being any necessity for cold working to obtain the desired shape. Indeed, in Egypt, copper remained as the principal material for a wide range of applications, though in Sumeria important new developments were under way.

Shortly before 3000 BC, the Sumerians (living in what is modern Iraq) imported tinstone from workings in Syria and Eastern Turkey and discovered that the addition of this ore to molten copper not only reduced its melting point but resulted in a new material with much superior metallurgical properties. The Bronze Age had begun. The use of bronze afforded three major advantages: firstly, its lower melting point significantly reduced the difficulty of managing the heating process; secondly, the molten metal was found to be less viscous, allowing greater refinement in casting; thirdly, it was possible to obtain harder (and therefore sharper) edges without the concomitant brittleness shown by work-hardened copper. By 2000 BC, a similar technology had spread to Egypt, but the next major advance came again from Mesopotamia when it was discovered how to make metallic tin. The addition of pure tin to molten copper, rather than the ore tinstone, allowed greater control over the alloy composition (typically 5–15% tin) and therefore its metallurgical properties, making it possible to *design* materials with properties appropriate to a wide range of applications. For example, large proportions of tin resulted in a whiter alloy well suited to making mirrors (though too brittle for tools) while smaller fractions could be used to make relatively soft rivets for use in bone handles where it was impossible to anneal any work-hardening.

Yet another advance was effected when craftsmen learned how to roast copper ores to drive off sulphur, thus eliminating the problem of gaseous inclusions in the finished metal. Then, soon after 2000 BC, bellows were invented to improve the efficiency with which air could be blown through the melt and this made it possible to form larger castings such as were needed for metallic door coverings. Both tools and weapons benefited from these developments, as well illustrated by the success of the Assyrian armies in the period prior to 1000 BC. Not only were their swords more effective in battle, but also the use of superior tools allowed their chariots to be better made, employing, for example, lighter and stronger spoked wheels. Also, about this time, it was found advantageous to add a small amount of lead to the melt, in the interest of obtaining a less viscous casting material. Thus, step by step, craftsmen were improving and optimising the properties of their materials to meet an ever-widening range of demands from both their military and civilian masters.

Versatile and manageable though the alloy bronze was, it fell somewhat short of being an ideal industrial material. One problem lay in the relative shortage of copper and tin in many parts of the developing world, another of its only modest performance in respect of hardness and toughness. Iron, as we now know, has more than a slight edge in both respects – not only is it widely available around the world but its many variations make it tremendously versatile. Bronze may still have a future in the world's art galleries but iron and steel have certainly stolen the industrial show. In fact, iron has almost as long a pedigree as copper, being used by both Sumerians and Egyptians as early as 4000 BC for such applications as spear points. However, these almost certainly depended on the discovery of meteorites, rather than the smelting of iron ores and represented only a minor contribution to the world's overall technological skills. Nevertheless, there is evidence for iron ore smelting in various parts of the Middle East from about 3000 BC onwards, together with a growing usage in tool making and weaponry. By about 1000 BC its use was widespread in Europe, the Middle East, China and India, growth continuing steadily towards the dominant position held by iron and steel in today's industrial economies.

The relatively slow take-up of ferrous technology resulted simply from the considerably higher melting point of iron (1535°C), compared with that of copper. It appears that no one succeeded in melting iron before the 9th century AD and the use of cast iron only became general from about the 15th century AD onwards. Prior to this, it was necessary to reduce the ores by heating them in intimate contact with charcoal to obtain a spongy, irregular mass which had then to be hammered at red heat to expel residual ore. In a similar manner, it was possible to combine several thicknesses of iron into a final billet and shape this to meet the application by further hammering on a suitable anvil. Indeed, it was often very necessary to do this in order to obtain iron with desirable properties. The difficulty lay in the poor control of carbon content in the starting material (which determines its hardness and ductility) and only by welding together several different samples was it possible to achieve the desired behaviour. Clumsy though this process may sound, in the hands of skilled practitioners it could be turned into a veritable art form. Famous examples are the 14th century 'damascene' swords made by folding several thin layers of iron over one another to produce beautiful patterns on the flat part of the blade, together with a very highly sharpened edge. A similar process was developed in Japan for making the 'katana' samurai swords which some authorities believe to be the finest swords ever made. Nevertheless, the overall process was slow and the composition of the resulting metal poorly controlled, the quality of the iron depending very much on the nature of the ores from which it was produced and on the craftsman's ability to generate sufficiently high temperatures. Furnace design improved only slowly. At the same time, a lengthy process of trial and error taught him how to control the material's properties on an empirical basis. In particular he found that 'wrought iron', the product of repeated heating and hammering, was soft and malleable, while material quenched abruptly from red heat took on characteristics of being extremely hard but undesirably brittle. Various intermediate processes yielded more appropriate combinations of behaviour which could be matched to specific requirements.

The Romans were efficient organisers who advanced the design of furnaces and considerably enlarged them so as to facilitate larger scale smelting but seemed to lack significant innovative flair – the overall process remained basically unchanged. This state of affairs continued into the Mediaeval period when, at last, cast iron (made in early blast furnaces) began to make an impact on the market place: cast iron firearms, for example, made an appearance in the 15th century. In parallel with this, however, bronze was still of major importance. During the 9th century a new and unusual cultural demand within Europe – for church bells – was met by bronze casting, while on the military front, cast bronze cannons and cannon balls were widely used between 1350 and 1450. A growing shortage of wood for charcoal led to the introduction of coal for heating furnaces, though charcoal was essential to the final stages of smelting in order to control the quality of the resulting metal – coal usually contained far too much sulphur which contaminated the iron.

And so we come to the 18th century and the Industrial Revolution which sparked an enormous increase in the use of iron and steel, witness, for example, the third Abraham Darby's first ever iron bridge which spans the River Severn at Coalbrookdale. This was built in 1769 and was followed in the 19th century by the coming of the railways, with their hundreds of miles of steel rail, and the subsequent development of iron steamships. A significant step in the evolution of the material technology was made, by Abraham Darby I, in 1709 when he successfully substituted coke for the ubiquitous charcoal and demonstrated much superior castings. There were still problems with impurities from the coke contaminating the iron and making the resulting pig iron too brittle but this could be avoided by a final stage in which the iron and coke did not come into contact. A measure of the overall success was provided by the quality of English cannons cast from Darby's metal. Unlike cannons made from conventional wrought iron, these showed no propensity for bursting in the heat of battle and led to a famous report by a French Brigadier, informing his Government that the bursting of French cannons was an accident so common that sailors 'fear the guns they are serving more than those of the enemy'!

The need for large scale castings for bridges, railways and steamships demanded larger furnaces and better air blasts, the latter being provided by the application of water (and later steam) power to driving 'blowing cylinders', rather than the old-fashioned bellows. The process was further improved by the use of a 'hot blast' whereby the incoming air was pre-heated by outgoing furnace gasses. The development of 'puddling' by Henry Cort in 1784 to achieve non-brittle wrought iron from coke-smelted pigs and Benjamin Huntsman's 1750 development of 'crucible steel' by heating the melt with suitable fluxes represented two further important advances. In both cases the essential requirement was to remove much of the carbon from the iron by oxidising it to carbon dioxide gas – only now were people beginning to appreciate the importance of the carbon content of iron and steel, perhaps the first step towards a proper scientific under-standing of such vital commercial processes. This was all very welcome, of course, but there remained yet another major drawback to the commer-cial application of steel on a grand scale – the existing processes were too slow and the resulting product too expensive.

A solution emerged just a hundred years later in the shape of what came to be called the 'Bessemer Converter'. The American William Kelly began experiments in 1851 to remove carbon from the melt by blowing air through it. The oxidation of carbon to carbon dioxide gas is exothermic (i.e. generates heat) so proceeds without any external heat source and the resulting reduction in carbon content and removal of other impurities such as manganese and silicon leads to much improved steel quality. The Englishman Henry Bessemer, thinking along similar lines, took out a patent for such a process in 1856 and described his ideas in a paper presented to the British Association for the Advancement of Science. Kelly countered with his own patent application in 1857 which, on being granted, acknowledged his priority over Bessemer. Unfortunately, Kelly went bankrupt later that year and Bessemer bought his patent in order to complete his own experiments without hindrance. His converter (a type of blast furnace) was mounted on trunnions which allowed it to be tilted for loading with a charge of iron and flux then again in order to pour the resulting steel into moulds – triumphantly, it yielded ten tons of steel in half-an-hour and the future of commercial steel-making was assured. There were still a few problems to be ironed out such as the need to remove phosphorus by lining the converter with dolomite but Bessemer's setting up on an industrial scale in Sheffield was the signal for that city to become pre-eminent in both steel production and steel products. Interest-ingly enough, the Bessemer process was soon replaced in Europe (though much less so in the United States) by the Siemens–Martin open hearth process, invented by the German, Frederick Siemens and improved by the Frenchman, Pierre Martin but the initial breakthrough by Bessemer was all-important in establishing the 'existence theorem'.

We are now very familiar, in the modern world, with a wide range of alloy steels made by introducing a few percent of various impurities such as tungsten, manganese, vanadium, chromium or titanium though what most of us may not realise is that Michael Faraday, working at the Royal Institution in London, was responsible for the first chromium steel as early as 1819. Faraday took seriously the need for science to contribute to engineering practice and, at the time, was working with an English cutlery manufacturer in efforts to unravel the mystery of how to make the famous Indian 'Wootz' steel. Several attempts had been made to replicate the Indian process in Europe but without success. Sadly, Faraday's scientific approach was no more successful than other people's empirical approaches but the application of scientific methodology in the 20th century certainly revolutionised the steel industry to the extent that it is now possible to design steels with desirable properties from basic molecular and structural models. The essential breakthrough consisted in the quantification of the so-called 'phase diagram' for the alloy formed between iron and modest amounts of carbon (typically up to about 7%). Steels contain less than 2.0% carbon while cast iron covers the range 2.0–6.67%, yet another recent advance being our ability to *measure* a material's carbon content with satisfactory accuracy. It was also recognised that iron can exist in different structural forms (i.e. different crystal structures) and that the carbon impurity atoms can take up various positions within the iron crystal lattice – also that iron and carbon can combine chemically to form a well defined compound Fe_3C (i.e. three iron atoms combined with each carbon atom) called 'cementite'. Thus, there exist: cementite with one crystal structure, ferrite (or α-iron) with a second structure, austenite (or γ-iron) with a third structure, martensite (a strained structure resulting from rapid quenching), pearlite (a mixture of ferrite and cementite), bainite (a plate-like structure of ferrite and cementite) and ledeburite (a mixture of austenite and cementite). Which of these forms occurs depends on the percentage of carbon and the temperature of the melt. Slow heating or cooling allows one structure to be transformed into another, whereas rapid quenching causes the high-temperature phase to be 'frozen-in', resulting in very different metallurgical behaviour. While the complexity of the system certainly works against ease of comprehension, at the same time, it allows considerable flexibility in our ability to 'design' material with any desired properties.

It would be quite inappropriate here to attempt any more detailed account of these many varieties of iron and steel but we might simply look at one specific example from recent work that illustrates the extent to which modern science has been able to influence the performance of steels for industrial use. In his 2002 John Player Lecture to the Institute of Mechanical Engineers, Professor Harry Bhadeshia enlightened his audience with the observation that 'iron is 10,000 times cheaper than an equivalent weight of potato crisps' but also described how it had been possible to tailor-make a special steel for railway lines which possessed unusual toughness, while minimising wear on carriage wheels. Careful study under the electron microscope showed that bainite contained platelets of ferrite approximately 10 microns long and 0.2 microns thick which played a major role in controlling both strength and toughness of the steel. However, it was found that the excellent properties of bainite were compromised by the apparently ubiquitous presence of cementite between the ferrite plates, causing poor resistance to fracture (poor toughness). However, further research showed that addition of silicon to the material prevented the formation of cementite and produced a steel with almost ideal characteristics for the problem in hand. Man's ability to 'design' materials has clearly taken on a very sophisticated swagger.

In summary, then, we have seen how technological progress determines man's ability to satisfy his many practical needs in ever more sophisticated ways, while it is fundamentally limited by his control (or lack of control!) over appropriate materials. First, it was necessary that he should appreciate their basic properties, then find adequate sources of the relevant ores and, finally learn how to extract the pure (or almost pure) metals. This, in turn, depended on his ability to make furnaces that were capable of the high temperatures necessary for melting these metals. However, ultimate success lay in his being able to understand exactly what made these materials tick – what factors, such as crystal structure and impurity content, influenced their mechanical properties and precisely how. In fact, the development of Materials Science as an academic subject grew out of such considerations and we should notice how important was the understanding of the micro-structure of the material, full appreciation of the atomic and molecular properties being an essential part of any understanding of macro-structural properties. Inherent in all this, of course, was the need for accurate measurement of these parameters – without measurement it would never be possible to achieve adequate control. Finally, a further lesson we may learn is the importance of developing cost-effective techniques for handling the materials – no degree of technical skill can be commercially justified if the resulting product is too expensive to grace the market place. Unsurprisingly, all these issues applied equally to the development of the materials which made possible the electronic wonders we briefly examined at the beginning of this chapter and now, at last(!), it is time to take a much closer look at them.

The word 'semiconductor' was coined in 1910 (in German!) by J. Weiss, a PhD student at the University of Freiburg im Breisgau. It occurred in his doctoral thesis which was concerned with the unusual electrical properties of a range of materials, typically oxides and sulphides such as ferric oxide Fe_2O_3 or lead sulphide PbS. His supervisor, Professor Johan Georg Koenigsberger was interested in the temperature-dependence of electrical conductivity in such materials and had propounded a theory to explain the fact that, unlike metals, they showed a marked increase in conductivity as the temperature was raised, an observation first made as early as 1833 by Michael Faraday. Faraday was concerned with the manner in which electrical conductivity changed when compounds melted and had studied the behaviour of silver sulphide Ag_2S as its temperature was increased toward the melting point. By the beginning of the 20th century, it was also well known that these materials were characterised by absolute values of conductivity which fell midway between those of metals and insulators – hence the name. To quantify this generalisation we note that typical values for the conductivities of metals (copper, aluminium, gold and silver, for example) and insulators (glass, silicon dioxide, mica, rubber) are 10^7–10^8 $(\Omega m)^{-1}$ and 10^{-14}–10^{-10} $(\Omega m)^{-1}$, respectively, covering a huge range of order 10^{20} times. Corresponding values for semiconductors lie in the range 10^{-2}–10^6 $(\Omega m)^{-1}$ which, in itself, represents a significantly large spread, making this a less than ideally precise way of defining them. However, as we shall see very soon, there are other criteria which allow us to reach firmer conclusions – though these were not available to Koenigsberger and Weiss in 1910. Only with the much improved insight provided by the development of quantum theory in the 1920s and 1930s could the electrical behaviour of semiconductors be properly understood and the rest of this introductory chapter is therefore devoted to providing an outline of such an understanding. Without it, the remainder of the book would be largely incomprehensible.

Probably the best starting point for our discussion is the distinction between electrical conduction in metals and the lack of it in insulators. The key discovery here must be seen as J.J. Thomson's discovery of the electron in 1897. Thomson was working in Cambridge, in the recently built Cavendish laboratory (set up by James Clerk Maxwell in 1874) where he concentrated on trying to understand the nature of the rays which emerged from the cathode in an evacuated gas-discharge tube, the so-called 'cathode rays'. These had earlier been studied by William Crookes and others but their precise nature was still unclear. Thomson demonstrated that they consisted of negatively charged particles with masses some 2000 times smaller than that of the hydrogen atom – the first time it was realised that such small sub-atomic particles existed. It was then a relatively simple step to assume that, if these electrons were responsible for electrical conduction in an evacuated tube, they might equally well account for conduction in metals. Two German theorists Riecke (1899) and Drude (1900) proposed that metals contained a 'gas' of free electrons which buzzed about in random directions within the metal, much as the molecules in a real gas did within their container. However, the 'electron gas' differed in so far as the application of an electric field to the metal (by connecting a battery voltage between opposite sides of the sample) caused the electrons to drift through the sample from one contact to the other and this moving charge represented the flow of electric current.

It was clear, then, that metals were good conductors because they contained a high density[1] of electrons, while insulators were poor conductors because they contained very few, while semiconductors were presumed to contain an intermediate density. It certainly constituted an important step forward, but still left unanswered the obvious question 'why?' – what were the differences between these various materials that led to such huge variation in electron density? Remember that we are talking about an overall factor of 10^{20}, while even the ratio between the densities in two different semiconductors could be as large as 10^8! There was clearly a need for better understanding of the physics of matter in the solid state, or, as it came to be called, 'solid state physics'. (Perversely, the name has changed yet again and we now speak of 'condensed matter physics', but the earlier title applied during most of the period covered by this book and I shall therefore adhere to it.)

The vital breakthroughs came in the period 1900–1931 with the advent of quantum theory. In 1913, Niels Bohr proposed his quantum model of the atom which, when it gave way to the more sophisticated formulations of wave mechanics (Erwin Schrodinger) and matrix mechanics (Werner Heisenberg) in the mid-1920s, came to be known as the 'old quantum theory'. It nevertheless provides a valuable introduction to these revolutionary new ideas in so far as it is conceptually easier to understand. From our viewpoint, the essential feature of the new theories was the proposal that electrons in an atom could occupy only a finite number of specific 'energy levels'. In other words, rather than being able to exist with a continuous distribution of energies, as predicted by classical mechanics, they were allowed only certain well defined energies which were characteristic of the particular atom in question. What was more, each of these energy levels was associated with a specific 'orbit'.

The picture of atomic structure which had emerged from Ernest Rutherford's work at the University of Manchester in 1911 depended on the existence of a tiny, but massive, nucleus, made up of positively charged protons and uncharged neutrons, surrounded by a cloud of negatively charged electrons, orbiting round the nucleus in a diminutive

---

[1] Use of the word 'density' to represent electron 'concentration' may possibly cause confusion, but it is a universal practice in semiconductor texts so I have continued with it here. Note that in this usage the units of density are $(\text{metre})^{-3}$ (or m^{-3}), not those of density as normally defined – kilograms $(\text{metre})^{-3}$.

reproduction of the planets in our solar system. Whereas the planets are held in their orbits by the gravitational attraction of the sun, electrons in their atomic orbits are attracted towards the nucleus by the electrostatic force which exists between oppositely charged particles. The size of the electron orbits (which determines the size of the atom) is typically about 1 Angstrom unit, or 0.1 nm (1 nm = 10^{-9} m), whereas the diameter of the nucleus is approximately 10^{-14} m, some ten thousand times smaller. In effect, the atom consists mainly of empty space, while it is the strong electrostatic forces between the particles which give the atom its substance. Overall, atoms are electrically neutral which requires the number of orbiting electrons to be equal to the number of protons within the nucleus so the complement of electrons, which determines the atom's chemical behaviour, is determined by the number of nuclear protons. The simplest atom, that of hydrogen, has a single proton and a single orbiting electron, the inert gas helium has two protons and two electrons, the alkali metal lithium has three protons and three electrons, and so on throughout the periodic table of the elements (see Table 1.1).

While the electric charges of the atomic particles were necessary to hold the atom together, it soon became apparent that these same charges implied a serious theoretical difficulty. Electrons following their curved orbits round the nucleus experience an acceleration (it is this which bends their trajectories from the straight and narrow!) and, as a result, classical electrodynamics predicted that they would radiate and therefore lose energy, thus gently spiralling inwards into the nucleus. In other words, atoms could not possibly be stable! (Note that the solar system neatly avoids this catastrophe because the planets are not electrically charged, but rely on gravity to control their orbits.) It was here that the 'quantisation' of energy inherent in Bohr's model came to the rescue – if electrons

Table 1.1 A simplified form of the periodic table of the elements

H^1								He^2
Li^3	Be^4		B^5	C^6	N^7	O^8	F^9	Ne^{10}
Na^{11}	Mg^{12}		Al^{13}	Si^{14}	P^{15}	S^{16}	Cl^{17}	Ar^{18}
K^{19}	Ca^{20}	[Iron]	Ga^{31}	Ge^{32}	As^{33}	Se^{34}	Br^{35}	Kr^{36}
Rb^{37}	Sr^{38}	[Pall.]	In^{49}	Sn^{50}	Sb^{51}	Te^{52}	I^{53}	Xe^{54}
Cs^{55}	Ba^{56}	[Plat.]	Tl^{81}	Pb^{82}	Bi^{83}	Po^{84}	At^{85}	Rn^{86}
		[RE]						
		[Act.]						

Iron = Iron Group
Pall. = Palladium Group
Plat. = Platinum Group
RE = Rare Earth Group
Act. = Actinide Group

could only exist in certain fixed energy states, this gradual loss of energy was impossible and the atom was stable, after all. And if this sounds a little like solving a problem by simply postulating that the problem doesn't exist, I should point out that Bohr's ideas were based on earlier work by Max Planck (which we shall examine in a moment) and further justified by later work of the Frenchman Louis de Broglie which attributed wave properties to the electron, this wave-like behaviour leading directly to the energy quantisation hypothesis. Much of quantum theory may seem like sleight of hand but it never quite descends to cheating!

The quantised energy levels of the orbiting electrons provided yet another vital link with reality. When atoms of a gas such as hydrogen were introduced into a glass discharge tube at moderately low pressure and an electric current passed through it, the gas atoms were found to emit a rather bewildering array of coloured lights, each with a well defined wavelength. In fact, more careful study showed that such 'emission lines' occurred in the ultra-violet and infra-red parts of the spectrum as well as in the visible, the details of these spectra depending on the nature of the gas atoms used. Each type of atom produced its own characteristic set of emission lines which was of considerable practical value in that it enabled the experimenter to identify components of a gaseous mixture but, from the theoretical point of view, it represented something of an enigma. It was only when the Bohr theory of the atom came on the scene that a plausible explanation emerged – each spectral line was generated by a process in which an electron moved from one atomic energy level into a second level of lower energy and, in doing so, it lost a fixed amount of energy. This energy was converted into radiation (i.e. light) whose wavelength bore a precise relation to the amount of energy lost and, because the energy levels of a particular atom were characteristic of that atom, it followed that the wavelengths of the corresponding emission lines were also characteristic of the atom.

Finally, in order to close the last remaining link in the argument, we must look at a slightly earlier development in quantum theory which goes by the rather impressive name of the 'Planck radiation hypothesis' and involves an esoteric bit of physics known as the 'ultra-violet catastrophe'. No matter how reminiscent this latter phrase may be of extra-terrestrial beings with ultra-violet ray guns, it simply represents another theoretical dilemma which faced the unwitting scientific community in the early years of the 20th century. It had been established experimentally that, when a body was heated it radiated a wide spectrum of wavelengths which spread smoothly, all the way from the infra-red to the ultra-violet but which showed a peak in its intensity at some intermediate wavelength and that this peak moved to shorter wavelengths as the body was made hotter. Attempts to explain this behaviour using classical theory came seriously to grief – instead of giving a peaked response the theory predicted that the intensity should increase steadily towards shorter wavelengths and actually tend to an infinite value at very short wavelengths, a totally unrealistic result. In 1900, Max Planck, professor of theoretical physics in Berlin, contrived to reconcile the theory with experiment but only by introducing a revolutionary new concept – light, instead of being smooth and continuous, as had originally been assumed, actually consisted of tiny packets of energy (which we now call 'photons') each carrying a quantum of energy given by the expression hv (where v is the frequency of the light and h is a small constant – now known as 'Planck's constant'). Planck himself was doubtful whether these quanta were anything more than a mathematical fiction but in 1905 Einstein showed that they did indeed exist and the quantum theory of light and matter was launched with a vengeance.

With regard to the atomic spectra, referred to above, this quantisation of light was of vital importance as it provided a precise interpretation of their origin. When an electron dropped from an upper energy level E_1 into a lower level E_2, a single quantum of light was emitted with a frequency v such that the energy quantum hv was equal to the energy difference ($E_1 - E_2$). In terms of the wavelength of the light, we can write this relationship in the form of a simple equation:

$$\lambda = 1.240/\Delta E$$

where λ is the wavelength of the light measured in microns (1 micron = 10^{-6} m) and $\Delta E = (E_1 - E_2)$ is the energy difference measured in units of the electron volt (the energy gained by an electron when it is accelerated through a potential difference of one volt). Note that we use the Greek letter Δ (delta) to mean 'a small difference between two larger quantities'. We shall make considerable use of this relation between energy and wavelength so it makes good sense to emphasise it by separating it from the text as we have done here. In practical terms, it is also worth noting that an energy difference of one electron volt corresponds to a wavelength of $1.24\,\mu\text{m}$, in the infra-red part of the electromagnetic spectrum (the visible part covers the rather small range from $0.4\,\mu\text{m}$ to $0.7\,\mu\text{m}$ or, in energy units, from $3.1\,\text{eV}$ to $1.8\,\text{eV}$). By way of comment, I should emphasise that this is the *only* equation in the book, so readers who love equations should savour it, while those who feel uneasy in their presence may take comfort from the certain knowledge that they will not be confronted with such embarrassment again.

This is all very well, you may say but are we any closer to understanding how semiconductors work? So far we have only considered some properties of single atoms in a low pressure gas discharge, whereas a sample of semiconductor material is made up of a large number of atoms arranged in a crystal lattice and these atoms are chemically bound to one another in order to hold the material together. This, of course, is what we mean by referring to it as a 'solid'. Not surprisingly, we can no longer describe the behaviour of a solid substance simply in terms of the properties of individual atoms – in the gaseous state, the atoms are far enough apart that we can ignore their mutual interactions, but in a solid these interactions are of major importance and we need a very different theoretical model to describe them. We still need to use quantum theory, of course, because we are concerned with the same (very small) atomic particles but it must be applied in a rather different way. In fact, it took until 1931 for a quantum theory of semiconductors to appear, in the shape of two papers in the *Proceedings of the Royal Society* by the Cambridge theorist A.H. Wilson. His work and that of others concerned with metals led to a very satisfying answer to our original question concerning the reason why metals, semiconductors and insulators contain such dramatically different densities of free electrons.

The crux of these theories can easily be stated. Whereas quantum mechanics, when applied to single atoms, predicts sharply defined energy levels for the orbiting electrons, in a solid these are replaced by 'energy bands', made up of a distribution of closely spaced levels. This, at first sight, may resemble the old classical model of the atom but it differs fundamentally in that the energy bands have a finite width – that is they cover a specific range of energies only – and are separated by regions which are completely devoid of electron states. In a sense, the sharp energy *levels* of the atom (with forbidden energy gaps between them) are replaced by energy *bands* (also with forbidden gaps between them). There is, however, a subtle and important difference between atomic energy levels and solid energy bands – the electron states which go to make up the band belong to the whole crystal, rather than to an individual atom – and this implies that electrons in a band are free to move spatially through the crystal. In other words, they can carry an electric current. But why should some solids (metals) contain lots of *free* electrons (electrons able so to move) while other solids (insulators) contain only infinitesimal numbers? The answer lies in the details of the energy bands, which differ from one material to another, but first we need to appreciate one vital point which is not immediately apparent. *An energy band which is completely filled with electrons cannot carry any current because there are no empty states into which an electron may move!* It also follows, rather more obviously, that a band which is completely empty of electrons is also unable to carry current. The existence of bands, therefore, is not sufficient to make a particular material into a conductor – we need something more – we need a band which is only *partially* filled with electrons.

We can now see the essential difference between metals and insulators – metals contain partly filled bands, while insulators contain only bands which are either completely full or completely empty. (The situation in metals involves overlapping bands but we need not delve into the details here!) What, though, of semiconductors? Like insulators, they contain either full or empty bands but the bands are spaced apart by relatively small forbidden gaps so that it is easy for an electron in the full band to jump up into the next higher empty band, a jump which occurs when the electron gains sufficient 'thermal energy' from the crystal lattice. At normal temperatures, the atoms which make up the crystal are vibrating about their notional lattice positions with amplitudes which increase as the temperature increases. Indeed, if the temperature is raised sufficiently, the vibrations become so violent that the whole crystal falls apart – in other words, it melts. We assume here that the temperature is well below the melting point but none-the-less there is still plenty of vibrational energy in the crystal, some of which may be transferred to an electron in the full band and, if this energy is sufficient, the electron may be 'excited' into the upper band where it is free to move through the crystal. This is an interesting concept because, clearly, the hotter the crystal, the more electrons are excited into the upper band and we have an immediate explanation for the fact that semiconductors show electrical conductivities which increase with increasing temperature, whereas metals, which depend on having partially filled bands, behave quite differently. (In fact, metals conduct less well at higher temperatures because the enhanced lattice vibrations make it more difficult for electrons to drift through the crystal – we say that they are more strongly 'scattered' – that is, deflected from their intended course.)

This represents a tremendous advance in our understanding but we need to consolidate by recognising two further features. Firstly, it must be apparent that the number of electrons thermally excited across the 'forbidden energy gap' will depend in some inverse fashion on the size of this gap. The larger the gap, the more energy the jumping electrons need, and therefore the smaller the probability of their acquiring it from the lattice vibrations. This explains the raison d'etre of insulators – these are materials which have forbidden energy gaps so large that very, very few electrons ever gain sufficient thermal energy to reach the upper band. By inference, semiconductors are characterised by moderate energy gaps – typically of order 0.2 eV to about 3 eV, while insulators may have gaps of 5 ev to 10 eV. We can also see now why semiconductors, themselves, vary in their electrical conductivities by quite large factors. As a final comment, we note that the density of electrons in the upper band depends *exponentially* on the ratio [(energy gap)/(temperature)] so it varies very strongly both with energy gap and with temperature, which explains the large variations of free electron density between different materials, referred to above.

The second point to grasp is that semiconductors possess a very unusual property in that they may show two quite different conduction mechanisms. So far, we have seen how free electrons excited into the upper band (known as the 'conduction band') result in electron conduction. Far less obvious, perhaps, is the fact that empty states in the lower band (the 'valence band') also give rise to conduction. When an electron is excited into the conduction band, it leaves an empty state behind, usually referred to as a 'hole' in the valence band and these holes are effectively free to move through the crystal. In reality, it is electrons which move but it is apparent (after a few moments or a perhaps few hours(?) of thought) that the movement of a negatively charged electron into the empty space, thus leaving behind another empty space, can be seen as the movement of a positively charged hole in the opposite direction. We talk of a current carried by 'positive holes'. Notice, too, that the density of free electrons in the conduction band must be equal to the density of holes in the valence band so the respective contributions of the two mechanisms are comparable to one another (they are not exactly equal because holes and electrons find it more or less easy to move through the crystal – usually electrons are more mobile than holes by a relatively small factor, of order two to ten times). Finally, we should introduce yet another technical word – we define this phenomenon of thermally induced conductivity as 'intrinsic conductivity' because it is a property of a pure semiconductor. (As we shall see in a moment, there is another kind!) While this oddity of semiconductor behaviour may seem like nothing more than just that, let me emphasise the point right away that it has enormous significance with regard to their practical applications. We shall certainly come back to it.

Interesting as all this may seem, there is a great deal more to semiconductors and we must now learn about one of their most important properties, the fact that it is possible to control the density of free electrons or holes by 'doping' them with certain specific impurities. In order to understand this, we shall consider the archetypal material silicon which is used in the manufacture of integrated circuits for computers and for many other applications (see Fig. 1.5). The silicon nucleus contains fourteen protons which means there are fourteen electrons circling round it. Of these, silicon uses just four with which to form chemical bonds with other atoms. Silicon crystallises in a form known as the diamond structure (for the simple reason that diamond also crystallises with this same structure) in which each silicon atom is chemically bonded to four other silicon atoms, arranged at the corners of a regular tetrahedron. In effect,

FIG. 1.5 Silicon wafers have grown rapidly in diameter since the introduction of the first integrated circuits in the 1960s. The more chips that can be accommodated on each slice, the less does each one cost. The six-inch wafer shown here is already dwarfed by its more recent counterparts. Courtesy of iStockphoto, © Alexander Gatsenko, image 7822588.

the silicon atoms share these bonding electrons in such a way that each atom has eight, a particularly stable arrangement. Now suppose that one silicon atom is replaced by a single phosphorus atom. Phosphorus is next to silicon in the periodic table and has a very similar electronic structure but with just one more bonding electron (five, instead of four) so, when substituted for silicon in the crystal lattice, it uses four of these electrons to bond with the surrounding silicon atoms, leaving one spare electron which can readily be detached, making it free to wander through the crystal. Putting it another way, the fifth electron is loosely bound and requires only a very small amount of energy (about 50 meV – or 0.05 eV) to excite it into the silicon conduction band. (This compares with the 1.12 eV required to excite an electron from the valence band – the band gap of silicon being 1.12 eV.) The phosphorus atom is called a 'donor' atom because it has donated an electron to the silicon crystal and the silicon is said to be doped 'n-type' (i.e. negative-type).

Let us look carefully at what this means. If we add a known amount of phosphorus to the molten silicon when we grow the crystal, we introduce a controlled number of free electrons into the silicon conduction band, each phosphorus atom yielding one such electron. At the same time, no holes are created in the valence band, so the resulting conductivity is purely electron-like. In this respect, this doping process differs from the thermal excitation process which we considered above, whereby electrons are raised directly from the valence band. Realising this, leads, fairly obviously, to the question of whether it might be possible to dope a silicon crystal to produce holes, rather than electrons and the answer is 'yes'. If, instead of doping our crystal with phosphorus, we choose aluminium (which sits immediately below silicon in the periodic table), this has the desired effect. Unlike phosphorus, aluminium has only three bonding electrons, which means, when it replaces a silicon atom, it uses all three in bonding but still leaves a missing bond and this hole in the bonding arrangement represents a hole in the valence band. We refer to the aluminium as an 'acceptor' atom because it accepts an electron from the valence band, leaving a hole behind and the silicon is said to be doped 'p-type' (positive-type). So, in summary of this wonderful sleight of technical hand, we now see that it is possible, by simple chemical doping, to control not only the density of 'free carriers' in the silicon crystal but also their type – that is whether they be electrons or holes. Notice, too, that we could introduce both donors and acceptors together, in which case the net doping effect depends on which species is present in greater number. If there are more donors than acceptors, the net doping will be n-type, whereas the opposite inequality will result in p-type doping. Finally, we refer to this form of conduction as 'extrinsic conductivity', to distinguish it from the intrinsic kind. There is absolutely no equivalent to this in the case of metals – any particular metal shows a characteristic conductivity which cannot realistically be varied – so semiconductors really are a dramatically different species. It is their remarkable flexibility which makes solid state electronics possible.

At this point in our discussion, it would be helpful to introduce a slightly more quantitative approach. To make clear the significance and power of doping we need to contemplate a few numbers, in particular the densities of electrons and holes in a number of practical cases. In Table 1.2, I have collected together some values for the densities

Table 1.2 Free carrier densities in some intrinsic semiconductors

Material	Formula	Band Gap (eV)	Density (m^{-3})
Indium Arsenide	InAs	0.35	1.3×10^{21}
Germanium	Ge	0.67	2.9×10^{19}
Silicon	Si	1.12	4.4×10^{15}
Gallium Arsenide	GaAs	1.43	1.1×10^{13}
Zinc Selenide	ZnSe	2.70	2.8×10^{2}
Copper	Cu	NA	8.5×10^{28}

of 'intrinsic' free carriers (i.e. carriers of electricity) resulting from thermal excitation across the band gap at room temperature for a few semiconductors, together with the value of electron density appropriate to copper, for comparison.

For guidance, I should emphasise that it is not the exact value which is important, rather the order of magnitude – we are concerned with powers of ten, rather than factors of two! The first point to notice is that, in a metal, such as copper, each copper atom supplies one electron to the 'electron gas' so the electron density is identical with the number of copper atoms per cubic metre. The second point to observe is that all the semiconductor values are orders of magnitude smaller than that for copper. The third point is the dramatic way in which intrinsic carrier densities depend upon the size of the forbidden band gap. The range of energy gaps is a factor of 7.7, while the resulting ratio of carrier densities is 5×10^{18}!

A vital conclusion we can draw from these figures concerns the densities of donor or acceptor atoms which have significance in modifying the electrical properties of any particular semiconductor. Clearly, these doping densities must be larger than the intrinsic carrier densities if they are to be useful. Take the case of silicon, for example where the intrinsic carrier density at room temperature is 4.4×10^{15} m^{-3} – if we were to replace 10^{13} m^{-3} silicon atoms by phosphorus atoms, it would make only a trivial difference to the conductivity but if we were to use 10^{17} m^{-3}, the resulting conductivity would be increased by about 20 times. In other words, the intrinsic carrier density sets a lower limit on the value of doping level which we could usefully use in making a practical transistor – in the case of silicon at about 10^{17} m^{-3}, in the case of germanium 10^{20} m^{-3}. There is, however, another aspect to these estimates based on the number of atoms per cubic metre in the crystal. This number is typically about 10^{29} which implies, in silicon, a minimum doping level of one part in 10^{12}, a very far cry indeed from the parts per hundred levels of interest to metallurgists in the steel industry! Practically speaking, such minute amounts of impurity are too small to control accurately and practical doping levels tend to range between about 10^{18} m^{-3} and 10^{27} m^{-3}. Even this represents an enormous range of conductivities and involves a quite unprecedented level of chemical control (bear in mind that even the best chemical analyses tend to be expressed in parts per billion – that is, 1 in 10^9).

All this has been heavy going so perhaps we should try to summarise the principal results.

1. Semiconductors exist because they possess an almost full valence band and an almost empty conduction band, separated by a relatively small energy gap – of order 1 eV.

2. Intrinsic conduction occurs as a result of thermal excitation of electrons from the valence band into the conduction band. Conduction takes place by equal numbers of free electrons in the conduction band and free holes in the valence band.

3. The density of free carriers in an intrinsic semiconductor depends strongly on the size of the energy gap and on the ambient temperature, increasing steeply as the temperature is raised.

4. It is possible to control the density of free electrons by doping the semiconductor with donor impurities. Each donor atom supplies one electron to the conduction band. Free holes may similarly be generated by doping with acceptor impurities.

5. The level of intrinsic free carriers sets a lower limit on the useful density of donor or acceptor impurities, but practical doping levels tend to range from about 10^{18} to 10^{27} m^{-3}. This represents fractional impurity levels ranging from 1 in 10^{11} to 1 in 10^{2} and constitutes an enormous range of conductivities.

6. There is no equivalent effect associated with electrical conduction in metals. Semiconductors are unique in showing this behaviour.

We have already made mention of several specific semiconductor materials but you, the reader, may be wondering just how widespread they really are. The answer is 'very'! In fact, something like 600 different semiconductor materials have been identified and that is probably not the end of the search. How many more may be discovered remains to be seen but it is well that we appreciate one important limitation to their practical usefulness – every single material so far commercialised has required a considerable investment in order to develop its technology. It follows that new materials emerge only when they are absolutely necessary for the solution of a specific problem and this has limited the number in commercial use to something less than thirty. A moment's thought will make clear why this is. We saw above how minute were the doping levels necessary to control free carrier densities, but we failed to emphasise the importance of purifying the semiconductor involved. Before the deliberately introduced donor or acceptor atoms can work in the desired manner, it is essential to reduce the level of unwanted impurities to something of the same magnitude as these doping levels – that is to one part in a billion or better – and this demands tremendous care and attention to detail when preparing the starting materials and growing the single crystals which are needed. (Contrast this with the levels of purity associated with making high quality steel – parts in a thousand, at best.) The semiconductor industry has faced challenges never before dreamed of in mastering material technology, challenges which could only be contemplated in the context of the huge commercial outlets that have resulted. The fact that modern electronics now permeates all our lives is an essential part of the commercial equation – the use of the absolute minimum number of semiconductor materials consistent with satisfying our demands lies on the other side of this equation.

In the following chapters we shall examine in some detail the difficulties which had to be overcome in order to bring several of these materials to the market place (Fig. 1.6 shows three important examples) – for the moment it is of interest to look in a general way at some of the similarities between the problems of developing semiconductors and those we examined earlier in the case of bronze and iron. There are obvious parallels between the two in the need to purify the starting material, though, as we saw in the previous paragraph, the absolute levels of purity are widely different. Similarly, it is essential to be able to control the proportion of impurity atoms which are incorporated, though again at vastly different levels. Structurally, too, the materials have to be in appropriate form. In the case of steel, for example, its metallurgical properties depend strongly on the particular phase and on the degree of strain present – we also noted that micro-structure, such as the combination of ferrite platelets interspersed with cementite, may have a strong influence. While metals are normally polycrystalline (made up of large numbers of tiny crystals misorientated with respect to one another), in the case of semiconductors, we are usually concerned with relatively perfect single crystal material which implies yet another technical challenge to the material technologist. The reason for this is the need for electrons to be able to move smoothly

GaAs   GaP   GaN

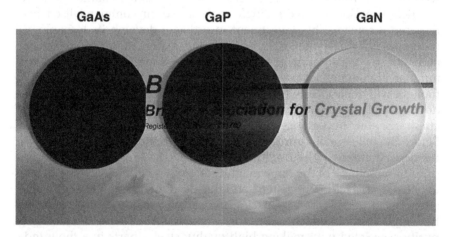

FIG. 1.6 Three compound semiconductor slices, gallium arsenide, gallium phosphide and gallium nitride, showing the sharp change in transparency associated with increasing band gap, from the arsenide to the nitride. Courtesy of Tom Foxon and Richard Campion.

through the sample – the irregular boundaries between crystallites in a polycrystalline sample result in scattering which seriously reduces the conductivity. Interestingly, however, as we shall discuss in Chapter 6, the use of carefully controlled microstructure can be of tremendous significance in modifying the electrical properties of semiconductors, another parallel between the two different regimes.

One feature of metal technology which made severe demands on man's ingenuity was the need for high temperature furnaces in order to melt the appropriate pure metals or alloys and very similar demands are made by the struggle to perfect semiconductor samples. Copper and bronze were tamed first by early man largely because they melted at considerably lower temperatures than those required for iron-based materials. A very similar situation arose in the first attempts to produce high quality synthetic semiconductor materials. Germanium was found to melt at significantly lower temperature that silicon so it was made first choice in the development race, even though it turned out to show certain undesirable properties. Though silicon proved much superior in its electrical behaviour, considerably greater effort was required in mastering its technology. Nevertheless, once the effort was made, silicon immediately triumphed in the transistor stakes, much as iron and steel had done in the realm of metals. The subsequent history of semiconductor materials has seen a series of struggles between silicon, the incumbent, and numerous young pretenders, with specific natural advantages, claiming preference in a variety of specialised applications for which silicon was less well equipped.

The final area of comparison between metal and semiconductor technology concerns the importance of reliable measurement. As we saw in the case of steel, it gradually became apparent that its metallurgical capabilities depended strongly on the fraction of carbon present. Too much carbon and the metal tended to became brittle. In this instance, the requirement was for techniques which provided measurement accuracy in the order of parts per thousand and similar modest challenges were thrown up by the need to monitor additives such as tungsten, manganese, chromium, etc. The demands made by semiconductor technology were very much greater, in many cases being beyond even the best available chemical analyses. Quite new techniques were needed, such as SIMS (secondary ion mass spectrometry) in which bits of material are chipped away with a beam of ions in an ultra-high vacuum apparatus and the resulting species detected in a 'mass analyser' (a sophisticated method for measuring atomic masses). In fact, a great many semiconductor measurements had to be made using clever electrical techniques – nothing else could match the level of sensitivity required – about which whole books have been written. In addition to such methods of establishing electrical properties, it was also important to apply a variety of structural measurements, such as X-ray diffraction and transmission electron microscopy. Knowing the detailed properties of the semiconductor materials was every bit as important as evolving accurate models of device behaviour.

If this Chapter has been a somewhat gruelling experience, take heart – we now know enough about semiconductors to appreciate the story which follows and are therefore ready to return to the early history of semiconductor development which preceded the dramatic events of Christmas 1947. It goes without saying, that the transistor did not happen by chance, and we are now in a position to appreciate the important sequence of events leading up to it.

CHAPTER 2

The First Hundred Years: From Faraday to Yesterday

We tend to think of scientific research as being very much a modern concept, on account of the intensity with which it is pursued these days, but the history of electrical conduction in solids certainly has a long pedigree. Perhaps the first such studies were those of a certain Stephen Gray who lived in London and Canterbury during the eighteenth century and who reported his findings to the secretary of the Royal Society in 1731. Gray was able to show that static electricity could be carried away by a range of materials and demonstrated the effect through an interconnected line of various components. The concept of *electrical conductor* was introduced in 1742 by a Frenchman, Jean Theophile Desagulliers, who was living at the time as a religious refugee in London. These early studies were hampered by the lack of a convenient source of electric current – that had to await the invention of the voltaic pile by Alessandro Volta in 1799, consisting, literally, of a *pile* of metal plates (silver and zinc) separated by sheets of cardboard, soaked in brine (the first of a long line of 'batteries'). This provided future experimentalists with a steady supply of current at a controlled voltage (depending on the number of plates employed), a big improvement on the previous methods relying on static electricity. However, it seems that Volta himself had performed experiments as early as 1782 in which he distinguished several different materials on the basis of their electrical conductivity – metals, semiconductors (actually he used the phrase *materials of a semiconducting nature*) and insulators.

Sir Humphry Davy, working at the recently established Royal Institution in London, made use of the Voltaic pile to study the temperature-dependence of conductivity of a range of metals, publishing his work through the Royal Society in 1821. He concluded that *the conducting power of metallic bodies varied with temperature, and was lower, in some inverse*

Semiconductors and the Information Revolution: Magic Crystals that made IT happen © 2009 Elsevier B.V.
DOI: 10.1016/B978-0-444-53240-4.00002-7

ratio, as the temperature was higher. (We saw in Chapter 1 that this was a consequence of the increased scattering of electrons by the increasing amplitude of thermal vibrations of the metal atoms, though no such explanation was available at the time – the electron itself was not discovered for another 78 years.) Davy's protégé, Michael Faraday had been taken on as his scientific assistant in 1813 and soon made a name for himself as a chemist. However, it was not until 1820 that Faraday began his highly significant experiments on electromagnetic induction which led to the development of the electric motor and, later, the electric dynamo. This work occupied him for the best part of 20 years but he also found time to explore several other electrical phenomena, such as the ionic conductivity of electrolytes (solutions of various salts in water) and the nature of conduction in a wide range of chemical compounds which he had prepared, himself. These included a number of oxides, sulphides, carbonates, sulphates, nitrates and halides, and he found quite different temperature dependences from those measured by his master for metals. A material of particular interest for Faraday was silver sulphide Ag_2S (which he knew as 'sulphuret of silver'). It conducted hardly at all at room temperature but at a temperature of about 175 °C there was a fairly rapid increase to nearly 'metallic' proportions. Faraday, not being mathematically inclined, did not record quantitative data, nor did he plot any graph of conductivity against temperature but it was clear that he had observed the positive temperature coefficient of electrical conductivity which we now regard as typical of semiconductor behaviour. His results were published in the Philosophical Transactions of the Royal Society in 1833.

As luck would have it, silver sulphide was an unfortunate material on which to concentrate because it shows a 'phase transition' at a temperature of 170 °C – that is to say, its crystal structure changes (even though its chemical composition remains unchanged) in much the same way as we saw for iron in Chapter 1. Its electrical behaviour is also very sensitive to the manner of its preparation – small departures from exact 'stoichiometry' (the ratio of silver atoms to sulphur atoms) have a marked effect on its conductivity. Interestingly enough, the precise nature of electrical conduction in silver sulphide was a subject for scientific debate until well into the twentieth century. For a long time, many people thought that conduction resulted from the movement of charged ions (Ag^+ and S^-) in just the way that had been established for liquid electrolytes (e.g., sodium and chlorine ions in the case of a solution of common salt) and it was only in 1929 that Hall effect measurements (which we shall meet formally in a moment) confirmed that conduction was due to electrons, rather than ions. Not that this example was unique – there were, as we shall see, many points of dispute concerning the electrical properties of semiconductors which lingered on right up to the Second World War.

However, in the interests of continuity, we must first return to the nineteenth century. The next discovery of interest to us was concerned with the effect of light in generating an electric current. The Parisian Becquerel family appear to have done for nineteenth century science what the Bach family did for eighteenth century music. In 1829 Antoine Cesar Becquerel constructed the first battery to give a constant supply of current (earlier versions suffered from a serious depolarisation effect which rapidly reduced the output) while in 1839 his son Alexander Edmund Becquerel observed that, when he shone light on one of the electrodes in a battery, the current increased. This was the first report of what we now call the 'photovoltaic effect' which was later to be associated with semiconductor-metal contacts. By far the most famous member of the family, however, was Antoine Cesar's grandson Antoine Henri Becquerel who was awarded a half share in the 1903 Nobel Prize for his discovery of radioactivity.

But, to return to our subject of semiconductors, we note that it was not until the 1870s that optical effects were recorded in relation to dry solids. Firstly, in 1873, Willoughby Smith observed that the resistance (the reciprocal of conductance) of a selenium sample decreased significantly when it was exposed to light. He was in process of testing a submarine cable and had need of a stable, high resistance standard resistor so was somewhat put out by the fact that his selenium bar showed apparently random fluctuations in its behaviour:

'Being desirous of obtaining a more suitable high resistance for use at the Shore Station in connection with my system of testing and signalling during the submersion of long submarine cables, I was induced to experiment with bars of selenium, a well-known metal [sic] of very high resistance. The early experiments did not place the selenium in a very favourable light [no pun intended?] for the purpose required, for although the resistance was all that could be desired – yet there was a great discrepancy in the tests, and seldom did different operators obtain the same results. While investigating the cause of such great differences in the resistance of the bars, it was found that the resistance altered materially according to the intensity of light to which it had been subjected'.

Note that Smith's observation differed from Becquerel's photovoltaic effect in that it represented a bulk property of the selenium, rather than being associated with a single electrode as in Becquerel's electrolytic cell. In fact, we can rather easily understand it in the light (still no pun intended) of our discussion of semiconductor properties in Chapter 1. There, we saw that thermal excitation of electrons across the forbidden energy gap resulted in conduction in both conduction and valence bands and we called this 'intrinsic conductivity'. In the case of Smith's selenium (which we now know to be a good example of a semiconductor!), electrons were excited across the gap as a result of light being absorbed by the material, this phenomenon being known as 'photoconductivity'. For each light quantum or 'photon' of light absorbed, one electron is excited into the conduction band and, as we can infer from Chapter 1, there is a simple relation between the photon energy hv and the band gap of selenium, 1.8 eV. Using the equation on page 20, we derive the fact that all wavelengths of light shorter than about 0.69 μm (690 nm) will be absorbed by the selenium and, as this includes most of the visible spectrum, we are not surprised that the incidence of daylight on Smith's standard resistor caused its conductance to increase by an amount proportional to the light intensity. Thus did a highly practical investigation of submarine cables lead to an important fundamental discovery in semiconductor physics (though the above *explanation* was not available for another 60 years).

The second discovery of an optical effect in a semiconductor followed just 3 years later when William Grylls Adams and his student Richard Evans Day at Kings College London observed a photovoltaic response from a sample of selenium. When they shone visible light on their sample, which was, in turn, connected to a galvanometer, it was found to cause a deflection in the galvanometer needle – in other words, it generated a 'photo-current'. This was a particularly significant observation because it represented the direct conversion of light into electrical energy, as pointed out in 1877 by Werner von Siemens (Charles Fritts in New York had made what was probably the first practical photovoltaic cell in 1876 and had sent samples to Siemens). Today, we know the device (usually made from silicon) as a 'solar cell', the mainstay power source for space research, including, of course, the first man on the moon project in 1969. Its application to the terrestrial generation of electricity, which we shall discuss in Chapter 10, could yet be of even greater importance. The conversion efficiency of Adams and Day's selenium cell was miniscule but their contribution to semiconductor physics was considerable – the essential point was that these strange materials showed altogether unusual properties and this was the period when they announced themselves to a rather incredulous scientific world.

Perhaps even more unexpected was the discovery in 1874 of the phenomenon of 'rectification'. A young German school teacher at the Gymnasium in Leipzig, Ferdinand Braun, (see Fig. 2.1) interested in the electrical properties of a range of minerals, particularly sulphides such as lead sulphide (PbS) and cuprous sulphide (Cu_2S), noticed that the current flowing through his samples differed according to the direction of the applied potential difference. He later moved to the University of Marburg, then to the University of Strasbourg but he kept worrying away at this surprising departure from Ohms Law (the law that says current

FIG. 2.1   Ferdinand Braun in middle age. Braun was active in the universities of Wurzburg, Leipzig, Marburg, Strasbourg, Karlsruhe, Tubingen and, again, Strasbourg during the late nineteenth and early twentieth centuries. He discovered the semiconductor rectifier, made important contributions to the early development of wireless transmission, invented the cathode ray tube and shared the Nobel Prize for physics with Marconi in 1909. Courtesy of WikiMedia.

is proportional to applied voltage, irrespective of direction), eventually concluding that the effect was related to the nature of the metal contacts. In designing a reliable method of contacting his semiconductor samples, he hit on a technique which involved a large area contact on one side, opposed by a wire spring contact on the other and he later reported that 'The anomalous phenomena occur most readily if at least one contact is small'. In effect, Braun had invented the 'cat's whisker' rectifier which played a significant role in the early development of radio and as a detector of microwave radiation in Second World War radar systems. Interestingly enough, there was another, independent observation of rectification during 1874 when Arthur Schuster reported the effect while studying the contact between copper and copper oxide. This also concerned the contact between a metal (copper) and a semiconductor (copper oxide) and provided a good example of how unexpected results frequently crop up in several widely separated locations. Once the electrical conductivity of semiconductors became an important scientific topic, it was always likely that similar phenomena would be discovered by independent investigators. By the same token, we might note that 2 years later Fritts also observes rectification in his selenium cells – clearly the effect was not confined to any one material system.

Mention of 'radio' brings our discourse to one of the most significant discoveries of the nineteenth (or any other) century. It was in 1888 that Heinrich Hertz demonstrated the existence of electromagnetic waves with wavelengths of the order of 5–10 m ('radio frequencies' of order 30–60 MHz), the end of a search which had intrigued many physicists, following theoretical work published by James Clerk Maxwell in 1864. Maxwell had developed a set of electromagnetic equations to explain the known properties of light but it was soon realised that these equations predicted the existence of electromagnetic waves at other frequencies and, in particular, at radio frequencies and the search for such waves then became a sort of scientific holy grail. If they didn't exist, grave doubts would be cast on Maxwell's superb interpretation of the properties of light and one of the great theoretical advances of the age might have to be abandoned. It was Hertz's Doctoral supervisor Hermann von Helmholtz who in 1879 suggested the project to Hertz but it was only when Hertz started work in Karlsruhe Technical College in 1885 that he was able to build the necessary apparatus. Three years later he used a spark gap transmitter and a similar spark gap detector to prove that electromagnetic waves could be transmitted through space across his laboratory. He then went on to show that these waves could be reflected and refracted in the same manner as light waves and also, most importantly, that they travelled at the same velocity as light. Maxwell, who died in 1879 and therefore did not live to see the day, was completely vindicated. Sadly, Hertz, too, died in 1894 at the age of 37 without ever imagining the wonders of radio-telegraphy which were to follow so rapidly on his work.

Indeed, one of the more interesting features of all this activity was the contrast between the reactions of pure physicists on the one hand and engineers on the other. Hertz, himself, and Oliver Lodge in England, who had been experimenting along similar lines, took great delight in the scientific proof of Maxwell's predictions, while Guglielmo Marconi immediately (1895) began studies with a view to using radio waves for communication and, following no more than modest success in this enterprise, made urgent approaches to first the Italian and then the British governments for the funds necessary to pursue his ideas further. Before this, in 1894, Lodge had even gone as far as to demonstrate radio wave transmission between two laboratories in Oxford some 50 m apart but still made no attempt to capitalise on it. Lodge could so easily have gone down to posterity as 'The Father of Radio' but that honour, as we all know, is usually accorded to Marconi. That it should not have been is argued vehemently by the many supporters of yet another candidate Nikola Tesla who in 1893 had described all the features of a wireless communication system in a meeting of the (American) National Electric Light Association. Tesla, a somewhat retiring character, was, in any case, concerned to demonstrate the wireless transmission of electric power, rather than of information so made little effort to pursue his claim to be the father of radio. Then again, if you happen to be Russian, you may well believe that the honour should have gone to Alexander Popov who clearly did demonstrate the transmission of information in St Petersburg in1896. Much earlier still, however, is the claim of an American dentist, one Marlon Loomis, who apparently conducted an experiment in radio transmission as early as 1866. Any attempt to pin the accolade to one particular chest is obviously fraught with difficulty and an adequate discussion requires considerably more space than is available here. We simply leave it at that. It is, after all, peripheral to our main purpose.

Incidentally, Braun played an important part in the success of radio-telegraphy, not only by using his cat's whisker to detect radio waves but also by advancing the design of both transmitter and receiver circuitry. In 1897, both Guglielmo Marconi and a German rival Adolph Slaby, using spark-gap transmitters, had reported wireless transmission over distances of about 20km but there appeared to be serious problems in extending the range. Larger aerial systems and more powerful sparks achieved disappointingly little improvement until Braun recognised the importance of (electrically) separating the aerial from the spark circuit. He developed an inductively coupled aerial circuit which could also be tuned to select a narrow band of radio frequencies (what we now know as 'selectivity') and demonstrated how the coupling could be optimised so that the electrical 'loading' of the aerial did not inhibit the spark. Marconi used a very similar circuit (without acknowl-edgement!) in his epoch-making transatlantic transmission experiments of 1901. Braun was also responsible for pioneering work in the achieve-ment of directional transmission of radio signals, a development of great interest to military users. As a result of these important technical advances Braun shared the 1909 Nobel Prize with Marconi – apparently perfectly amicably, in spite of Marconi's unauthorised 'borrowing' of Braun's innovations! Though he was far more interested in such techno-logical improvements than in their commercial exploitation, Braun was persuaded to join a small start-up company Funkentelegraphie GmbH in 1898 which, via several intermediate stages, metamorphosed into the powerful German company Telefunken, founded in 1903.

It was also in 1897 that Braun made yet another highly significant invention, that of the cathode ray oscilloscope, still known in Germany as the Braun tube. It made possible the pictorial representation of an electrical waveform and proved of immediate value for the proper understanding of spark-gap radio wave generation. This was the time when cathode rays were very much in the scientific news – in Wurtzburg, Rontgen had discovered X-rays in 1895 and in Cambridge, J J Thomson the electron in 1897 so perhaps Braun's invention was overlooked to a considerable extent. It certainly received far less publicity but there can be no doubt of its long-term importance, leading, as it did in the twentieth century, to the ubiquitous television tube.

Ferdinand Braun's many talents have led us into a diversion, from which we must, however reluctantly, retrace our steps. Our interest is in the history of semiconductors and it is to this which we must now return. The key facts about semiconductors to be noted here are the following: (1) the increase of conductance with increasing temperature had been observed as early as 1833, (2) the increase of conductance as a function of light absorption (photoconductivity) had been observed in 1873, (3) the phenomenon of rectification (asymmetric current vs voltage characteristics) had been observed in 1874 and (4) the direct conversion of light into electric current (the photovoltaic effect) had been observed in 1876. So, well before the nineteenth century had come to a close, four important characteristics of semiconducting materials had been established experimentally. There was still much to learn before it would be possible to exploit their unusual features to the full but these four discoveries provided a framework within which semiconductor behaviour could be usefully classified. Two important factors which were still lacking were those of an acceptable theory to explain the above phenomena and an appreciation of just how critical was the degree of impurity and structural control to the detailed electrical properties. The former had to await the application of quantum theory in the 1930s, while the latter gradually came to be recognised during the Second World War as part of the development of microwave radar detectors.

   The early years of the twentieth century saw considerable activity in semiconductor research. Not only was it widely realised that the cat's whisker rectifier could be used to detect radio waves but there were also exciting developments in semiconductor physics. Ferdinand Braun reported his experiments with the cat's whisker in 1901 and, within a very few years, a variety of patent applications and publications in learned journals followed (from other authors, that is). The original form of detector used by Hertz, a second spark gap, was fairly rapidly replaced by a more sensitive technique in the form of a 'coherer'. This, the brainchild of a Frenchman, Edouard Branly, was invented in 1890 and consisted of a glass tube filled with metal particles which, under the influence of radio waves, cohered so as to provide a good conducting path but, on being given a gentle tap, parted company again to offer a comparatively low conductance. For some years the cat's whisker and the coherer competed for the affections of the burgeoning cohort of wireless enthusiasts but both were suddenly overtaken by an alien species – a vacuum tube!

It was in 1904 that John Ambrose Fleming patented his 'Fleming valve' or 'thermionic valve', or 'vacuum diode'. In 1881 Fleming was appointed to the London branch of the Edison Electric Light Company and one of his preoccupations was the blackening of the glass bulb due to evaporation of the filament material. In attempting to prevent this, he experimented with a metal plate within the glass envelope and discovered more or less by accident that a positive voltage on this plate (or 'anode') drew a current from the filament, while a negative voltage produced no current – in other words, he had made a rectifier. By this time, Fleming had taken up an appointment as Professor of Electrical Engineering at University College London and was also a consultant to the Marconi Company so it was no surprise when his valve formed the basis of the Marconi-Fleming valve receiver, one of the first of its kind and a forerunner of what became the vacuum tube revolution. The coherer suffered from the fact that it had to be reset after every pulse of radio wave energy and was therefore extremely slow in its response. The cat's whisker was a big improvement in this respect but proved frustratingly difficult to use – for optimum performance, it was necessary to find just the right spot for making contact to the semiconductor crystal and any undue vibration could be sufficient to upset this carefully sought sweet spot. In contrast, the rectifier diode proved simple and reliable, and rapidly became the preferred method of detection. Its only drawback was the relatively minor inconvenience of the need for a battery to provide current for heating the filament to generate a source of free electrons in the vacuum space.

However, this was only the first stage of the vacuum tube revolution. Even more important was the invention of the 'triode valve' by Lee de Forrest in 1906 (see Fig. 2.2) (he called it the 'audion' but the term triode soon became widespread). The addition of a third electrode, in the form of a wire grid between filament (cathode) and anode, allowed the anode current to be independently controlled by the application of a small 'grid voltage' and gave the triode the priceless capability of 'amplification'. A small alternating voltage at the grid could be reproduced in the anode circuit with much enhanced amplitude. Here was a device with a huge potential for commercial exploitation. For example, it was taken up by

FIG. 2.2 A selection of triode thermionic valves in a valve amplifier. The triode was invented by Lee de Forest in 1906 and was adopted by Bell engineers for use in long-distance telephone communications in the early years of the twentieth century. It enjoyed a long history until competition from the transistor and integrated circuit reduced sales to fairly modest proportions during the 1970s. Courtesy of iStockphoto, © Valeriy Novikov, image 8117713.

Bell Telephone engineers in 1912 as a means of amplifying telephone signals on long distance telephone lines and it was soon seen as a vital component in radio receivers designed for loud-speaker output, rather than the less convenient earphones which had been intrinsic to earlier sets. So by the onset of the First World War, it was clear that the vacuum tube had won an important electronic battle but, as we shall see, the semiconductor device would eventually win the war. Not that the valve yielded pride of place without a fight - the invention of the transistor in 1947 may have been the first nail in the vacuum tube coffin but it was not until the mid 1970s that sales of vacuum tubes reached their peak and there are still numerous specialised applications even today.

In the first quarter of the twentieth century the role of semiconductor devices may have been in the doldrums but the physics of semiconductors was burgeoning. More experimentalists took up the challenge to quantify their detailed properties and new techniques came on stream. The most important of these was based on what was already well established as the 'Hall effect'. It had been discovered in 1879 by a research student Edwin Herbert Hall, working at the Johns Hopkins University in Baltimore. His supervisor Professor H A Rowland (better known for his work on diffraction gratings and their use in spectroscopy) was concerned with a problem originally raised by Clerk Maxwell, concerning the effect of a magnetic field on an electric current. Rowland felt that a magnetic field at right angles to the direction of current flow should result in a tendency for the charge carriers to be deflected to one side of the sample, thereby setting up a potential difference across it. He had attempted, unsuccessfully, to measure this in a metal sample but Hall later demonstrated the correctness of the hypothesis by measuring such a potential in an evaporated gold film. Though Rowland may have been unduly modest in not attaching his own name to it, it has come to be known as the Hall effect and is still very much alive today – indeed, no less than four Nobel prizes have been won as a result of recent and novel ramifications of the effect.

We should, perhaps, remind ourselves that, at the time of its discovery, the Hall effect could not be properly interpreted because the nature of the charge carriers was still unknown. The mechanism of ionic conduction in electrolytes was reasonably well established and there were those who favoured a similar mechanism in solids but the idea of the electron gas had, inevitably, to await J J Thomson's discovery of the electron in 1897. However, one important feature of Hall's discovery was readily appreciated – the *sign* of the charges could be ascertained from the sign of the transverse voltage (or Hall voltage). Both positive and negative charge carriers were deflected to the same side of the sample, resulting in opposite polarities for the Hall voltage. In Hall's original experiment with gold the sign turned out to be negative and several other metals followed this behaviour but then confusion set in when it was found that yet others gave positive signs – confusion which could only be resolved by the advent of a quantum theory of metals in the late 1920s.

From the viewpoint of our story, the Hall effect turned out to be of supreme importance as a tool for understanding the behaviour of semiconductors and is still widely used today. The fact that the *sign* of the Hall voltage gives us the sign of the relevant charge carriers has proved of inestimable value but, of even greater significance was the recognition that the *magnitude* of the Hall voltage can also be related to the density of free electrons in the conduction band (or holes in the valence band). Bearing in mind that most practical applications of semiconductors depend on appropriate doping to make samples either n-type or p-type, it is clear that any tool which provides a ready method of measuring both the type and the density of free carriers must be of paramount importance to anyone faced with the task of analysing semiconductor behaviour.

In the early years of the twentieth century, however, such clear cut advantages were less readily appreciated. While several research groups seized the opportunity to apply the technique to their semiconductor samples, they were scarcely in a position fully to understand the results they obtained. As with metals, there was puzzlement over the discovery of both positive and negative Hall voltages. By this time, the electron was well established as a negative charge carrier but there could be no explanation for positive carriers which were equally mobile (positive *ions* were known to be very much less mobile). What was even worse, the variation of Hall voltage with temperature threw up some very peculiar results, including in some cases a change of sign which was extremely difficult to understand. Then, to rub technological salt in the scientific wound, several materials showed apparently random variations between different samples and led the more sceptical physicists to write off semiconductor research as a complete waste of time:

> One shouldn't work on semiconductors, that is a filthy mess; who knows if they really exist – Wolfgang Pauli (letter to Rudolph Peierls, 1931).

> Working on semiconductors means scientific suicide – friends of Georg Busch, 1938.

> What are semiconductors good for? They are good for nothing. They are erratic and not reproducible – anonymous Professor of Applied Physics, during the Second World War.

With the benefit of nearly 70 years of hindsight, we may be tempted to look somewhat patronisingly at such pessimism but I have little doubt that it could reasonably be understood at the time. The rapid theoretical and experimental advances which cleared up this 'filthy mess' in time for the discovery of transistor action in 1947 must have been very hard to anticipate.

The first application of Hall effect measurements to semiconductor materials seems to have been by Koenigsberger's group in Freiburg, in the period 1907–1920, during which time many of the experimental incongruities could be seen to materialise. Then numerous other groups took up the challenge for, in spite of the obvious uncertainties of interpretation, the subject held undoubted fascination and, indeed, some positive advances were made. One such was the realisation that the density of free carriers in semiconductors was much lower than that appropriate to metals. Metals were characterised by roughly one free electron per metal atom, semiconductors by densities several orders of magnitude smaller. Another such was the discovery by Karl Baedeker (son of Fritz Baedeker, editor of the world-famous travel guides) at the University of Jena that the density of free carriers in copper iodide (CuI) could be controlled over a wide range by the expedient of exposing it to iodine vapour (or to a solution of iodine in alcohol). The Hall voltage indicated conduction by positive carriers and their density increased dramatically on exposure to iodine. Cuprous oxide (Cu_2O) was found by Carl Wagner to show similar behaviour, the free carrier density increasing strongly on exposure to an oxygen atmosphere. By contrast, zinc oxide (ZnO) showed a negative Hall voltage, the density of free electrons *decreasing* with exposure to oxygen. In all three cases, the explanation can be found by considering the precise stoichiometry of the compound. CuI and Cu_2O came to be known as 'defect' semiconductors – there being a small deficiency of copper, compared to that required for perfect match to the iodine or oxygen. ZnO, on the other hand, was recognised to be an 'excess' semiconductor, the small excess of zinc resulting in a corresponding density of free electrons. These departures from stoichiometry can be seen in the same light as the doping by impurity atoms which we described in Chapter 1 and, perhaps more importantly at the time, provided an explanation for the large random differences measured between different samples of the same material. Here was the first hint of understanding that stoichiometry and purity were of paramount importance in determining the electrical properties of semiconductors. Careful studies of these properties over a wide temperature range also revealed the distinction between intrinsic and extrinsic behaviour – at moderate and low temperatures, large variations in carrier densities were measured for many materials (the result of uncontrolled impurity or stoichiometric doping) whereas, at high temperatures, all samples of the same material tended to show (at the same temperature) the same free carrier density.

Thus, very gradually, some degree of order appeared in the vast array of experimental data which had accrued but the vital semiconductor breakthrough came when Allan Wilson published his two classic papers on the 'band theory' of solids in 1931. Not only did he explain the origin of conduction and valence bands in semiconductors, but also the mechanism of impurity doping which provided a basis for the detailed understanding of extrinsic semiconductor properties. From this point, it is easy to trace a path to the work at Bell Telephone Laboratories in the post-Second World War years which produced the point-contact transistor. Amusingly enough though, Wilson was convinced that silicon, the future semiconductor material par excellence, was a metal, yet another illustration of the confusion still dominating the subject – the likelihood is that the only samples of silicon then available were so impure that their intrinsic semiconducting nature was totally obscured! At the time, Wilson was a student of R.H Fowler in Cambridge but, in 1931, he had chosen to spend a year in Leipzig where the Institute for Theoretical Physics was run by Werner Heisenberg, one of the fathers of quantum theory and immortalised by his 'uncertainty principle', putting limits on the degree of accuracy to which we can hope to measure any microscopic physical quantity. Heisenberg also had a hand in introducing the concept of positive holes and it was largely a result of the stimulating atmosphere in Leipzig that semiconductors could begin to challenge their own uncertainty principle – did they really exist?

According to Wilson, there could be little doubt that semiconductors really did exist but we should not be too surprised to learn that his conclusion was not universally accepted, not, at least, in the short term. Indeed, there were sceptics so influential as to frighten away those who showed even a glimmer of faith. Professor B Gudden at the University of Gottingen held extremely pessimistic views concerning the pedigree of semiconductors and published a series of review articles between 1924 and 1939 in which he consistently doubted the existence of intrinsic semiconductors (in this respect he probably had a point – very few semiconductor samples existed with a degree of purity sufficient to show intrinsic conduction at room temperature). Like Wilson, he was convinced that silicon was a metal and so great was Gudden's influence that when, as late as 1943, another German physicist, Karl Seiler, measured the conductivity of silicon as a function of temperature and found clear evidence of both extrinsic and intrinsic regions, he was too scared of Gudden's reputation to publish his findings. Yet again, in 1944 a Gottingen graduate student, J. Stuke observed similar results for germanium but was not allowed to publish them. Let there be no mistake – science has its prima donnas too!

No matter what traumas were experienced by those earnest seekers after semiconducting truth, the practitioners of semiconductor device development were by no means prepared to await the outcome. The 1930s were a time when semiconductor science lagged far behind empiricism, driven, as it was, by an emerging commercialism. We have already referred to the use of the cat's whisker rectifier as a detector of wireless signals, (see Fig. 2.3) an activity which first registered the importance of silicon as a practical semiconductor material. Much empirical searching through the 'known' (if that be an acceptable term?) semiconductors had shown silicon to offer the most stable, if not quite the most sensitive performance. However, there were other, much more pressing applications for semiconductor rectifiers. Whereas the invention of the thermionic valve took much of the wind from the detector sails, it blew strongly behind the development of battery chargers and AC–DC converters. A

FIG. 2.3 An example of a cat's whisker detector used for detection of radio signals in an early crystal set. The crystal, in this case, was galena (lead sulphide) such as used by Ferdinand Braun in his original studies of rectification. Courtesy of Maurice Woodhead.

triode valve demanded a steady DC voltage of typically 100–200 v on its anode, whereas the purveyors of electric power to the domestic market had chosen (for very good reasons) to use alternating current (AC) at a frequency of 50 or 60 Hz. This meant that the domestic radio required either a hefty battery or (preferably) a device which could convert AC to DC – a rectifier. At the same time, the directly heated cathode which supplied free electrons to the vacuum required a DC current supply at a voltage of about 2 v, frequently provided by a rechargeable accumulator. This, in turn, demanded a low voltage battery charger which, again, made use of rectified AC.

The cat's whisker was in no position to cater for these two markets because it was, by definition, a very small area device and quite incapable of dealing with the currents involved. A very different technology was called for, based on a much larger area rectifying contact and it was here that Schuster's (1874) discovery of rectifying action at a contact between copper and copper oxide and Fritts' (1876) observation of similar behaviour in selenium photo-cells came into their own. The first copper-copper oxide rectifier was demonstrated in America by Grondahl and Geiger in 1927 and the first selenium rectifier by E. Presser in 1925. Most of the major electrical firms took up further development of one or the other device and commercial rectifiers proliferated. We see here one of the first examples of *directed* research – if fundamental understanding of the properties of selenium or copper oxide could be seen to produce improvements in performance or a reduction in manufacturing costs, there was clearly good reason to pursue it – and it is interesting to note that this kind of activity was taken up much more rapidly in the United States where the culture of the practical probably held greater sway than in Europe. Research in Germany or Great Britain at that time concentrated on understanding Nature, rather than trying to make money out of her.

It is worth looking briefly at the technology for making these rectifiers. The copper oxide version consisted of a disc of copper, about 1 mm thick which was oxidised on its upper surface by heating in air at a temperature of 1000 °C, then annealing at a temperature of 500 °C and finally applying a 'counter electrode' in the form of a soft metal such as lead which could be pressed onto the oxide (with, possibly a layer of graphite in between). Selenium devices were made by forming a layer of selenium (by melting selenium powder) on a metal washer which might take the form of steel, cleaned, sandblasted and plated with nickel or possibly it might be aluminium. Again a counter electrode was formed using a mixture of lead, tin, bismuth and cadmium, deposited in a spray process. Manufacturing details were often obscured in the interests of commercial secrecy and, as they were inevitably based on empiricism, rather than scientific understanding, there were wide variations in practice. One feature common to both types was the need to stack several units together in order to achieve satisfactory 'reverse breakdown' characteristics. The rectification process involved large currents flowing in the 'forward' direction compared with negligibly small currents in 'reverse'. However, reverse current tended to increase dramatically when the reverse voltage exceeded about 20 v and, in order to supply DC voltages of say 150 v for typical valve circuits, rectifiers had to achieve reverse voltages of about 250 v and this implied that at least 10 units must be connected in series. This was facilitated by making each rectifier in the form of a circular washer and stacking them on an insulating rod.

These rectifiers were incorporated in a wide range of electronic equipments, including radios, measuring instruments and controls. Military communications and fire control systems came to rely on them and the lack of understanding of their detailed functioning left the end user extremely vulnerable to unexpected 'wobbles' in manufacturing procedures. A classic example of this was known as the 'thallium catastrophe' which hit the German war machine during the Second World War. Solder used in the manufacture of selenium rectifiers, ordered from a new supplier, led to a catastrophic decline in rectifier characteristics. The problem was traced to the presence of thallium in the solder and a change of solder soon put things to rights but nobody was any the wiser as to the precise nature of the problem. This was only one of a range of similar difficulties caused by uncontrolled trace impurities and simply served to emphasise the lack of proper understanding of the processes employed. With our superior knowledge of the importance of minute impurity levels in semiconductor behaviour, we can readily understand that the relatively crude manufacturing techniques in use were never likely to provide adequate control. Nevertheless, it is worth noting one or two features of the technology – we have here an early example of the use of semiconductor material synthesised *in situ* rather than the reliance on natural samples dug from the ground (such as the lead sulphide crystals used to make cat's whisker detectors) and, again, the use of high temperature processing to form copper oxide can be seen as a parallel to the kind of processing used in the iron and steel industry and, more importantly, as a forerunner to the much more sophisticated processing developed in the latter half of the twentieth century for the manufacture of integrated circuits.

Given the relative crudity of the technology available to rectifier manufactures, the wonder is that success was as widespread as it certainly was. By 1939, for example, copper oxide rectifiers were capable of supplying currents of 30,000 amps at 6 v for electrolytic coating or 1 amp at 5000 v for radio transmitters. Such performance is extremely impressive when one remembers that almost nothing was known about the scientific basis of the rectification process involved. Not only were bulk semiconductors little understood, but the properties of the metal–semiconductor interfaces responsible for the rectifying behaviour were, until 1939, totally without scientific explanation.

For many years there had been speculation about the origin of rectifying characteristics (the shape of the current vs voltage plots). As we saw earlier, Braun believed that his cat's whisker devices depended on the contact between the pointed wire and the semiconductor crystal but not everyone was convinced. There was also doubt as to whether the mechanism was electronic or thermal, a controversy put to rest by G.W. Pierce in 1909 in favour of the former. In copper oxide devices, rectification occurred at the copper–copper oxide interface but in selenium devices much work went into proving that the important interface was that between the selenium and the metal base on which the selenium was deposited. However, the crucial question concerned the electronic nature of these interfaces and it is remarkable that, after years of controversy, no less than three separate papers should appear in 1939 proposing an acceptable theoretical model. Their authors were an Englishman (Neville Mott), a German (Walter Schottky) and a Russian (B. Davydov). The crucial factor in all these explanations was the concept of a 'surface barrier', which arises because the semiconductor and its contact metal have different 'work functions'. (The work function is the amount of energy required to remove an electron from the material and take it to infinity – an abstract concept if ever there was one but very useful to physicists!) This implies that, in general, there will be an energy barrier at the interface, meaning that electrons require a certain amount of energy to pass from the metal into the semiconductor or vice versa. In recognition of his major contribution to the understanding of rectifiers, the barrier is known as a 'Schottky barrier' (no doubt 'Mott–Schottky–Davydov barrier' would have been just that little bit too clumsy!) The exciting point, however, was the realisation that if an external voltage is applied in the forward direction, the effect is to lower the height of this barrier, thus making it easier for electrons to pass over it, whereas a reverse voltage has no effect. In other words, we should expect the forward current to increase rapidly as the forward voltage is increased but the reverse current must saturate at a rather small value and remain independent of applied voltage – a close approximation to the ideal rectifier characteristic. (The increase of reverse current which we referred to above represents a 'breakdown' effect which occurs at some fairly well defined reverse voltage and is secondary to the main predictions of the barrier theory, nothing ever being quite as straightforward as one would like it to be!)

We need to add two further points to this brief summary of the barrier theory. Firstly, we note that the calculated current-voltage characteristic provided no more than a rough account of those measured on real rectifiers. Clearly, real devices were a good deal more complex than the simplified model on which the theory was based and there was no practical possibility of unravelling the details because there were no means of acquiring the detailed information appropriate to actual devices. This was something of a disappointment but simply represented an acknowledgment of reality – real devices made with uncharacterised materials and ill-defined structures could never be expected to yield electrical characteristics in line with an idealised model. Secondly, and much more encouragingly, the barrier model provided a good qualitative explanation for the photovoltaic effect which had been observed on quite a wide range of rectifying structures. Implicit in the barrier model is the prediction that within the barrier region, just below the surface of the semiconductor there exist a high electric field. It was then possible to understand that when light was absorbed near the semiconductor surface, generating both holes and electrons, these particles would move rapidly in opposite directions as a result of this surface electric field. Holes moving one way and electrons the other both represent electric currents *in the same direction* so the barrier theory was able to explain that the photovoltaic effect, along with rectification, was also a surface effect.

The concept of the Schottky barrier represented a second major step in the theoretical understanding of semiconductor behaviour (the first being Wilson's band theory to explain their bulk properties) and it is significant, if a trifle fortuitous, that it should appear right at the start of the Second World War when the cat's whisker rectifier was effectively reborn as a detector of microwave radiation. Radar systems existed well before the War (see Fig. 2.4) but they made use of relatively long wavelength electromagnetic waves and this severely limited their accuracy. (It is a fundamental property of any imaging system that the smallest spot to which 'light' can be focussed is of the order of the light wavelength – this limits the resolution of an optical microscope, for example, and similarly limits the amount of information that can be squeezed onto a compact disc. By a reciprocal relationship, the requirement for the narrow beam demanded by a radar system implies that the aerial should be much larger than the wavelength so, particularly for airborne radars, this implied the use of short wavelengths.) Early in the War, it became

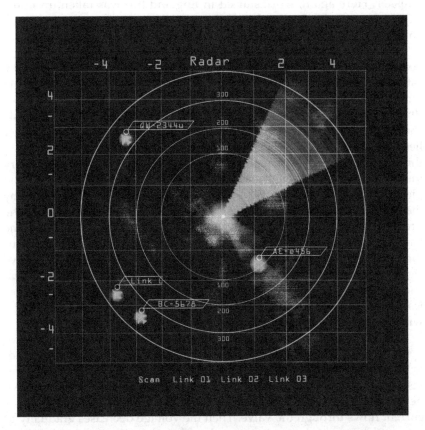

FIG. 2.4 A typical airport radar screen. As the aerial is scanned across the sky, the scanned area is represented by the rotating segment, the distance to the target being represented by its distance from the centre of the screen. Courtesy of iStockphoto, © George Cairns, image 607770.

apparent that truly effective radar systems must operate at wavelengths of the order of 10 cm, though at the time this so-called 'microwave' spectral region had been relatively poorly explored. In particular, there was a serious lack of microwave sources (i.e., generators) and of radiation detectors. The need for a high power microwave source to generate pulses of energy was paramount if such systems were to have a useful operating range and this need was splendidly satisfied by the invention of the cavity magnetron, by Randall and Boot, working in Birmingham University, which was subsequently developed by the General Electric Company (GEC) to produce powers of up to 10 kw and further developed at the MIT Radiation Laboratory, following the Tizard mission to the USA in September 1940. Meanwhile, the concomitant need for a sensitive and reliable detector was satisfied by our old friend, the cat's whisker. Here again, work started in England but was taken up in the USA, in particular by Purdue University, University of Pennsylvania and Bell Telephone Laboratories.

After Braun's pioneering efforts and its initially widespread take up on both sides of the Atlantic, the emergence of the thermionic valve made the cat's whisker detector more or less redundant. It could scarcely compete with the valve's greater reliability and much improved 'noise' performance (the level of random background interference). Why, then, one may ask, should the device return to life in the guise of a radar detector? The answer lies again in the realm of operating wavelength (or, more conveniently, in this case, frequency). Thermionic valves had proved themselves to be very much at home in the frequency range of 50 kHz–1 MHz used by the radio pioneers but they suffered from a serious limitation when confronted by the challenge of microwave frequencies (typically 3 kMHz or 3 GHz). The problem was associated with the time it took electrons to traverse the vacuum space between cathode and anode (about 1 mm) in a detector diode, when compared with the time scale on which such a microwave signal varied. The former time was typically 10^{-8} s, whereas the signal changed from one peak to the next in a time of 3×10^{-10} s, some 30 times less. In effect, the valve simply couldn't respond rapidly enough to cope with the signal it was supposed to rectify.

Let us work this through by doing a 'thought experiment'. Suppose we apply an alternating voltage between anode and cathode of our valve and, first of all we consider it to be at a low frequency so that the valve can respond. Imagine the situation which holds when the signal is at a positive peak – electrons from the cathode will be attracted to the anode and a current flows through the valve. Then the voltage decreases gradually to zero so the current decays in sympathy. Then the voltage goes negative – electrons are now repelled from the anode so no current flows until the signal becomes positive again. Clearly, current only flows during positive half-cycles of the applied voltage and, if we follow the current amplitude during this time, it will increase from zero to a peak, then decrease again in a symmetrical manner back to zero. The overall result is a series of current pulses, all with positive amplitude. The *average* current is a steady DC level which is the rectified (or detected) signal which we can use to drive some kind of indicator device. Now suppose that the signal is varying at a high frequency, too high for the valve to respond. Again, imagine a positive peak of the signal – electrons are attracted towards the anode and start to move. However, before they have moved any significant distance, the signal swings down to zero and then it rises to negative peak which repels electrons away from the anode. The electrons which had started to move towards the anode now move back towards the cathode until the signal returns to a positive sense, when they try again to move towards the anode. The sequence repeats but at no time do any electrons reach the anode so no current flows – in other words, there is no rectified output and the diode simply fails to perform its desired function.

It was this palpable failure of the thermionic valve that led experimentalists to turn again to the semiconductor rectifier and, in particular, to the cat's whisker. Why though should this be superior? The answer is far from simple but one aspect of the explanation concerns a 'transit time' of a similar nature to that we have just discussed for the diode valve. In the case of a Schottky barrier rectifier, the distance which charge carriers have to move in order to generate a current corresponds to the width of the barrier region within the semiconductor and this distance is typically about $1\,\mu$ (10^{-6} m), 1000 times less than that of the millimetre appropriate to the valve. Thus, it is easy for electrons to make the transit in a time less than the separation of adjacent peaks in the microwave signal. This argument would apply to almost any semiconductor rectifier so why was it the cat's whisker that won the day? It is here that the going gets complicated but it depends on the fact that the metal–semiconductor contact looks like a combination of a capacitor and two resistors and this capacitance–resistance combination has an associated 'time constant' which represents the time it takes electric charge to flow into and out of the capacitor through the resistors. It represents a limit to the response time of the rectifier in much the same way the transit time does, so it is important that this RC (resistance–capacitance) time constant should be as small as possible. Without delving into any more detail, let us just say that the very small contact area associated with the cat's whisker is ideal for achieving this and, as things worked out, it proved just adequate to provide rectification at the necessary high frequency. Any large area rectifier would have been incapable of doing so. Fatefully, the newly formulated Schottky barrier theory came just in time to provide a proper understanding of this important truth.

The story of the microwave crystal detector began in England, following the invention of the cavity magnetron in 1940 which made possible the development of microwave radars, offering the improved resolution necessary for many wartime demands. So, at the very beginning of the War, pressure was placed on researchers to come up with a working microwave detector. Initially, the 'crystal valve', as it was somewhat oddly called, was regarded as nothing more than a short-term stopgap prior to the development of a suitable thermionic valve. In the event, no such development occurred and the crystal rectifier gradually took on the characteristics of stability and ruggedness which ensured its successful long-term use in a wide range of radar applications. The British work was summarised in a paper published in 1946 from the three participating organisations: the Clarendon Laboratory, Oxford; the British Thomson Houston Company and the General Electric Company. It makes interesting reading.

One of the first problems to be solved was the choice of semiconductor material; the second was that of whisker material. Various naturally occurring crystals were tried but only galena (lead sulphide) and silicon showed acceptable promise. Galena required only a very light contact pressure from the whisker and this made it susceptible to vibration which all too easily destroyed the carefully acquired 'good contact'. It was decided, therefore, to concentrate on the combination of silicon with a tungsten whisker which was incorporated into a standard capsule for ease of mounting into the waveguides in common use at these frequencies. (Because of the relatively short wavelengths involved, microwave signals could conveniently be propagated inside rectangular section metal pipes, rather than being carried along copper wires, as was standard practice at lower frequencies. We shall discuss another example of this in connection with fibre optical communications in Chapter 8.) The whisker was pressed against the silicon sample by means of a suitable screw arrangement until the desired electrical characteristic was obtained, then a locking screw was tightened to stabilise the contact. An important discovery, that vigorous 'tapping' of the crystal could result in much improved performance, led to its incorporation in the standard production process! The British Thomson Houston Company set up the first production line and slightly later GEC followed suit. When large scale production of detectors began in the United States, the capsule employed was made interchangeable with the British version, a vote of confidence in the merit of the original approach.

Perhaps the key discovery was concerned with the quality of silicon used. It was found that different batches of 'commercially pure' silicon yielded very variable performance (as we saw in Chapter 1, mere chemical standards of purity were far from adequate!) so it was decided to re-purify the material by melting it in vacuo, a difficult procedure on account of silicon's propensity to react chemically with almost everything! Eventually, the use of beryllia crucibles was found to result in the purest silicon and this became standard practice, though it was then found necessary to dope the silicon with some 0.25% of aluminium to realise material with the right resistivity and appropriate stability, this latter property being determined, it was thought, by the degree of surface oxidation. Indeed, the development of so-called 'high burn-out' crystals (those which could withstand high power microwave pulses) was achieved by adding aluminium, beryllium or magnesium because these elements formed suitable oxide layers on the silicon surface.

It is interesting from a number of viewpoints to look back at this British work. We see some tentative recognition of the importance of purity and of attempts to achieve the necessary purity by re-melting the silicon under vacuum but there was no reference to Wilson's band theory or his clear account of n-type or p-type doping to control the free carrier density. Even though aluminium was added to the melt, no mention was made of the fact that this would be expected to result in p-type conduction. No measurements of the Hall effect were made in order to confirm such ideas nor was there any attempt to grow single crystal material. Interestingly, no reference to the Schottky barrier model of rectification appears – the principal concern was with oxide surface barriers, rather than with the barrier intrinsic to the semiconductor–metal interface. The work was certainly very successful in so far as it produced working detectors with reasonably reproducible characteristics and, where required, high back voltage behaviour, and, in wartime, this must surely be regarded as paramount. It is fascinating, though, to appreciate the lack of any real semiconductor background in the approach adopted, a reflection, I believe, of the lack of experimental semiconductor research undertaken in Great Britain between the wars. As we have seen already, semiconductors had a very dubious reputation scientifically and British scientists had a tendency to look for research in the 'pure sciences', unsullied by 'dirty' problems related to poorly controlled materials. (It probably came as quite a shock to the scientists involved to find themselves press-ganged into working in such an alien environment!) It is even more fascinating to note that, while the vast majority of pioneering work on semiconductors had been pursued in Germany, German radar developments lagged well behind those in the United Kingdom (though this probably represented a lack of respect for things scientific within the Nazi high command, rather than any fault of German science).

In comparison with the modest size of the British effort, the Americans were able to mount a much larger programme. By the end of 1940, the MIT Radiation Laboratory in Boston was already staffed by 40 scientists, while at its peak it employed 1200 physicists and nearly 4000 people in total. One of its roles was to award and monitor research contracts in other laboratories such as the Universities, Pennsylvania and Purdue. Frederick Seitz was in process of building a new research group at Pennsylvania when he was asked to take a contract on silicon. A little later, the physics group at Purdue, under Karl Lark-Horovitz was awarded a second, and concentrated their effort on germanium. Lark-Horovitz was convinced, as early as mid-1942 that both silicon and germanium were true intrinsic semiconductors, clear evidence of both intrinsic and extrinsic conduction regimes having been established by the two groups. The key to this long-awaited advance in semiconductor physics was, of course, the recognition that high purity material was essential to establishing fundamental semiconductor properties and the American workers devoted considerable effort to this aspect. They also made use of Hall effect measurements to gain a better understanding of electron or hole behaviour, including their scattering by both lattice vibrations and by 'ionised impurity atoms' (e.g., when a donor atom gives up its spare electron to the conduction band, it becomes positively charged and the resulting electrostatic force between the atom and free electrons causes 'ionised impurity scattering', the trajectory of free electrons being thus deflected). On a strictly practical point, the Purdue group were able to demonstrate that detectors with high back voltage performance could be made from very lightly doped germanium and recognised that this correlated with the existence of a thick Schottky barrier at the metal–semiconductor interface. They understood the significance of the barrier theory and were therefore able to apply it.

With due respect to the considerable amount of earlier work, it is probably true to say that this wartime activity saw the birth of modern semiconductor physics. It established the band gaps of these two important materials, it confirmed the use of donor and acceptor atoms to control conductivity, it recognised the importance of highly perfect crystals (though, in this case, only in the form of small crystallites in polycrystalline samples), it yielded an understanding of free carrier mobility (determined by scattering processes) and it proved once and for all that material purity was an essential pre-requisite for both good devices and good physics. We shall meet again and again this interdependence of material quality, device performance and physical understanding – if we wish to make effective semiconductor devices with well controlled properties, we also need to understand the physics behind their performance, while both these requirements depend on the availability of highly pure and structurally perfect semiconductor materials. And the fact that none of these things is achieved easily explains the need for the kind of effort employed on crystal valves during the Second World War. It also explains the nature of the Bell Laboratory programme which led to the realisation of the transistor in 1947 but that is quite another story and deserves a chapter to itself. We shall conclude this chapter by looking back over some of the developments which shaped the post-war scientific world as a suitable nursery for the nurture of this precocious semiconductor child.

Science, let us never forget, is a human activity and suffers from human foibles in much the same way as any other human creative act. In many ways it depends on the character of the individual scientist and this is particularly true in respect of his or her attitude towards the 'pure science' versus 'commercial exploitation' dichotomy which we have touched on already in connection with the early history of radio. Many scientists regard the satisfaction to be gained from discovery and analysis as reward enough in itself, others regard such developments as merely stepping stones to commercial success, while some (perhaps most?) take a position between such extremes and can combine their instinct for discovery with an ability to seek innovative applications for their work. Some scientists, like many artists, work best on their own, creating a unique output which they can truly call their own – but often overlooking the fact that their personal contribution has been built on foundations provided by others (compare Newton's famous remark that, if he had seen further than other men, it was only because he had had the good fortune to sit on the shoulders of giants). At the other extreme are people who prefer the stimulus of interaction with members of a team and who gain their personal satisfaction from the team's success. We are all different and any account of scientific advance must recognise the fact. What is certain, however, is the fact that the *organisation* of science changed considerably over the period covered by the present chapter.

In the eighteenth century, the typical scientist was a 'gentleman' of means who was able to finance his research from his own fortune and to pursue it in his own private laboratory. The Honorable Henry Cavendish was an excellent example of the type, producing a wealth of 'Electrical Researches' in the period 1771–1781. Almost by definition, he pursued his research purely out of interest and looked no further than personal satisfaction (and the approbation of his peers?) as an adequate reward. In England, this began to change with the foundation of the Royal Institution in 1799 where Humphry Davy and his protégé Michael Faraday performed their life's work, and it should not be overlooked that both of them regarded research as having a practical outcome. (Indeed, this was in line with the Institution's original brief in favour of 'Diffusing the knowledge and facilitating the general introduction of useful mechanical inventions and improvements'.) Davy is famous, of course, for the invention of the miner's safety lamp and Faraday took on a number of commissions from the British Government and from Industry which have tended to be somewhat eclipsed by his electromagnetic researches. There seemed to be no great hurry to replicate this innovative step elsewhere – most laboratories at that time were seen merely as teaching adjuncts – and the first American venture into the field, in the shape of the Franklin Institute, had to wait until 1844. The Clarendon Laboratory in Oxford and its counterpart in Cambridge, the Cavendish, were inaugurated in the 1870s and the Jefferson Physical Laboratory at Harvard followed in 1884 but these were university laboratories with interests in pure, rather than applied research. (The story is told at the Cavendish, that, when J.J. Thomson discovered the electron, the resulting celebrations included a toast 'may it never be useful to anyone!'). The world's first *industrial* laboratory was probably that established about 1870 by Thomas Edison in Memlo Park but Edison's mode of research might not even be recognised as such today. Trial and error, rather than logical thinking, took pride of place. Though a basic research laboratory was established by the American GE company as early as 1900, it was not until the 1920s that the concept of industrial research became widely accepted – the GEC laboratory in Wembley and the first formally established Siemens laboratory were opened in 1914, the Philips laboratory in Eindhoven in 1919 and Bell Telephone Laboratories was established as a separate entity in 1925 – nevertheless, by the beginning of the Second World War, industrial research organisations were generally available to meet the demands of the appropriate military authorities.

Perhaps the more remarkable wartime innovation was the drafting of university researchers into *organised* research in the interests of the war effort. Not only were they obliged to work on subjects remote from their personal interests, but they found themselves thrown together with colleagues and strangers, alike, to form teams dedicated to specific outcomes. The radar research to which we have already alluded was only one example, though an excellent example of its kind. It is ironic, perhaps, to see a reversal of roles in so far as the American empiricism which had been dominant during the nineteenth century now played second fiddle to fundamental science, while the British love of purity in research appeared to be moving in quite the opposite direction. No doubt the driving forces were those of urgency and availability of resources, rather than any change in leopard's spots but it is interesting to note the American movement into basic research following the war – once the concept of large scale funding for fundamental research took hold, America's greater resources would always play a dominant role.

In summary, we note that the very nature of research had changed out of all recognition from that of the 'gentleman's club' approach typified by the early days of the Royal Society. It was now taken for granted that research could be organised with specific ends in view, people could be directed and expected to concentrate on such ends, rather than on blue-skied horizons. They were also accustomed to working in teams, rather than as brilliant individuals. This was not to say that scientific fundamentals were no longer important – quite the contrary – but they were no longer seen as the ultimate end, so much as important means to some practical outcome. And, having said this, we are now in position to follow the fascinating story of the transistor.

CHAPTER 3

Birth of the Transistor: Point Contacts and P-N Junctions

The post-war years were inevitably going to see fundamental change. Europe, Russia and Japan were in dire need of physical reconstruction, political institutions were poised for a major effort in an attempt to prevent any similar slide into armed conflict. The European Union was born, a socialist government was voted into power in England with a brief to introduce free social services, thereby rescuing the poorer sections of society from the fear of unemployment and the financial shock of ill health. German and Japanese military might was reigned in. Hope for a better life was widespread and it was natural that the application of technology to improved lifestyles should play an important role. If science could be harnessed effectively to the successful pursuit of war, surely it should be possible to employ it in the interests of domestic peace. Indeed, much of this came to pass, but very few people could have imagined, even in their wildest dreams, just how far-reaching were to be the ramifications of the electronic revolution which sprang from the immediate post-war activity in the Bell Telephone Laboratories in Murray Hill, New Jersey. The invention of the transistor shortly before Christmas in 1947 must rank as one of the most traumatic of man's technological innovations, certainly to be bracketed with the discovery of the wheel in its long-term effect on his way of life. Fittingly, the three principal participants were awarded the Nobel Prize in 1956 but even they must have been surprised by the manner in which their invention spread into so very many unlikely spheres of application. How, exactly, did it happen? Let us forget, for the moment, the long-term effects – it makes a fascinating story, in itself.

The context within which the transistor came into being was that of long-distance communication. It all began quite a long time ago and, if we are fully to appreciate the background, we must first retrace our steps to

Semiconductors and the Information Revolution: Magic Crystals that made IT happen © 2009 Elsevier B.V.
DOI: 10.1016/B978-0-444-53240-4.00003-9

the nineteenth century, in order to pick up yet another strand of the fabric, forming the elaborate tapestry of post-war technological progress. One of the first successful attempts to speed the delivery of messages over distance can be associated with Napoleon's need to communicate with his far flung armies. It took the form of a line-of-sight semaphore system invented by Claude Chappe and his four brothers in the 1790s. Chappe, of aristocratic descent, was destined for the Church but these plans fell foul of the anti-clerical philosophy of the French Revolution and the family turned to invention for its livelihood instead. They demonstrated a working system in 1792 and, helped, no doubt, by the fact that Ignace Chappe was a member of the Revolutionary Legislative Assembly, the system was adopted for an experimental link between Paris and Lille, some 120 miles distant. This having proved successful, further links were set up across France and were widely used for Government and military communication, though later attempts to involve commercial interests seem to have fallen on relatively deaf ears. The system was widely copied – the British Admiralty, for example, established a similar connection between London and Portsmouth in 1795 – but, by the middle of the nineteenth century, was largely superseded by the electric telegraph. It suffered from being relatively slow, as each message had to be received and re-transmitted at each station, and was available only during daylight hours when visibility was adequate.

An early suggestion for an electric telegraph came from the French physicist Andre-Marie Ampere, in 1820. This was based on his observation that an electric current flowing along a wire caused a magnetic needle to be deflected. The initial version of the Ampere telegraph made use of one wire for each letter in the alphabet and was, in consequence more than a little clumsy in operation (not to mention expensive in realisation). However, the idea was later taken up in 1833 by a duo of German physicists, Wilhelm Eduard Weber and Karl Friedrich Gauss, who set up a communication link via their local church tower in Gottingen, then by a duo of English physicists, William Cooke and Charles Wheatstone. They took out a patent in 1837 for a five-wire system, then, in 1841 they set up a two-wire system which operated between Paddington and Slough and provided a public service. However, bearing in mind the fact that the transistor was a purely American affair, we should concentrate, here, on the development of American communications. Joseph Henry (1830) demonstrated the use of an electromagnet to activate a bell at the remote end of a one mile cable and, in 1835 Samuel Morse established the concept of pulsed current to transmit the dots and dashes of his famous code and recorded them by activating an ink marker on a strip of paper. In 1843, an experimental telegraph line was under construction between Washington and Baltimore, some forty miles distant, and news items were successfully transmitted between the two cities in the following year. The Western Union Telegraph Company came into being in 1851 and, by 1861, had established the first trans-continental telegraph line, by utilising that otherwise useless bit of land alongside fast-developing railroad tracks. It is of interest from the viewpoint of much later digital communication systems, that Western Union also introduced the concept of multiplexing (interleaving groups of current pulses) as a means of increasing the number of signals which could be accommodated on a single telegraph line – four each way on its inception in 1913. Not that telegraphic communication was anything less than an international activity – the first transatlantic cable was laid as early as 1857, and, though this was a failure, reliable, if rather slow, links were established between Ireland and Newfoundland in the mid-1860s. Clearly, nineteenth century entrepreneurship was driving communication technology ahead at a brisk pace and there was now no shortage of interest in its application to commerce.

Valuable as the electric telegraph had been in stimulating both national and international communication, there can be little doubt that the introduction of the telephone represented a giant step forward and probably made much greater inroads into domestic usage of appropriate transmission lines. As everyone knows, the telephone was invented by a Scotsman, Alexander Graham Bell, (see Fig. 3.1) who was granted the first patent in 1876 and there can be no doubt that Bell was greatly influential in its rapid commercial take-up. (Bell did for the telephone what Edison

FIG. 3.1 Alexander Graham Bell speaking into a prototype model of his telephone, c.1876. This version made use of a liquid microphone which was soon replaced by a more reliable carbon microphone. Bell's patent was challenged by several rivals but he successfully fought them off and went on to commercialise the telephone, making a large fortune in the process. Courtesy of WikiMedia.

did for the electric light bulb, he commercialised it.) The fact remains, however, that an Italian émigré, Antonio Meucci demonstrated a working telephone system as early as 1855 but, sadly, was too poor to afford the $250 fee involved in taking out a patent. He attempted to challenge Bell's application but, again, was probably too poor to employ effective legal representation and lost his case. Interestingly, Bell fought no less than 600 court cases (the majority being trivial) in defence of his priority and won them all! We might draw the conclusion from this just how important the patent was seen to be – as, indeed, it certainly was (in a 1939 U.S. Congressional Report it was described as 'the most valuable single patent ever granted') – but it is interesting to record one instance of a failure to appreciate the fact by the then President of Western Union. Bell and his backers were prepared to sell the patent to Western Union in 1876 for the miniscule sum of $100,000 but were turned down, with the comment that the telephone was 'nothing but a toy'! Two years later, Western Union would gladly have paid $25 million but, by then, Bell was no longer interested in selling.

The Bell Telephone Company was founded in 1877 and a key patent acquired, 2 years later, from the Western Union for the use of a carbon microphone (originally invented by Thomas Edison). This replaced the far less robust fluid microphone employed in Bell's original experiments and ensured a future that was based on a truly practical technology. From that point, the company never looked back. The first telephone exchange opened in New Haven, Connecticut in 1878 and, within a few years, most large American cities were similarly serviced. An important development in 1882 saw Bell acquire control of the Western Electric Company, which became its manufacturing arm and this was followed in 1885 by the incorporation of the American Telephone and Telegraph Company (AT&T) with a brief to build and operate a long distance telephone network. This it did successfully, reaching Chicago in 1892 and San Francisco in 1915. In a company restructuring programme of 1899, AT&T acquired the assets of the Bell Company and thereby became the parent company, expanding its brief onto the international scene in the early years of the twentieth century, selling a wide range of telephone equipment round the world. However, in 1925, the International Western Electric Company was sold to the newly formed ITT Company and AT&T retired to fulfilling its original brief of providing long distance communications within the American mainland. This, nevertheless, was taken to include providing its subscribers with international connection and the first transatlantic telephone service, based on wireless transmission, came into service in 1927. Reliability problems, finally, led to this service being transferred to the first telephone cable link (TAT-1) in 1956, a similar transpacific link being completed in 1964. Various microwave connections were also set up about this time, culminating in the first transatlantic satellite transmissions via Telstar in 1964. Altogether this was a truly impressive achievement by a remarkably vigorous and technologically innovative company. We should look in more detail at some important aspects.

From our point of view, the first item of interest concerns the development of long distance telephone links and we should recognise an essential difference between telegraph and telephone signals. The former were generated by some kind of generator which allowed the amplitude of the current pulse to be fully under the sender's control, whereas the amplitude of a telephone signal was limited by the performance of the microphone. Within reasonable limits, a telegraph signal could be set at whatever level was appropriate to the length of the line – any loss introduced by a long line could be compensated by an appropriate increase in amplitude at the transmitter. (Though there clearly were limits – an early transatlantic cable was actually burned out by one over-enthusiastic telegraphist!) This was not the case for telephone transmissions. Long lines demanded that the relatively weak signal be boosted at regular intervals to compensate for losses along the line and, in the 1890s, no suitable technology was available. The necessary amplifier simply did not exist until the development of the thermionic valve (or vacuum tube) in the early years of the twentieth century, all very frustrating for AT&T, struggling to meet its rather ambitious challenge of a trans-America telephone service. However, it did have some long-term benefit for the company, as we shall shortly see.

Like so many scientific developments of the time, the advent of the triode valve depended on a certain degree of serendipity, as well as the application of imagination. The story began in Thomas Edison's famous laboratory at Menlow Park in New York State. In 1883, Edison was desperate to develop a suitably long-lived incandescent lamp for his future lighting systems and was experimenting with the use of a metal plate within the lamp's vacuum envelope. He hoped to trap atoms which were being vapourised from the carbon filament, and thus to avoid blackening of the glass, but, in passing, noted that a positive potential applied to the plate caused a small electric current to flow in the plate circuit. The electron was yet to be discovered (in 1897, by J.J. Thomson, in Cambridge), so Edison had no means of interpreting his observation and was, in any case far too busy to take it seriously – as was his wont, he simply patented it and moved on.

Remarkably, the phenomenon lay dormant for 21 years until resurrected by an English physicist, John Ambrose Fleming who happened to be working as a consultant for the Edison Electric Light Company. (Fleming had a long and distinguished career in academia, being, for

1 year only, the first professor of Physics and Mathematics at my own University of Nottingham.) He was intrigued by the original Edison observation and, being now able to understand the effect, following Thomson's discovery of the electron, realised that it could be used to make a rectifier (a device which allowed current to pass in only one direction). He took out a patent for the device in 1904. As luck would have it, he was also acting as consultant to the Marconi Company and was able to provide Marconi with an improved wireless detector in the form of what came to be called a diode valve (a device with two electrodes). This proved of enormous value to Marconi but, perhaps the really important breakthrough came from the American side of the Atlantic, when Lee de Forest added a third electrode in the form of a wire grid between the filament (cathode) and plate (anode) which could be used to control the current through the valve. He took out a patent in 1908, calling the resulting device an Audion but subsequent usage has universally favoured the title of triode. A small voltage swing at the grid electrode effected a large change in anode current and provided considerable power gain. Here was the device which the Bell telephone engineers desperately needed to fulfil their dreams of a transcontinental telephone system and fate was certainly on their side. De Forest was in serious financial difficulty as a result of a series of law suits and was happy to sell his patent to Bell for a very modest sum. The only problem was that the device didn't work!

Perhaps we might look briefly at the reason. Its operation depended on the fact that a thin wire filament, heated to a high temperature by passing an electric current through it, emitted electrons into the vacuum space (the process of thermionic emission). This loss of negatively charged electrons tended to leave the filament, or cathode, positively charged, thus attracting the electrons back to it, the net effect being a so-called negative space-charge close to the filament. Application of a positive voltage to the anode, or plate, electrode drew electrons away from the cathode and generated a current in the anode circuit. The grid electrode was normally biased negatively with respect to the cathode, which had the effect of repelling electrons back to the cathode and therefore reducing the anode current. When used as an amplifier, the alternating signal voltage was superimposed on the negative grid voltage, giving rise to a corresponding fluctuation in anode current. However, if the vacuum within the valve envelope was inadequate, some of these travelling electrons collided with gas atoms and ionised them – that is, they knocked secondary electrons from them – producing positive ions, which were attracted towards the negative grid to produce an ion current in the grid circuit. This represented power dissipation in the grid circuit and implied a serious reduction in power gain for the amplifier. The larger the anode current, the greater was the probability of ionising collisions and the worse became the effect (revealed by the presence of a faint blue glow, emanating from the ionised gas atoms). The device would amplify at very low signal levels but with ever-depreciating efficiency as the signals increased – a body blow for its application in telephone repeater stations.

It was at this point that the Bell Company made an important commitment. Following the forceful advice of Frank B. Jewett, a senior member of staff, working for Western Electric, it was decided, as a matter of policy, to start employing university graduates specifically to do research, with a view to solving the company's technological problems. It was also seen as important to generate new patents, Bells original telephone patents having expired in 1894, leaving the company vulnerable to innovative rivals. Jewett was given the go-ahead to recruit suitable staff and, by 1911, he had put together an appropriate research team. One of their first tasks was to improve the performance of the thermionic triode valve for application to long distance telephone communication and they made a good job of it – by improving the vacuum within the valve envelope and redesigning the electrode geometry they achieved amplifier performance suitable for the necessary repeater stations, thus enabling the successful development of a transcontinental telephone service – in January 1915, the first calls were put through from New York to San Francisco. This and other similar successes ensured that the arrangement became permanent, and in 1925, The Bell Telephone Laboratories was formally established. It grew rapidly into one of the most powerful research organisations in the western world, with laboratories situated in many locations about the United States. The Murray Hill establishment, where the transistor first saw the light, was built in 1941 (an industrial contribution to the American war effort?) and was to grow considerably in size after the war. For many years, it was the headquarters of Bell research.

It is here that we take up our story again, in the shape of AT&T's participation in the programme to develop effective radar detectors. In particular the Bell scientists were beginning to appreciate just how important was the quality of the semiconductor materials employed and to take their first faltering steps towards its accurate control. Bearing in mind the status of Bell research, it was scarcely surprising that they had been working on semiconductor devices well before the war. In fact, Walter Brattain, one of the triumvirate who were to be credited with inventing the transistor, spent several years studying the properties of copper oxide rectifiers and had even been involved in a crude attempt to make a solid state amplifier. He and his colleagues had recognised that the rectifying behaviour was probably associated with some kind of barrier to electron flow between the copper metal and the adjacent oxide film and had argued that, if a control electrode could be introduced in order to vary the height of this barrier, it should be possible to make an amplifier working on similar lines to the triode valve. The problem was that, not only had they little idea of the physical scale of this barrier, but they lacked the necessary technology to make a structure of the appropriate size. No amount of experimental trial and error showed even the smallest effect. Interestingly, William Shockley, who was eventually to be the team leader in their later transistor work, came to Brattain's lab, in 1939, with a very similar idea. He had been studying the theoretical papers by Mott and by Schottky which predicted the barrier between a semiconductor and a metal contact and saw the same possibility as Brattain but, once again, practical realisation remained chimerical.

It is interesting to note at this point that the pre-war Bell scientists were not alone in hankering after a solid state amplifier. It has recently come to light that a remarkable Russian scientist Oleg Vladimirovich Losev also had the idea to replicate the triode valve in semiconductor form. Working in Leningrad during the 1930s, he made a detailed study of cat's whisker radio detectors which, when appropriately biased, could be used as amplifiers and oscillators. He also made use of these devices in a Crystadine radio, a heterodyne system based on the unusual properties of silicon carbide and zinc oxide detectors. What is more, he is said to have developed a three-terminal silicon carbide device which demonstrated current control by way of a third electrode, though not showing net gain. In this respect, he was clearly ahead of his American contemporaries but unfortunately, the paper in which he reported these results was written during the Nazi siege of Leningrad and not only failed to reach its intended journal editor but has never been found. Losev himself became a victim of the siege so we shall probably never know how close he came to discovering the transistor several years before it finally happened in Murray Hill.

It is important for us to recognise two aspects of the situation within Bell. Firstly, the laboratory management (in the person of Mervin Kelly) had already made the decision to concentrate scientific effort on solid state work and, secondly, Bell scientists were actively seeking means of building a solid state amplifier, even before the war. There may have been only the vaguest idea as to how this might be achieved, but it was important enough to attract laboratory funding. Summing up the Bell philosophy: if research could improve communications technology, as it had done prior to 1940, there was even more likelihood that further benefit would accrue from the study of the new field of solid state physics. Such thinking was of particular significance in so far as it represented a fundamental change from the old cut-and-try methods that had characterised much American work in the late nineteenth and early twentieth centuries. Needless to say, the war would interfere in these plans but, as things turned out, by no means negatively. Crucially, Bell scientists were required for work on microwave radar and this was to include the development of detector crystals. It was an entirely appropriate decision because they already had considerable experience with silicon.

For several years before the outbreak of hostilities, one of Bell's long-standing researchers, Russell Ohl had been using cat's whisker silicon rectifiers to detect short wave radio signals and he had built up a valuable expertise in their use. But one overriding fact had emerged from this work; the silicon then available was characterised, to a frustrating degree, by apparently random variation in behaviour – even a single sample could show rectification in both senses, depending exactly where the cat's whisker point contact was made and the variations he observed between different samples could also be quite dramatic. It was this kind of experience which had led many European scientists to eschew work on semiconductors altogether, but Ohl realised that his difficulties were probably the result of inadequate purity and set out to try and improve the quality of his material by melting the silicon in a vacuum furnace. But progress was slow. Silicon, being a highly reactive element, tended to dissolve bits of the crucible in which it was contained, added to which, it was difficult to maintain an adequate vacuum. Ohl persisted, however, and persuaded various colleagues to help, including two metallurgists, Jack Scaff and Henry Theuerer, who possessed an electric furnace, containing an inert atmosphere of helium gas. They heated Ohl's silicon in quartz tubes and produced samples with significantly improved uniformity. Even so, some samples still rectified in one direction, others in the opposite sense. Clearly, there were two distinct types of silicon (what we recognised in Chapter 1 as n-type and p-type) and even this more sophisticated apparatus was not capable of controlling which would be produced in any particular experiment. Nevertheless, this certainly represented progress, as it was now more than ever certain that unknown impurities dominated the silicon properties – in principle, at least, the way ahead was clear; purify the silicon to an adequate extent and its properties could yet be tamed.

An unexpected bonus came Ohl's way, too. One particular sample showed a strong photovoltaic effect (shining light on it generated a voltage of about half-a-volt between its ends) which Brattain recognised as similar to the effect he had observed with his copper/copper oxide rectifiers. The important difference, though, was that, in the silicon sample, it was a bulk effect, rather than the surface effect with which he was familiar. This silicon sample contained what we now routinely refer to as a p–n junction – the top half of the crystal was n-type, while the bottom was p-type – and the position of the junction could be clearly seen after subjecting the sample to a chemical etch (in this case nitric acid). In fact, Ohl had, in 1939, produced the prototype of the now widely used silicon solar cell. Kelly saw how important such a device might be in powering telephone repeater stations in the far-flung wastes of the American

Mid-West and swore those concerned to secrecy. No matter how freely the Bell scientists might collaborate in the wartime effort to perfect silicon cat's whisker radar detectors, the p–n junction solar cell was to remain under wraps for the duration of the conflict.

This, the first, recognition of a p–n junction may have owed much to serendipity but it was, without doubt, a very important step forward. Such junctions were to become omni-present in the world's electronic future, so it behoves us to enquire briefly into their modus operandi. Bringing together regions of n-type and p-type material results in an interface, on one side of which exists a high density of free electrons, while, on the other, is a high density of free holes. Such sharp changes in density are unstable and these respective free carriers inevitably diffuse across the interface, in Nature's attempt to achieve parity. (This is paralleled by the urgent rush of gas molecules into an evacuated container, on its being opened to the atmosphere – Nature not only abhors a vacuum but, more generally, any sharp change in density, be it of molecules in a gas, molecules in a fluid or free electrons in a semiconductor.) But electrons differ from gas molecules in that they carry an electric charge, and their diffusing across the interface sets up an electric field which tends to draw them back again. In another way of saying this, it sets up a potential difference across the interface and this represents just the same kind of barrier that Mott and Schottky pointed out must exist at a metal/semiconductor interface. If we now ask what happens when light is incident on the sample, close to the junction, you may remember from Chapter 1 that light with photon energy greater than the semiconductor band gap is absorbed, thereby exciting electrons from the valence band into the conduction band (and leaving positive holes behind in the valence band). The electric field at the p–n interface then sweeps these newly generated free carriers across the interface, electrons from the p-side into the n-side and holes from the n-side into the p-side. Such a flow of free carriers constitutes an electric current (both components of *current* being in the same direction, even though the *carriers* move in opposite directions!). If we connect a piece of copper wire between the ends of the semiconductor sample, this current will flow round the external circuit, and this rather simple (though extremely clever) device can be seen to have converted light energy into electrical energy, without the intervention of any moving parts. (We shall meet p–n junctions again in various guises, so it would be well to feel comfortable with the above explanation before moving on.)

Wartime activities at the Murray Hill laboratory took many scientists into new and unfamiliar fields and into very new and unfamiliar modes of working. It was all very stimulating and driven by a strong sense of national pride but there were many who felt relief at the prospect of a return to normal civilian life. Not that anything was going to be quite the same as in pre-war days. Kelly had been planning a major shake-up of his research organisation for much of the interim period and in 1945 he sprung his ideas on a not-altogether-unsuspecting world. There was to be a new emphasis on fundamental science as the basis for better under-standing of materials and phenomena, which would lead to commercially viable innovation in communication technologies. Kelly was aware of a U.S. Government move to stimulate the application of basic science in many fields of activity and recognised that industries, in general, and the communications industry, in particular, would become ever more competitive. He firmly believed that Bell must be at the forefront of these developments and placed a clear emphasis on the specific area of solid state physics. This new Group was to be headed by an outstanding physicist, in the person of William Shockley (returning from a distinguished wartime advisory role) and a chemist, Stan Morgan, a relaxed individual who provided the man-management skills to complement Shockley's somewhat sharp-edged technical contribution. Walter Brattain, also back from a wartime struggle to counteract the threat of German U-boats to allied Atlantic convoys, was a key experimentalist member and, in line with the philosophy of establishing a firm base in fundamental science, a theoretician, John Bardeen, was recruited from the University of Minnesota. Bardeen was to distinguish himself by winning not one but two Nobel Prizes in physical sciences, an achievement that well illustrates the standard of competence inherent within the team.

Shockley kick-started their programme, as it were, by proposing a study on a new type of semiconductor amplifier, which later came to be known as a field-effect transistor. As we saw earlier, Shockley was convinced that he could make a solid state amplifier to replace the well established thermionic valve and his strength of personality made it almost inevitable that this was the way in which the programme would develop. The idea was to make a thin 'brick' of semiconducting material, with metal contacts at its ends, so that a current could be passed along it, between the 'source' at one end and the 'drain at the other end (a terminology taken from the analogous situation of water flowing along a pipe) and to control this current by applying a voltage to a metal plate, which formed a capacitor with the top surface of the brick. Applying a positive voltage to this 'gate' electrode should induce an equal, but opposite, negative charge on the semiconductor, in the form of free electrons, which would enhance the source-drain current. Then, superimposing an alternating signal voltage onto the gate would produce a corresponding, but much larger AC voltage in the drain circuit. The nature of the capacitor meant that very little current could flow in the gate circuit so very little power would be dissipated there, whereas much larger currents could appear in the drain circuit and considerable power gain should be available. At least, that was what Shockley's calculations told him! Alas, when a suitable structure was made by depositing a film of silicon or germanium on a ceramic substrate and applying an electric field to the surface by way of a gate electrode, nothing happened! The effect was well over 1000 times smaller than Shockley's prediction. Even worse, when Shockley tried to patent his idea, the Bell patent agents discovered an earlier patent by a Polish academic, Julius Lilienfeld, taken out, remarkably enough in 1930! Fate, at this juncture, was hardly being kind to poor Shockley.

Bardeen was called in to try to explain this negative result and spent some time working out a theory of so-called surface states, quantum states on the surface of a semiconductor which could trap free electrons. He argued, correctly, that, when a gate voltage was applied, the electron charge induced in the semiconductor was immobilised in surface states, rather than existing as the hoped-for free electrons which would enhance the source-drain current. Once this was realised, it became almost mandatory for the team to investigate the nature of these surface states – could a way be found to overcome their nefarious influence? Oddly, however, and significantly, Shockley seems to have opted out. Whether he was smarting from the fact that it was Bardeen who came up with the explanation, whether he felt disillusioned at the failure of his device ideas, or whether he was simply exercising his will-o'-the-wisp need to be involved in everything about him, he appears to have left the semiconductor group to get on with it, whilst he took up a number of other solid state topics. In defence, it must be said that he was responsible for the whole solid state activity at Murray Hill and had every right to devote time to non-semiconductor aspects. In the circumstance, though, it was an unfortunate decision on his part.

It was natural for the work to concentrate on the two semiconductor materials which were most highly developed. As we saw in the previous chapter, the war years witnessed significant advances in the understanding of silicon and germanium, particularly with regard to their purification and controlled doping. In particular, the Purdue work on high-back-voltage germanium detectors had demonstrated the importance of using high purity starting material, which could then be doped with aluminium to make it p-type. What was more, samples of such material were made available to the Murray Hill laboratory through their participation in the national programme to optimise detector performance and Walter Brattain was in a position to produce suitable samples from his 'goodies' drawer when the need arose. This proved to be a vital part of the Group's success in discovering transistor action and contrasted with Shockley's earlier attempts to make a field-effect transistor, which were performed on evaporated films. These were of poor quality, in terms of their crystal structure, consisting, as they did, of fine-grained polycrystalline material, which was characterised by very low electron and hole mobilities. The interfaces between these grains acted to scatter free carriers – that is to deflect, or even, reflect them, thus hindering their paths through the sample and resulting in a very high electrical resistance (i.e. very small current flow). It was, perhaps, characteristic of Shockley that he found it difficult to accept the vital importance of crystal quality in determining device performance – he was usually in a hurry to have his many ideas put to experimental test and was intolerant of what he saw as other scientist's time-wasting efforts to perfect material properties. He suffered from a similar blind spot where device technology was concerned – if a device could be drawn on paper, he assumed it could be made next day! He was not alone in such beliefs, of course, but, with the well known benefit of hind-sight, we can now see just how dangerous this cavalier approach could be. It was to haunt him in his later attempt to set up his own semiconductor company.

But we must return to the technical story. The question immediately arose as to how to demonstrate the actual presence of surface states on silicon and germanium samples. Without this verification, Bardeen's theory was no more than that – a plausible explanation of the difficulty experienced in making a field-effect amplifier. The new emphasis on fundamentals demanded clear experimental proof before it could be regarded as sound. Bardeen and Brattain argued that, if electric charge could exist at the semiconductor surface, it should be possible to detect it using a measurement of surface photovoltage. This required a metal electrode to be brought up close to it and caused to vibrate in a direction normal to the surface, while shining a light on the semiconductor surface. Photons absorbed close to the surface generated electrons and holes which altered the amount of surface charge and caused a change in the voltage between the metal and the semiconductor – this could be measured with the vibrating probe. In effect, they were using the field effect as a measurement tool, rather than trying to make an amplifier with it. It was a clever and potent weapon to wield against the marauding surface states and immediately showed its worth by proving unequivo-cally that these states certainly did exist on both silicon and germanium surfaces. What was more, they were able to make reasonably accurate estimates of their density – and there were more than enough to explain the lack of an observable field effect. This was a wonderful step forward and, alone, put the Bell scientists well ahead of the international field. The next question was: could they be either removed or, in some way, neutralised?

But Bardeen was more immediately concerned to improve their understanding still further and encouraged Brattain to repeat his experi-ments over a range of temperatures (a frequently used technique to test new theoretical ideas). Brattain was only too willing and proceeded to cool his test gear in a Dewar flask. However, the experiments promptly ran into trouble due to moisture condensing on the sample surface, giving a series of spurious readings. Brattain knew very well what he should do to eliminate such effects – it would be necessary to revamp the equipment so as to include the measuring head in a vacuum enclosure, thereby excluding the moist air which was causing the difficulty. The problem was that this would probably take several weeks to engineer – and Brattain was in a hurry! He decided to compromise. If he were to immerse the sample, with its vibrating metal plate, in a suitable liquid, this would also have the effect of excluding the air and providing a stable environment – and it would be very much quicker. It was bad practice, in the sense that it introduced an unknown variable into the experiment but he was prepared to live with that – he was, in any case, well aware of what he was doing. He first tried several organic liquids, such as ethyl alcohol, acetone and toluene and, to his surprise, observed significantly larger photovoltages than before but the major surprise came when he used distilled water. The voltage shot up to something like a volt, comparable, in fact, to the effect seen in Ohl's famous p–n junction photocell. This was certainly dramatic but even more exciting results followed. The group electro-chemist, Robert Gibney realised that the hydrogen and oxygen ions in the water were modifying the amount of charge on the semiconductor surface and proposed that Brattain should try changing the voltage applied to his vibrating probe. This had an immediate effect – it was possible to enhance or reduce the size of the photo-effect, depending on the direction of this change, and, to everyone's amazement, it could even be reduced to zero. In this condition, the system was behaving as though there were no charge in the surface states at all! It clearly opened the way to the demonstration of the long-sought field effect. Brattain's risky throw of the experimental dice had produced a spectacular success.

All that was needed now was an electrical contact to the semiconductor surface in the region immediately beneath the water and this they very neatly achieved by using a small water drop on the silicon surface, through which a metal point was lowered onto the surface, while being insulated from the water by means of a layer of wax. A contact to the water drop was made by a second wire which could be used to control the surface state charge, as in the earlier experiment. This electrode acted as the gate, while the metal point served as the drain contact, a source electrode being made on the back of the sample. Hey presto, it worked! To their great delight, they achieved net gain with a silicon device, then obtained even better results with germanium. There was, however, a problem – gain was limited to frequencies below 10 Hz, on account of the very low mobility of the ions in water. To understand this, we need to appreciate that, when an alternating signal voltage was applied to the gate (the water), any change in semiconductor surface state charge had to wait whilst water ions moved through the water to the semiconductor surface. Because the experiment was concerned with heavy ions, rather than electrons, the response was extremely sluggish.

The next step was obvious – the liquid had to be replaced by some other medium which would, hopefully, have the same property of modulating the semiconductor surface charge but which would not rely on ionic conduction to supply charge to the interface. Because germanium had shown much greater gain in their earlier experiments, they decided to stay with it and to use germanium oxide as the liquid-replacement. Once again Gibney was called in – he oxidised the germanium surface, producing a thin layer of oxide through which the insulated metal point (the drain contact) could be pressed, to make contact with the germanium below. Again the source contact was on the back of the germanium, while the gate contact was in the form of an evaporated gold dot on top of the oxide. Success was immediate – the device gave gain at frequencies up to about 10 kHz – utterly vindicating their arguments. But it was not the complete answer, because, though there was voltage gain, there was no power gain, a consequence, Brattain realised, of the device geometry, the gate electrode being too large. Bardeen made an estimate that the separation between the two contacts should be no more than 50 µm (about the thickness of one of the pages of this book!), so Brattain set about redesigning the electrode structure. He achieved the very small separation by covering a plastic wedge with gold foil which he then slit along its apex with a razor blade, bringing the wedge down onto the germanium surface with pressure from a crude spring, fashioned from a paper clip (see Fig. 3.2). At last, this wonderfully Heath Robinson (or, since it was clearly American, Rube Goldberg) arrangement gave both voltage and power gain at a frequency of 1 kHz, while, feeding some of the output back to the input, produced an oscillator, the final proof that they had, at last, made a solid-state amplifier. Celebrations were very much in order. It had taken less than 2 years of co-ordinated effort to go from complete puzzlement to working device, a hugely creditable outcome and a wonderful Christmas present for the Bell management which had put its faith in the virtues of fundamental research.

Two minor qualifications appeared to cloud the euphoria, one purely technical, and easily solved, one of a personal nature, and much less tractable. Considering the technical problem first, we come to recognise yet another amazing piece of serendipity. When Brattain was preparing his germanium slice for contacting, he had thoroughly washed it and, quite unintentionally, dissolved the oxide film which Gibney had so carefully prepared on its surface. The two gold electrodes, therefore, made contact

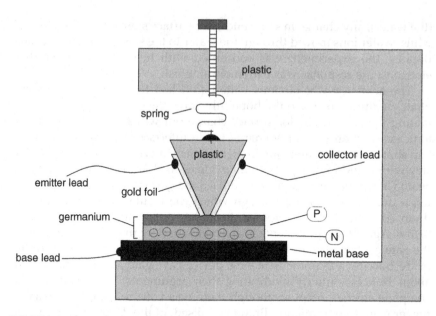

FIG. 3.2 A schematic reconstruction of Brattain's paper clip, point contact technology which he used to demonstrate transistor action for the first time. The plastic wedge was covered with metal foil, cut through with a razor blade to effect the emitter and collector contacts and pressed down onto a germanium crystal by means of the paper clip spring. Courtesy of Wikimedia.

directly with the germanium and there was no oxide film to control the amount of charge in surface states. Clearly, the device was functioning in a different mode from the field-effect mechanism which had been antici-pated. Brattain already had a clue to this, having noticed that the sign of the voltages needed to obtain the desired performance on germanium was opposite from that which worked for silicon. It took some discussion and a fair amount of clarifying experimentation to prove that the surface of the n-type germanium was 'inverted' – a thin layer of holes was present just below the surface. We shall discuss this further in a minute but, for the moment, let us simply accept it and consider the consequences. Applica-tion of a positive voltage to the gate caused these holes to be pushed towards the drain contact while a negative voltage on the drain resulted in their being swept into the drain circuit. Varying the gate voltage there-fore produced a variation in drain current in much the same way that varying the grid voltage on a thermionic valve varied the anode current. The essential point was that the device depended on the existence of an inversion layer at the germanium surface and this layer resulted from the presence of surface states, as had been predicted in Bardeen's original theory, its precise nature depending on the details of surface treatment –

etching, anodising, abraiding, etc. It was now clear that the point-contact transistor did not depend for its operation on the field effect but on the presence of this inversion layer and, once the group recognised this, they decided on a completely new set of labels for the various electrodes – the point contacts became known respectively as emitter and collector, while the back contact on the bulk germanium was called the base (the emitter injected holes into the base region, while the collector collected them from the base) and such terminology has been used ever since.

The personal problem concerned the relationship between Shockley, on the one hand, and Brattain and Bardeen, on the other. As we noted earlier, Shockley had taken the opportunity to further his interests in one or two very different solid state topics but, none-the-less, as leader of the Group, he tended to assume responsibility for the Group's success. It was he who took charge of demonstrating the new device and took prime position in the numerous photographs taken of the three of them. In one famous example, he is shown peering into a microscope, when, in practice, this had clearly been Brattain's role, and Brattain could hardly be expected to relish being sidelined in this manner by his boss's craving for the limelight. The sad truth was that Shockley, pleased though he was at the team's success, was secretly frustrated that it had been achieved without his involvement. Of course, his original suggestion was that they should explore the concept of a field-effect device which had determined the direction of the research but, not only was he unable to patent the idea, it turned out, ironically, that the transistor which had emerged depended on a quite different principle, that of 'minority carrier' flow. Shockley's burning personal ambition was yet to have even greater influence on the future direction of semiconductor device research but, for the moment, we shall leave him fuming quietly to himself, whilst we look at the concept of minority carriers in semiconductors and the manner in which they can affect device behaviour.

In Chapter 1, we saw how semiconductors could be doped with donors to generate controlled densities of free electrons (or with acceptors to generate positive holes) and we assumed that, in an n-type material, the density of holes was zero. While this is very nearly true, the hole density is actually just finite. Thermodynamic arguments show that the product of the electron and hole densities $(n \times p)$ must equal the square of the intrinsic free carrier density n_i^2 and we can use this relationship to estimate the density of holes in an n-type sample. For example, in silicon at room temperature, the intrinsic free carrier density is about $4 \times 10^{15} \, \text{m}^{-3}$ so, if a silicon crystal is doped with $4 \times 10^{22} \, \text{m}^{-3}$ donors, thus generating $4 \times 10^{22} \, \text{m}^{-3}$ free electrons in the conduction band, it is easy to see that it will also contain $4 \times 10^{8} \, \text{m}^{-3}$ positive holes in the valence band. As we said, this is a very small density, but, nevertheless, finite. For obvious reasons, these free holes are referred to as minority carriers and, in many cases, it is entirely safe to forget about them, but there are special circumstances in which they become important – the transistor being a good example. When the surface of the semiconductor contains charge in surface states, this sets up an electric field which, if it has the appropriate sign, may attract more minority carriers to the surface. The above relationship between free electron and free hole densities still applies, so a surface layer exists in which the minority carrier density is greater than in the bulk, while the electron density is correspondingly lower. In an extreme case, such as we met in the germanium transistor, above, it may happen that the surface region may actually have more minorities than majorities – the surface is said to have been inverted. In other words, for the case of the n-type germanium, used by the Bell team, the surface was actually p-type and, in the absence of any applied voltage, this represented a perfectly acceptable thermodynamic equilibrium.

Another interesting point emerges from an understanding of the inversion layer under the emitter contact and the resulting flow of minority holes to the collector. Such an inversion of the semiconductor surface occurs more easily in semiconductor materials having a small band gap and germanium, with its smaller gap, happened to be an ideal choice from the viewpoint of achieving an inversion layer, though it was chosen for those particular experiments largely because Brattain had a high quality sample in his drawer and this probably followed from the fact that germanium melts at a lower temperature than silicon and is chemically less reactive – it is therefore easier to purify. Once again, we recognise how lucky Brattain was – firstly, he used a liquid gate to circumvent an experimental inconvenience and thereby discovered how to control the occupation of surface states; secondly, he accidentally removed the oxide layer from the germanium surface and thereby obtained a direct contact between the emitter probe and the inversion layer; and, thirdly, he chose germanium, rather than silicon for these crucial experiments, which made the inversion layer possible. If this sounds like sour grapes, let me make clear my admiration for the wonderful fusion of theoretical and experimental skills demonstrated by Bardeen and Brattain – success in such an enterprise frequently depends on riding one's luck and appreciating the point of an experimental result, even when it might be totally unexpected. This they certainly did, learning immediately from each new discovery and building on it to design the next move. Far be it from me (or, I suspect, anyone) to denigrate their achievement but it is salutary to analyse the manner in which progress is actually made (rather than the way it may have been intended) – we can often learn much more from an honest appraisal.

One final step was necessary to turn the somewhat ramshackle experimental set-up into a practical commercial device – and that proved far from trivial. Brattain's razor blade technology had to be replaced by a carefully controlled positioning of a pair of true point contacts, sealed in an appropriate encapsulation, resulting in the tiny, three-legged capsule which became familiar to millions of professional and amateur electronic enthusiasts during the 1950s and 1960s. The legs were actually fine wires, projecting from the capsule, enabling the circuit engineer to make connection to the emitter, base and collector terminals, respectively, in much the same way as was done in the case of a triode valve, but it was necessary to learn new circuit techniques, as the impedances and voltage levels of the transistor differed considerably from those of the valve. Bell hastened the international spread of the new solid state amplifier by offering licences to anyone prepared to pay the necessary fee but it should not altogether surprise us to realise that the take-up was cautious, rather than wildly enthusiastic. Undoubtedly, the new device was very much smaller than the valve, it worked with much smaller, and more convenient voltages, it turned on more or less instantaneously (the valve needed time for the filament to heat up before it could function properly) and, not having a hot filament, it was more efficient. Against these advantages, would-be users were faced with redesigning their circuits and they were obliged to cope with a device whose electrical characteristics varied somewhat alarmingly and which was undeniably noisy in operation (the wanted signal had to compete with an array of spurious background signals generated within the device itself). It was also considerably more expensive. Hearing aid manufacturers were certainly interested but the rush of interest remained just that – interest!

In this context, we meet a well known characteristic of any new invention which is seen as a replacement for an established device – in this case the triode valve. The very people who were given the task of developing it were those who already had the 'appropriate' experience, and they were the people who were most reluctant to see its benefits. After all, if successful, it bid fair to put them out of their jobs! They also felt vulnerable on account of their lack of understanding of the new device which came from a completely different scientific stable from the vacuum environment in which they felt comfortable. Being experts in the detailed application of the valve, they naturally tended to try to use transistors as direct replacements and to see any lack of success as a weakness of the upstart device, rather than of the inappropriate circuit. Early transistors had defects enough, without having to contend with unsympathetic applications engineers. At all events, progress was disappointingly slow, though, when one appreciates all these factors, understandably so. It was perhaps significant that the most rapid progress was made by small, new companies who had none of the baggage carried by the large firms and who were therefore much quicker to realise the importance of designing customised transistor circuits, rather than simply trying to modify well known valve circuitry.

We learned enough about point contact technology in Chapter 2 to realise that any device dependent upon it could expect teething troubles, and we should remember that, in this case, not only were there two such contacts, which doubled the chance of failure, but they had to be placed extremely close together on the germanium surface, whilst their separation was critical to transistor performance. Hardly surprisingly, Bell found it far from easy to engineer a reliable commercial version and it was not until 1952 that the point contact germanium transistor was first introduced into telephone switching equipment. This and a small number of military applications seem to sum up its commercial impact. Gradually, the device resulting from Bardeen and Brattain's sterling research efforts came to be seen as the proof of an existence theorem, rather than the prototype of a commercially successful amplifier, a judgement that may seem unfeeling and cruel but one which, in view of subsequent developments, has an undeniable ring of truth. That this came on top of a restructuring of the solid state personnel had the most unfortunate effect of leaving the two principal inventors of the transistor out in the cold and only emphasised the growing split with Shockley, to which we have already referred. The time has come to return to this aspect of the story.

In the weeks following the first demonstration of transistor action, Shockley was desperate to come up with some contribution which could clearly be seen as his, alone. He needed to be recognised as the man who was personally responsible for the future of solid state electronics in a way that was never likely to emerge from the point contact device work. He was convinced that Brattain and Bardeen's transistor could never be satisfactorily engineered into a commercial device (in this he was probably right) and he had an idea for an alternative structure which might overcome some of its major disadvantages. He shut himself away from his colleagues and worked feverishly to develop it into a plausible amplifier, keeping it very much to himself until he felt confident enough to reveal it to them as a fait accompli – a Shockley fait accompli. Once again, his ideas were based on a kind of solid state version of the triode valve, consisting of three semiconductor layers of alternating conductivity type; for example, p–n–p. The central n-layer would act like a triode grid to control current flowing between the p-type emitter and collector layers and he gradually realised that such a structure involved the flow of minority carriers (holes in this case) through the base region in the middle. This was a little like the case of the point contact transistor but here the minorities flowed through the bulk of the base layer, rather than along the semiconductor surface. An obvious advantage, of course, was the absence of any metal point contacts and it therefore promised to be easier to fabricate a reliable commercial version.

In the event, Shockley was forced into a somewhat premature revelation by developments in the laboratory. Bell management had been quick to draft new staff into the Solid State Group to build on the startling breakthrough already made, and there had been considerable discussion within the Group concerning the precise mechanism whereby minority carriers made their way from emitter to collector in the Brattain structure, in particular, whether the holes flowed in the bulk of the germanium or only along its surface. The controversy was abruptly concluded by a clever experiment conducted by a new member of the Group, John Shive. Shive made a sample of germanium in the form of a thin wedge, placed point contacts on either side of it and obtained transistor action through the bulk of the material. Clearly, holes could flow through the bulk and, when Shive presented his results at a Group meeting, Shockley was obliged to describe his revolutionary ideas for a new transistor structure, in case the experiment triggered a similar idea in Bardeen's fertile brain. Needless to say, Bardeen and Brattain were taken aback by this account of ideas which they might have expected to be privy to – the chasm between them and Shockley was widening rapidly. It was further emphasised by a management decision to ask Bardeen and Brattain to continue with their surface studies, while Shockley was given some new assistants to help him develop his three-layer transistor. Feeling unfairly sidelined, the pair soldiered on but Bardeen eventually decided to leave semiconductor physics behind and concentrate his interest on the theory of superconductivity. Sadly, even this didn't work for him and in 1951 he accepted an offer from the University of Illinois and left Bell Labs altogether. It was at Illinois that he made an important contribution to the understanding of superconductivity – in the famous BCS theory, B stands for Bardeen and he, together with Leon Cooper and John Shrieffer, was awarded the 1972 Nobel Prize for this work. It was a bizarre coincidence that BCS theory finally came together in 1956, the same year that Bardeen (together with Brattain and Shockley) went to Stockholm to receive his award for inventing the transistor. The Murray Hill laboratory was clearly the loser in this particular transfer of loyalties but the world's understanding of superconductivity certainly benefited greatly. It is, as they say, an ill wind that brings nobody any good.

Returning to Shockley, it was now clear that the future of solid state electronics lay firmly in his scientifically very capable hands and he was determined to seize his opportunity with both of them. It was already accepted within the Group that minority carriers could diffuse through bulk semiconductor material and this appeared to be more than a little relevant to the operation of his (still hypothetical) three layer (p–n–p) device. It gradually became clear that this should be seen as a pair of p–n junctions, separated by an intermediate base layer and the new team concentrated on understanding the function of each. We have already met the p–n junction as a photocell but, in Shockley's device, it was the way in which current depended on applied voltage (the so-called current-voltage characteristic) which was of greater significance. We need to think now of the fact that a p–n junction can act as a rectifier – that is, in one direction current flows easily, while in the opposite direction, the junction presents a very high resistance to current flow – and examine the way in which this comes about.

Firstly, we consider a forward biased p–n junction, where the p-side of the junction is made positive with respect to the n-side. In this case, the applied voltage has the effect of reducing the barrier to current flow, making this the easy direction. Positive holes are pushed across the junction into the n-region, while negative electrons are drawn from the n-side into the p-side. In other words, the junction acts as an injector of minority carriers. If we now imagine a junction in which the p-side is heavily doped and the n-side relatively lightly doped (often written as a p^+–n junction), this results, under forward bias, in current which is carried almost entirely by holes – a large density of minority holes flow into the n-side while only very few electrons are injected into the p-side. Though the mechanism is quite different, it mimics the action of the emitter point contact in Brattain's transistor but with one important difference. Whereas in the point-contact device the holes are flowing in a p-type inversion layer where there are very few electrons present, here, the injected minorities diffuse away from the junction, through an n-type region containing a relatively high density of electrons. Inevitably, therefore, they recombine with electrons as they cross the base and, if the base region is relatively thick, very few holes will actually reach the collector. Once again emitter and collector must be close together if the device is to provide useful gain, though for a somewhat different reason. This, then, would form the emitter junction in Shockley's transistor, but what of the collector?

The collector consists of an n–p junction, lightly doped on both sides. A reverse bias (making the p-side negative with respect to the n-side) has the effect of increasing the barrier to current flow and actually builds up a high electric field in the junction region. Such a junction acts, not as an injector of minority carriers, but as a collector, the high field sweeping them from the n-region into the p-region. The combination of a p^+–n emitter and an n–p collector acts in a very similar way to a point contact transistor but does it while avoiding the complications associated with those problematic points. (There is still a need to make metal contacts to each of the three regions – the p^+ emitter, the n base and the p collector – but it turns out to be much easier to do this reliably.) Finally, we note one further feature of the collector – because it represents a high barrier to current flow, it acts like a very high electrical resistance. This means that it is possible to place a moderately high resistance in series with the collector (and through which the collector current flows) without reducing collector current significantly. Thus, when a fluctuating signal voltage is applied between emitter and base, (the input voltage) it results in a corresponding fluctuation of collector current and, since this current flows through the series resistor, it produces a relatively large voltage across it. This is the output voltage, which may be as much as 100 times the amplitude of the input voltage.

This, then, was Shockley's transistor and, because it was essentially made from p–n junctions, it became known as the junction transistor, to distinguish it from the point contact device. As we shall see, it was the junction transistor which eventually took the electronic industry by storm and in this sense Shockley has an unassailable claim to being the father of modern electronics, his doubts about the original point contact transistor being very well founded. It was, perhaps, unfortunate that he should have gone about it in the way he did but there can be no doubt of his excellent scientific insight, driven, as it was by an intense personal ambition. Once the idea was clarified, all that remained was to make it! His first instinct was to use a sequence of evaporated layers, each doped appropriately, but this was never going to be successful – as explained earlier, such material is polycrystalline and suffers from having very low electron and hole mobilities. (One unavoidable consequence is that minority holes would diffuse only slowly across the base region and most of them would recombine before reaching the collector.) Something much more sophisticated was required and many members of the new Group would be involved before a successful device eventually emerged.

Perhaps the first contribution of note was an experiment undertaken by Richard Haynes to measure the average lifetime of minority holes in n-type germanium. He took a sample of high back voltage germanium and cut from it a thin filament, only a few hundred microns thick, then formed a series of emitter contacts along the length and a collector contact at one end. The experiment consisted of measuring the time taken for a pulse of injected holes to drift along the filament and the manner of its decay along its length. It was possible, from this, to derive values for both the mobility and the average lifetime of the holes (i.e. how long they survived, before recombining with the sea of electrons). Typical lifetimes were found to be of order 10 µs. The technology Haynes developed for making his filaments was then seen as eminently adaptable to making a transistor and, by August 1948, the Group was able to demonstrate a filamentary transistor using only one point contact. Spurred on by strong expressions of interest from the U.S. military, they finally eliminated both points and produced the first junction transistor early in 1951. This success, however, owed much to several new names – improved material quality resulted from the work of Gordon Teal and William Pfann, junction formation from that of Morgan Sparks, filament technology Gerald Pearson and the addition of ultra-fine contact wires, again, William Pfann. We need to look in detail at some of these contributions.

The most remarkable development was undoubtedly that of single crystal growth by Teal, if only because it started as a personal bit of research, done in his spare time (often at night) and completely unauthorised by Bell management! Teal was convinced that semiconductor devices should be fabricated from single crystal material, containing no grain boundaries to scatter electrons or holes but, being unable to convince Shockley or any of the others, he was obliged to pursue his ideas on his own. After some time, he was given a minimal amount of financial support to build a crystal puller (explanation to follow) and used it to grow germanium crystals of a quality that even Shockley had to agree was far superior to anything previously available. Teal took as his model a growth technique originally described by an obscure Polish scientist Jan Czochralski in 1916, as part of a study of the crystallisation of metals. The idea was to melt germanium in a graphite crucible by RF heating, dip a small seed crystal into the melt and slowly withdraw it (the whole apparatus being surrounded by hydrogen gas contained in a quartz vessel). The seed grew into a large single crystal, from which a great many suitable filaments might be cut. A fortunate aspect of the method concerned a phenomenon known as impurity segregation – most of the impurity atoms in the germanium remained in the liquid, thus producing crystals, significantly purer than the starting material. Then, by recycling some of the grown crystals, even greater purity was achieved, resulting in minority carrier lifetimes as long as 100 μs, 10 times longer than Haynes had measured on high-back-voltage material. Here was proof, indeed, of the benefit to be obtained from careful attention to material quality.

What was even more exciting, crystals containing p–n junctions could be produced by doping the germanium ingot during growth. Sparks worked alongside Teal to demonstrate this. They grew an n-type crystal for a certain time then Sparks dropped a small pill of germanium doped with gallium (a group III acceptor) into the melt, so that the crystal continued to grow with p-type conductivity. The current-voltage characteristics of the resulting junction were close to those predicted theoretically by Shockley – a great boost for the crystal growers but also for Shockley, who became an enthusiastic supporter of the new approach. It was an obvious further step to add a second doping pill to counter-dope the later stages of the growth, thereby making it n-type again, and thus forming an n–p–n structure – a junction transistor, which showed power gain at the first time of asking. In the first instance, the base region was some 500 μm wide, which limited the frequency response of the device to rather modest values – a long way short of those then being achieved with point contact transistors – but Sparks was able to speed up the counter-doping stage and increase this to a respectable 1 MHz. (The base region was now down to about 50 μm wide and it required a touch of genius from Pfann to attach very fine gold wires to it.) The real good news, however, was that these filamentary junction transistors were very much quieter than their point contact predecessors and now looked capable of challenging the thermionic valve in many applications. This was March 1951. Just over 3 years since the initial breakthrough by Bardeen and Brattain, Shockley's bloody-mindedness was finally beginning to pay dividends. It is important to appreciate that the attainment of operating frequencies in the region of 1 MHz was significant because this approximated to the well-known Medium Wave Radio Band between 0.53 MHz and 1.61 MHz. The ubiquitous transistor radio (see Fig. 3.3) followed soon afterwards.

However, Bell did not have things all their own way. The policy of selling licences meant that quite a number of rivals came rapidly up to

FIG. 3.3 A typical transistor radio from the 1950s. The first such radio, the Regency, using TI germanium transistors, appeared in 1954 and was followed by a host of competitors from all over the world. The name 'Sony' was first used in connection with that company's first pocket radio. Courtesy of iStockphoto, © Dan Herrick, image 2565026.

steam (in research, at least) and the next important step forward came from the (American) General Electric Company, in the person of John Saby, a member of research staff at their Schenectady laboratory. In fact, his announcement came slightly before Bell revealed their success with grown-junction devices and, in a sense, it rather took the wind out of the Bell sails. Saby had demonstrated a new method of making junction transistors based on an alloying technique which proved quite considerably easier in manufacture. The idea was to take a thin slice of n-type germanium and place two pellets of the group III metal indium, one either side of it, then alloy these into the germanium by heating to about 500 °C. The indium diffused towards the centre of the slice, making both sides p-type, while leaving a thin n-type base region in the middle, thus forming a p–n–p transistor. Control of the base thickness was far from easy but this structure had the advantage that it was far easier to attach a wire contact to the base than was possible with the Bell structure. The technique was rapidly taken up by RCA and by Philco, who further improved it, using a carefully controlled etch to thin the germanium from both sides before alloying in the indium. Operating frequencies up to 10 MHz were achieved and the way was clear for possible application to the RF end of a standard radio receiver – the transistor radio was now a technical possibility, all that was needed was someone bold enough to take the commercial plunge.

More of this in a moment but other applications for the new pretender were coming into prominence. In 1952, Bell were already starting to use point contact transistors in telephone switching circuits, where their noisy performance was less of a disadvantage. As with any digital technique, it was only necessary to distinguish between on and off states, much easier to do than to register the precise signal amplitude required in analogue methods. Spurious noise signals might cause serious interference in analogue circuits, while being no more than a minor nuisance in digital equipment and it needed only modest insight to appreciate that transistors might have a major commercial future in computing. The digital computer and digital processing of information were very much in their infancies in the 1950s but the principles were already well established. Digital computers were already in existence and, while this is certainly not the place for a detailed survey of computer history, we might usefully mention one or two early examples. The Atanasoff-Berry computer at Iowa State University was built in 1941, using 280 valves. It used digital signal processing and was fully electronic in operation, though it was not strictly programmable. As is now widely known, Colossus (probably the first programmable, all-electronic digital computer) played an important part in the deciphering of Nazi secret codes at Bletchley Park during the

Second World War. The first such machines were delivered in 1943, and, by the end of the war there were 10 of them in use, each one employing 2400 valves. (A transistorised descendant known as ERNIE – Electronic Random Number Indicator Equipment- was later used to select winners in Prime Minister Harold McMillan's Premium Bond scheme.) In 1946, the ENIAC (electronic numerical integrator and calculator) computer at Penn State University contained 18,000 valves and dissipated 150 kW! Other computers came to light in the universities of Manchester and Cambridge (ca. 1949), the Cambridge EDSAC (electronic digital storage automatic calculator) apparently being a serious rival to ENIAC. Two salient points are of concern to us. First of all, these machines used digital techniques and this meant that their circuitry had to deal with short pulses of current, which, in turn, implied that they had to respond at high speed. For example, the handling of audio signals using standard analogue methods requires circuits to operate at frequencies up to about 10 kHz, while performing the same function digitally demands frequencies up to about 10 MHz – a 1000 times faster. Digital computers operated faster still. Valves could operate at these speeds but transistors were obviously close to their limit. On the other hand, valves were close to their limit when it came to heat dissipation and reliability. There was clearly a need for a solid state device having low power dissipation, which did not require frequent replacement as a result of failure, but it would have to be capable of working at high frequencies. Though progress in transistor technology had been impressive, it was clear that there was more still to be done.

The US military, in particular, was keen to see rapid developments in this direction and, importantly, had money to invest. Not only did they see important applications in digital processing, such as, for example, that needed to display readily interpretable radar pictures, but there was an obvious requirement for solid state devices in missile guidance systems. Here, it was not so much a question of the speed of operation as of the speed of flight. Accelerating thermionic valves, having fragile filaments, to high velocities led to many filament failures, with consequent wastage of expensive missiles – transistors might be more expensive than valves but that would be a small price to pay for saving even a single missile. Readily available military money was just the spur to which the fledgling solid state electronics industry was happy to respond and this injection of funds had a major influence on its development, an important aspect being the growth of effective competition. Large, well established electronics companies such as Raytheon, RCA, Philco and Sylvania were galvanised into action by the thought of juicy contracts which took much of the risk out of their accompanying commercial ventures. Nor was Bell averse to accepting government money – it has been suggested that, in the period 1953–1955, nearly half of Bell's R&D funding came from government sources – and in 1954, their engineers built the first all-solid-state digital computer, TRADIC (transistorised digital computer) on a military contract. It was based on germanium point contact transistors, 700 of them, together with over 10,000 germanium diodes. This was in marked contrast to the situation in Europe and Japan. Neither Germany nor Japan was allowed to rearm and other European nations were in no financial state to contemplate such profligacy. There can be little doubt that military funding stimulated transistor development in the USA through the whole of the 1950s – it was further stimulated in 1957 by the Russian launch of the first Sputnik and kept very much alive by President Kennedy's man-on-the-moon programme throughout the 1960s – the rest of the world being left to play catch-up.

It is worth emphasising the use of germanium diodes here because they actually dominated digital computer technology throughout the 1950s. For example, the Minuteman Missile programme used diode, rather than transistor logic and germanium, rather than silicon diodes. Transistors may have been fairly readily available at this point but logic circuitry could equally well be designed using diodes, and diodes were much cheaper and more reliable. In 1955, diode sales were almost 10 times greater than those of transistors and, even in 1959, they still exceeded transistor sales by about 50%. It was only when the integrated circuit came on the scene in the 1960s that the use of diode logic finally came to an end – and, as we shall see, that had much to do with the development of the field-effect transistor.

But, we must leave the heady world of international finance and return to the technical battle being waged at the research bench. The year 1952 was significant for a number of reasons, not least because Gordon Teal, one of Bell Labs' most go-ahead materials scientists relocated to Texas in the interests of his wife's health. In doing so, he joined a relatively small and unheard-of outfit known as Texas Instruments, where he found greater freedom to pursue his personal whims. In particular, he was determined to press ahead with Czochralski crystal growth of silicon, being convinced that it, rather than germanium, was the material of the future. But silicon presented far harder materials problems than germanium had done – it possessed a significantly higher melting point (1412 °C as against 937 °C) and was more highly reactive, tending to pick up impurities from anything it touched, when in the molten state. There clearly had to be a reason why Teal was prepared to tackle them and this lay in germanium's relatively poor thermal behaviour. Because of its smaller band gap (0.67 eV, compared with 1.12 eV for silicon) it suffered from a phenomenon known as thermal runaway. When germanium devices operated in particularly warm environments, the additional heat generated by their electronic functioning in turn generated additional (intrinsic) electron-hole pairs in the way described in Chapter 1. These extra free carriers resulted in increased current flow, which generated more heat, which generated yet more carriers, which – caused the device to overheat and cease to function. A glance at Table 1.2 in Chapter 1 shows that, at room temperature, the intrinsic free carrier density in silicon is nearly four orders of magnitude smaller than that of germanium, implying that silicon is sensibly free from runaway problems, a feature which, incidentally, appealed particularly strongly to our friends the U.S. military – they with the relaxed purse strings – much military equipment being called upon to function in extreme climatic conditions. Teal was not alone in recognising this situation but he was more than averagely determined to do something about it. Having set up a successful silicon crystal puller at Murray Hill, he was now about to repeat the trick in Dallas.

Those Teal left behind at Murray Hill also made significant progress in 1952. Even further improvement in germanium purity was achieved by Pfann, using a sophisticated application of the impurity segregation effect. This employed an apparatus called a zone refiner, in which a number of thin molten zones were moved slowly along a germanium ingot, contained in a horizontal boat. After several passes, the bulk of the ingot was highly purified while most of the impurity atoms had been swept to the end, which could then be discarded. Impurity levels in such material reached unbelievably minute values, as low as one impurity atom to 10 billion (10^{10}) germanium atoms and minority carrier lifetimes were correspondingly improved – values as long as 1 ms could be achieved routinely. The technique, conceived by Pfann in 1950 and applied by him to purifying both germanium and silicon, has since been applied quite widely to the purification of metals. Combined with a similar process known as zone levelling, which allows a desired impurity (a dopant) to be spread uniformly throughout an ingot, it revolutionised material preparation and established standards for the whole future of semiconductor technology. We shall see later that it also made a huge impact on semiconductor physics, finally allowing high quality measurements to be made on material of known provenance. Gone were the days when learned scientists could warn their colleagues off semiconductor physics as an area of professional suicide. We shall say something about semiconductor physics towards the end of this chapter – for the moment it must suffice to note the major contribution made by Pfann, who joined Bell as a technical assistant without a degree but who proved beyond doubt that paper qualifications are not always essential to career success.

Hard work and dedication are also necessary, as well illustrated by Teal's progress at TI. It took him until about the middle of 1954 to overcome silicon's reluctance to cooperate but he was then able to announce to an incredulous world that TI had silicon grown junction transistors in production. It was a significant turning point – from that moment on, germanium's future as first choice was numbered – it was a breakthrough which was to have serious ramifications for numerous competitors. Philips, in Europe, for example, suffered badly from having invested heavily in germanium, as did companies in Japan. But, TI had clearly surprised their American competitors, too. The large, established companies were slow to react to new ideas in the way that a small, start-up outfit like Texas was able to do. And this was not the only surprise sprung by TI. 1954 was the year of the transistor radio, in the shape of the Regency TR1 radio which came onto the market in October, sporting four germanium grown-junction transistors, supplied by TI. It was something of a gamble from start to finish and probably made very little money for anyone but it certainly started a trend and gave TI a head start in the business of supplying transistors to many of the big manufacturers. They were destined not to remain small and start-up for very long.

Back at Murray Hill, the Bell scientists (with or without PhDs) had not stood still. 1954 also saw the introduction of another new transistor process technology – the double-diffused mesa structure. Bell first demonstrated it using germanium but followed up with a silicon version in the following year. The basic idea was to place the semiconductor sample in a heated quartz tube, through which flowed suitable dopant gases. Gas atoms, striking the semiconductor surface, may be incorporated into the solid, then diffuse into the bulk, doping the semiconductor either n-type or p-type according to whether they are from group V or group III of the periodic table. The trick used to make a transistor structure depended on the differing diffusion rates of different dopants. For example, to make an n–p–n silicon transistor, a lightly n-doped silicon slice was simultaneously diffused with two dopants, aluminium acceptors and antimony donors but the lighter aluminium atoms diffused faster than the heavier antimony, forming the p-type base, while the antimony donors, present in greater quantity, formed the n-type emitter. The cross-sectional area of the device was then defined by etching a mesa shape, that is, a flat-topped, steep-sided bump, much like similar naturally occurring land masses, also known as mesas (from the Spanish or Portuguese words for 'table-top'). The method was adopted in the interest of gaining better control over the base width, values less than 5 μm being routinely obtained, resulting in cut-off frequencies greater than 100 MHz, the magic figure which allowed transistors to be used in FM radio circuits. Bell standardised on this structure for several years though it had one fairly serious drawback – contact to the base region had to be made through the emitter and there was always a danger that the contact might short-circuit the emitter junction.

The gradually increasing emphasis on silicon as the semiconductor of choice also demanded an innovative step towards improving its purity. As we said earlier, one of the principal difficulties lay in silicon's chemical reactivity and its tendency to incorporate impurities from any container used to hold it when molten. There seemed no answer to this – molten silicon surely had to be held in some kind of container? Well, no! It was Henry Theuerer, who had worked with Gordon Teal, before he left Murray Hill, who came up with an ingenious solution, known as floating-zone refinement. A vertical silicon rod was held at both ends while a narrow molten zone was moved through its length, using a narrow heating coil. Being stabilised by surface tension, the molten region was never in contact with its surrounding quartz tubing, so the danger of contamination was minimised. Only a single zone could be supported so it was necessary to make several more passes than in Pfann's original horizontal process but this was no more than a minor inconvenience in pursuit of the highest possible purity. From this point onward, silicon could match germanium at every turn, while showing no sign of thermal runaway.

Any account of the early years of solid state electronics bumps frequently into the person of William Shockley. Having stimulated the discovery of transistor action in 1947, invented the junction transistor in 1948, seen it demonstrated in 1950, and been an influential figure behind almost all the new developments in the early 1950s, including the double-diffused transistor structure, Shockley had also published the first definitive book on the new field 'Electrons and Holes in Semiconductors'. It was a tour de force and written, amazingly, whilst he was deep in the birth throes of the junction transistor. It provided, for the first time, an ordered account of the interaction between quantum mechanics and semiconductor device performance, illustrating dramatically just how far man's understanding of these still new materials had progressed in the half-decade since the end of World War II. It also illustrated the vital symbiosis between basic research and device development which sat centrally in the Bell philosophy. In a foreword to the book, Ralph Bown, then Research Director at Murray Hill, commented: 'If there be any lingering doubts as to the wisdom of doing deeply fundamental research in an industrial research laboratory, this book should dissipate them'. The scientific world could recognise both Bell Research and William Shockley at the height of their powers. Bell was to remain there for decades but Shockley, alas, was not. Things started to go wrong for him in 1956, when his all-consuming restlessness persuaded him to leave Bell and set up his own company, Shockley Semiconductors on an industrial park near Stanford, California. The change of location coincided with another major change in Shockley's life – he took the opportunity to change his wife – but that is another story.

The new venture began with high spirits. Shockley used his reputation to recruit a body of young high-flying technocrats and Arnold Beckman, of Beckman Instruments guaranteed financial backing. It should have been a tremendous success. The fact that it wasn't must be laid largely at Shockley's door – his undoubted technical ability belied his lack of management skills and the tendency, already demonstrated at Murray Hill, to put his own hobby horses before the collective good eventually brought the whole edifice crashing down. Commercial success in such an enterprise depended on bringing a suitable product to market as quickly as possible, in the interest of establishing a financial return, and this demanded careful choice of a device which could be manufactured well within the technical skills and equipment then available. This was clearly seen as having top priority by most of the staff but not, alas, by Shockley himself. He preferred to concentrate on another of his own ideas, a four-layer p–n–p–n switching device which he believed should have a future in digital processing. The technical problem was that it was very difficult to make – the managerial problem was that it split the organisation right down the middle. Frustration grew rapidly and rebellion followed. Beckman was initially sympathetic to the rebels' cause but, in the event, found himself unable to deal firmly with Shockley. Being a technical man himself, perhaps he couldn't quite overcome his admiration for Shockley's genius. The result was that, on 18 September 1957, just 18 months after the official start-up, eight of the brightest members of staff resigned en bloc. It was a significant turning point in the history of the semiconductor industry – these 8 people promptly set up the Fairchild Semiconductor company which was to play a major role in the future of the industry. It was little more than a mile down the road from Shockley's enterprise and can be seen as the beginning of a rapidly expanding and fiercely competitive solid state electronics nursery, now known as Silicon Valley. The names Robert Noyce and Gordon Moore, in particular, would resonate down the annals of twentieth century electronics. Poor Shockley was to end up mired in controversy connected with the vexed question of race and intelligence, a story taken up in Joel Shurkin's book, 'Broken Genius'. I shall say no more here.

The trend may have been towards an ever widening take up of transistor electronics, both in America and in other parts of the world – Philips made its first transistor within a week of the first announcement from Bell in 1948 and its English subsidiary, Mullard was manufacturing germanium alloy transistors as early as 1955, so too were their rivals GEC, while another small start-up company in Tokyo produced its version of the transistor radio in this same year (the name Sony being invented specifically to sell this particular product!) – but Murray Hill was still the source of important new technologies. It was also in 1955 that Bell scientists made one of the most significant discoveries ever to hit the headlines – except that it was altogether a low key affair, stumbled upon more or less by accident and not immediately recognised as significant at all. This was the discovery that the surface of silicon could be oxidised to produce a highly stable and highly insulating film of silicon dioxide, or silica (SiO_2). It arose from experiments designed to diffuse appropriate dopant atoms into silicon, which had to be done at high temperatures (above 1000 °C) and in an atmosphere of flowing hydrogen. A chemist, Carl Frosch, accidentally let a small amount of oxygen into the hydrogen stream, forming water vapour, which oxidised the silicon surface, producing a smooth, hard, insoluble layer, which clung tenaciously to its silicon parent, giving it (by way of optical interference) a uniform green appearance. One's mind inevitably goes back to 1947, when Brattain formed a similar layer on germanium, only to find that it was water-soluble and washed off when he tried to clean it. Serendipity had struck again but in reverse.

It wasn't immediately obvious just how important this particular discovery was to be, 4 years passing before the question was finally answered, and answered not at Bell but at Fairchild. In fact, two vital properties of silicon oxide came to be used in semiconductor technology, one purely chemical the other electronic. Because the oxide was chemically stable and resistant to attack, it could be used as a mask to define regions that were to be diffused. One chemical which did attack it was the powerful (and dangerous!) acid, hydrofluoric acid (HF) and this could be used to dissolve a small area of oxide, leaving a hole through which dopant atoms could be diffused into the silicon below. This provided an alternative (and superior) method from mesa etching to define the active area of a diffused transistor. It was then realised that, having diffused one region of the silicon, the surface could be oxidised again and a second region doped with another dopant. Suppose, for instance, that the original silicon slice was lightly n-type and an initial area was diffused with aluminium or boron to make it p-type, then a second area could be defined *within the p-type region* and diffused with an n-type dopant such as phosphorus or arsenic. This second diffusion was arranged to penetrate to a depth slightly less than the p-type diffusion, thus forming an n–p–n transistor structure in which all the metal contacts could be made *on top of the silicon slice*. It was not until 1959 that this so-called planar process was announced and, by this time, the scientists involved had left Bell and were working for Fairchild Semiconductor. The Swiss physicist Jean Hoerni, one of the original eight, was the principal protagonist but Robert Noyce was also seriously involved. It gave Fairchild a significant advantage in the competitive business of making reliable and reproducible transistors and was to lead, as we shall see, to the planar integrated circuit, which has formed the basis of almost every new development in solid state electronics from that time forward.

The use of silicon oxide as a diffusion mask had one other advantage. When the dopant atoms were diffused into the silicon, they also diffused a small distance sideways underneath the oxide mask and this meant that the interface with the surface of the resulting p–n junction was covered by the oxide, which formed an ideal passivating layer. At last, the reliability problem that all transistor manufacturers were struggling to overcome was well and truly under control. No longer was device performance subject to the vagaries of random dust particles or minute amounts of condensation – the oxide film effectively excluded all such intervention. It was a wonderful example of one new 'stone' accounting for two highly exotic 'birds'. Incidentally, if I may be allowed to change the metaphor, it also hammered the final nail into the germanium coffin. If thermal runaway problems were not sufficient to bring it down, its lack of a stable oxide surely sealed its fate for all but a very few specialised applications. Germanium, having set the stage in dramatic style in 1947, was now about to take its final curtain call. (But see Chapter 10 for a possible reprieve!)

There was yet another property of the interface between silicon and its oxide, which was exploited by Bell Labs, and which was to prove of at least equal importance. It resulted, in 1960, in the long-sought field-effect transistor finally coming into being but we shall consider this in the next chapter. For the moment, it is worth spending a few moments looking at a very different aspect of this mad rush into man's electronic future. We saw earlier that the tremendous progress towards perfecting semiconductor materials not only made possible their control in complex device technology but also enabled the development of worthwhile semiconductor physics. The 1950s saw a remarkable surge in our understanding of semiconductor properties, so we shall conclude this chapter with a brief account of these developments, building on what we learned in Chapter 1. We shall look at two aspects: on the one hand, details of electrical conductivity and, on the other, some considerations of optical properties.

In a very general way, electrical conductivity depends on two parameters, those of free carrier density and free carrier mobility. The former is determined by either thermal excitation of carriers across the band gap (i.e. between valence and conduction bands) or by doping with impurity atoms, the latter by scattering of free carriers due to imperfections in the crystal lattice. A better understanding of both can be obtained by measuring electrical conductivity and the Hall effect as a function of temperature and this was a feature of post-war semiconductor physics. As we saw previously, the Hall effect gives us a measure of the free carrier concentration, together with carrier type – combining this with conductivity measurements yields, in addition, values for free carrier mobility (a measure of how easily the carriers move through the crystal). Looking first at free carrier density, it is easy to distinguish intrinsic from extrinsic densities by their dependence on temperature. The intrinsic density depends very strongly on temperature and on the band gap of the semiconductor concerned (0.67 eV for germanium, 1.12 eV for silicon). Knowing the band gap, it is straightforward to calculate the intrinsic density and its temperature dependence. On the other hand, the extrinsic density depends on the density of donor, or acceptor, impurity atoms, which, in high quality material can be accurately controlled in the growth process (or, possibly in a diffusion process). It also depends on temperature but very much less strongly than the intrinsic density. In this case, the relevant energy is the ionisation energy of the donors (or acceptors) which, in silicon and germanium tends to lie round about 50 meV (this is the energy required to excite an electron from a donor atom into the conduction band, or a hole from an acceptor into the valence band). At high temperatures, the extrinsic density saturates at a value equal to the donor (or acceptor) density, whereas the intrinsic density continues to increase rapidly. In this manner, the predictions of quantum theory were accurately confirmed for the first time, giving physicists much greater confidence to pursue more sophisticated investigations.

Two scattering mechanisms dominated the temperature-dependence of mobility in both these materials. At temperatures near room temperature, mobility is usually limited by the thermal vibration of semiconductor atoms, which effectively disturb their regular crystal lattice positions – an electron, for example, travelling through the crystal, suddenly bumps into a silicon atom because it is displaced from its nominal lattice point. As the temperature is raised, these lattice vibrations become larger, so the scattering effect increases – carrier mobility therefore decreases. As the temperature is lowered below room temperature, mobility increases but not indefinitely because, at low temperatures, it is limited by another mechanism known as ionised impurity scattering. When an electron leaves a donor atom, the atom is left with a positive electric charge and this ion interacts with free electrons by way of the resulting electric field. A theoretical calculation shows that the mobility should increase with increasing temperature and this was found experimentally. Overall, it means that mobility goes through a maximum value at some temperature in the region of 100 K (−170 °C), decreasing both above and below this temperature. Once again, detailed comparison between experiment and theory gave encouraging agreement.

A particular feature of free electron behaviour in a semiconductor is the rather surprising effect of the crystal lattice in enhancing mobility – remarkably, the electrons behave as though they were considerably lighter than their ordinary free-space mass, which means they are accelerated by an electric field more easily than would be expected. This is a peculiar quantum mechanical effect which cannot possibly be understood on a classical model. It originates in the idea that a free electron in a crystal belongs not to any particular atom but to the crystal as a whole and, in this sense, it is completely free to move anywhere in the lattice – almost as though it had zero mass. When theorists looked in more detail at the interaction between the electron and the lattice, they found that the zero-mass idea was not quite correct but, nevertheless, the effective mass should be significantly less than the free-space mass m_0 – typical values in silicon and germanium are in the region of 0.1 m_0–0.5 m_0 and, interestingly, they vary with the direction of motion with respect to the crystal lattice planes. Measurement of these effective masses, using a technique known as cyclotron resonance bore out such strange ideas and allowed theorists to calculate electron and hole mobilities for comparison with experimental data.

Following these weird concepts leads to another feature of semiconductor behaviour which concerns their optical properties. In Chapter 1, we pointed out that light with photon energy greater than the semiconductor band gap could be absorbed by the semiconductor, generating hole-electron pairs in the process but quantum mechanics suggests that this is not the whole story. In practice, materials should be divided into two classes, depending on the details of their conduction and valence bands. As was well understood in the 1950s, silicon and germanium both belong to the class of materials whose band gaps are indirect. This means that in order to absorb a photon, they must simultaneously absorb a quantum of lattice vibrational energy (called a phonon) and this makes the process less probable. (It depends on the likelihood of the two particles coming together at the instant of absorption.) Two practical consequences follow; absorption of light is correspondingly weaker and recombination of electrons and holes tends to generate heat, rather than light. This contrasts with the class of semiconductors which have direct energy gaps, such as gallium arsenide, gallium nitride, zinc selenide and many others. In these materials optical absorption is strong and recombination has a far greater probability of generating light, indeed, in favourable cases, light may be produced with efficiencies approaching one hundred percent (see Chapter 7).

The 15 years from 1945 to 1960 saw a quite dramatic change in both our understanding of semiconductor properties and of their application to solid state electronics. A whole new industry had developed, new companies had come into existence (one had already crashed!) and the whole ethos surrounding these materials had changed. How exactly had it come about? In a word – investment! By the end of the period, large numbers of scientists were working in the field and they cost money, money which companies obviously believed would earn an excellent return (though, as we have seen, military contracts also made a vital contribution). Several factors combined to stimulate these developments, in the first instance the determination of Bell to maintain its stranglehold on U.S. communications by offering more and better service to its customers, the recognition that basic research had an essential part to play in generating new products, the personal belief of Mervin Kelly that solid state devices could revolutionise communications and the concept of research as an organised activity with specific aims in view, rather than merely a blue skies activity. Clearly, the war had played a role in encouraging the idea that scientific research could be organised and that scientists should work in teams, rather than as inspired individuals but nor should one overlook the contribution made by fiercely ambitious individuals such as William Shockley – many strands went into making the tapestry. But, if one were to single out a specific development which was essential to the whole activity, it must surely be the realisation that material quality was of paramount importance. The discovery of transistor action in germanium point contact devices depended, in the first instance, on the availability of relatively high purity material left over from wartime work on microwave radar detectors. It was this which made plain the need for pure material but the concomitant requirement for high quality single crystals became apparent only gradually. In fact, it was only Gordon Teal's bloody-mindedness and determination to follow his convictions, in the face of widespread scepticism that saved the day. His resurrection of the Czochralski crystal growth technique proved the point and opened the way for others to follow. Pfann's invention of zone refining then completed the battery of methods available to both the device engineer and the semiconductor physicist. The vital synergy between material quality, device operation and physical understanding was, at last, widely appreciated and, once the lesson had been learned in relation to germanium and silicon, it could be readily applied to other semiconductors. Not that anyone was in doubt that such material research was both technically and fiscally challenging (all the more so when applied to compound semiconductors, such as gallium arsenide) but the experience that had accrued during the 1950s proved unequivocally that it was essential to any successful device programme. The other lesson which had been learned from many attempts to obtain high frequency transistors for FM radio and digital computers concerned the need for well controlled device technology. The early days, when a few good devices could be selected from a broad spectrum of random specimens would never form the basis of a successful commercial enterprise. Device yield came to be seen as a key parameter in any well run production line.

At this point, we leave the problems of the transistor in favour of those of the integrated circuit, the next and even more significant development in the solid state story, significant enough, without a doubt, to merit a new chapter.

CHAPTER 4

Micro and Macro: The Integrated Circuit and Power Electronics

The initial take up of transistor electronics may have been a trifle sluggish but by the middle of the 1950s things were changing fast. In 1951, there were 4 companies marketing transistors, 8 in 1952, 15 in 1953 and as many as 26 by 1956. Stimulated by military contracts, silicon diffused devices were being manufactured in considerable numbers and, by 1960, there were no less than 30 US companies making transistors to a total value of some $300 million. Compared to the size of the industry today, which is of order 10^{12}, this may appear tiny but it nevertheless represented dramatic progress over the single decade which had elapsed since the first devices went into production. In addition to Bell, the list included well known companies, such as GE, RCA, Sylvania, Raytheon, Philco, CBS and Westinghouse, but there was also a clutch of new firms, having no previous experience of electronic manufacture. These included TI, Fairchild, Germanium Products, Hughes and Transitron. Interestingly, the newcomers gradually increased their market share throughout the 1950s, until, in 1957, they claimed no less than 64%. At first sight this seems quite remarkable, until one remembers that the big boys were principally concerned to make and sell valves – transistors being no more than a promising sideline. In 1955, the annual market for valves was about $500 million (at about 70 cents apiece), while the total market for electronic components came to over $3000 million, compared to some $30 million for transistors (at $2 each). Indeed, the valve market was still increasing up to 1955 and only showed significant fall-off towards the end of the 1960s. For the start-up companies, of course, transistors were their life and they could be forgiven for a degree of aggressive marketing.

In addition to military applications, transistors were beginning to appear in domestic products such as portable radios, car radios, tape

Semiconductors and the Information Revolution: Magic Crystals that made IT happen © 2009 Elsevier B.V.
DOI: 10.1016/B978-0-444-53240-4.00004-0

recorders, hearing aids, etc. These latter applications made only modest demands on solid state circuitry, employing no more than five or six transistors each but the military demand for digital processing and computing led the way into considerably more complex circuits and already in 1955 IBM had produced a commercial computer employing 2000 transistors. There was still some way to go before solid state computers reached even the size of the ENIAC computer, which employed 18,000 valves, but it was clear that this must be a major market for solid state devices in the not far distant future, and the size of some circuits was already beginning to cause concern. Though transistors offered huge reductions in size, weight and heat dissipation, there was a cloud on the horizon. It was relatively easy to envisage very large numbers of transistors enclosed in a moderately sized container but those who had given much thought to the matter recognised that each and every one had to be linked by a complicated mesh of wires. It soon became apparent that the wiring might take up even more room than the silicon and, much more seriously, that, as the size of circuits increased, the likelihood of faulty interconnections (each individually fashioned) must increase with it. Added to this was the fact that typical yields of working transistors could be as low as 20% and, even towards the end of the decade, yields of between 60 and 90% were normal. Little by little, the idea grew through the industry that solid state electronics was about to suffer a 'tyranny of numbers'. Little by little, it took on a significance which overshadowed all else – purification, diffusion, alloying, contacting, etching, passivation all paled into insignificance by comparison. It was like having a tourniquet tightly restricting ones capacity for future growth and the whole industry was beginning to feel the pressure.

Looked at critically, towards the end of the 1950s circuit manufacture was in a mess – transistors were being made in batches on slices of silicon, chopped up to produce individual devices, then reassembled by armies of young women with tweezers and microscopes, mounted on printed circuit boards, connected to an appropriate selection of passive components by minute wires, intricately soldered with miniscule soldering irons, then tested to confirm (hopefully!) that all was well. While this may have been acceptable in the interests of building a transistor radio it was clearly incompatible with the needs of computer manufacture. Something better just had to be thought out – but what? The attempts being made currently showed little sign of producing a solution.

Already by 1950, a miniaturisation programme was in hand at the American National Bureau of Standards, funded by the Navy. It rejoiced in the name of Tinkertoy and involved a sort of three-dimensional stack of components formed by screen printing on ceramic substrates, with valves at the top of the stack but it took no recognition of the fact that the transistor had been invented 3 years previously. It appears to have ground to a halt in 1953 but was resuscitated by the Army in 1957 in the form of its Micro-Module plan which did include transistors. The idea was to form individual components, each on identical small ceramic chips which could be clipped together to make any desired circuit. It was an interesting idea but came to grief in 1964, having consumed something like $26 million of Government money – mainly at RCA. Events had, by then, overtaken it. A related two-dimensional miniaturisation programme was undertaken at the Diamond Ordnance Fuze Laboratories (a military research organisation) and this may have seen the first use of photolithographic techniques to define components and circuit layout. It was started in 1957 but, again, ran out of steam, under competition from the integrated circuit which came on the scene in the following year.

Interestingly, the first serious proposal to develop truly integrated circuits came from an English scientist, Geoffrey Dummer of the Royal Radar Establishment in Malvern. Dummer joined the Telecommunications Research Establishment, TRE (as it then was) at the beginning of the Second World War and led a small team devoted to improving the reliability of British radar systems. He became a much respected figure in the field of reliability physics, being awarded the MBE in 1945, and included in his brief some aspects of solid state electronics. In 1952, he presented a paper at a Washington DC conference specialising in reliability and in which he proposed the idea of making an electronic circuit in a single piece of silicon or similar material – 'with the advent of the transistor and the work in semiconductors generally, it now seems possible to envisage electronic equipment in a solid block with no connecting wires'. He referred to rectifying layers, amplifying layers, resistive layers, etc., and suggested that the various elements could be inter-connected by cutting away appropriate areas of material. It was perhaps unfortunate that he should choose this particular conference because his ideas appear to have been largely overlooked by the solid state electronics community. Sadly, he was also a prophet crying in the wilderness in his own country. Though he did manage to persuade his management to award a small contract to the Plessey Company with the aim of building a prototype flip-flop circuit, their initial failure was immediately used as an excuse to shut down the whole activity. It seems that the Plessey efforts were inevitably doomed to failure because they were using grown-junction transistor technology which was scarcely compatible with Dummer's proposal. Apparently, the only worthwhile outcome of the work was a large-scale model of the proposed circuit but even this generated far more interest amongst visiting American scientists than it did within the home team. It was, perhaps, an idea before its time. We should remember that the junction transistor had only been realised in 1950 and in 1952 the industry was still struggling to acquire basic technologies but it was certainly a sad failure on the part of British management – just think how different the world might have been if the fundamental patent had been a British one.

And just think how often has this particular cry been heard in British scientific and engineering circles!

It was not until 1959 that the first patent applications were filed and, as so often happens in such cases, there were two of them, one from Jack Kilby at TI and one from Robert Noyce at Fairchild. Kilby was first to make his proposal, and in somewhat bizarre circumstances. In mid 1958, he had only very recently joined the TI research activity from his previous employment at Centralab, where he had learned something of current miniaturisation projects and whence he had spent time on a Bell transistor course. The rest of his colleagues took their vacations at this juncture, leaving Kilby, who had not yet built up any leave entitlement, to muse silent musings in an empty laboratory. He used the time to think himself out of the job which was probably lined up for him (about which he was far from enthusiastic!) by dreaming up the concept of the very first practical integrated circuit. He realised that not only transistors could be made from silicon or germanium – so could resistors and so could capacitors. The former, he proposed to make by defining a narrow bar of semiconductor and using the natural resistivity of the material, the latter by using a simple p–n diode. This is an aspect of the p–n junction that we have so far overlooked – quite simply, because the barrier field at the junction sweeps free electrons and holes away from the junction, this region is empty of carriers and is therefore highly insulating. The junction looks like two conducting regions (a p-region and an n-region) separated by an insulator, which is just what a capacitor looks like, too! While the standard capacitor consists of two metal plates separated by a thin film of insulator, the junction consists of two semiconductor 'plates' separated by a thin film of insulator. However, the junction enjoys an important advantage – the value of the capacitor can be controlled by varying the doping levels on either side. It can also be varied by applying a reverse bias voltage across the junction.

On his boss's return at the end of July, Kilby presented him with a specific proposal for making a crude integrated circuit in the form of a phase-shift oscillator (a circuit in which an amplifier is arranged to have a portion of its output returned to its input in the correct phase to make the circuit oscillate, the necessary phase shift being obtained via a resistor-capacitor network). All the components were to be fashioned from silicon, though they had to be connected together with pieces of fine wire. As it turned out, Kilby was unable to lay hands on a suitable sample of silicon so the test circuit was made from germanium, employing a ready-made transistor, but nothing else was changed as shown in Fig. 4.1 – it looked rather like an elongated bird's nest but it oscillated at 1.3 MHz, the very first functional integrated circuit. For a good measure he also made a

FIG. 4.1   The very first integrated circuit. It was built by Jack Kilby at Texas Instruments in 1958, demonstrating that the functions of resistance, capacitance and gain could all be achieved within a single crystal of germanium. Kilby's circuit lacked an integrated technology for connecting the various components together, making use, instead, of fine copper wires. Courtesy of Texas Instruments Inc.

digital circuit, a so-called flip-flop which had two stable states, one representing a digital zero, the other a digital one. This worked too. Here was Dummer's purely conceptual idea realised in down-to-earth practical form (though at this juncture Jack Kilby knew nothing of Dummer's work). It may not have looked very sophisticated but it proved the existence theorem beyond any shadow of doubt. Real circuits could be made without the need for extraneous resistors and capacitors. This was September 1958.

Immediately the idea was taken up within TI research – silicon replaced germanium and people began to design and build circuits from scratch, using purpose-made semiconductor samples, rather than the cobbled together bits and pieces Kilby had been able to glean from helpful colleagues. Improved definition of circuit elements was achieved by the introduction of photolithography (a technique we shall examine in more detail later) as replacement for the crude black wax technology which had been widely used up to that point. There was also some speculation about the best method of interconnecting the various components. The aim was, of course, to make circuits for demonstration purposes and to form the basis for a patent application. However, the question of interconnection was still not properly resolved when they found themselves hurried into applying before they were quite ready, by a rumour (false, as it happens) that RCA were about to disclose a similar invention. So, in March 1959, TI went public with its Solid Circuit, a flip-flop which outperformed similar circuits ten or more times larger. One year later the first device to be commercially available was on offer for evaluation by a somewhat sceptical industry.

TI's panic over patent rights could well have been justified, not by any announcement from RCA but from its rival start-up company Fairchild. The key invention in this case was that of planar technology by Jean Hoerni and it was a direct result of this patent application that Noyce was stimulated into proposing his version of the integrated circuit. The Fairchild patent agent was doing his job in nagging Noyce for other ways in which planar technology might be exploited, when the idea came to him. If photolithography could be used to define transistors by way of an oxide surface film, then it might equally well be used to define evaporated metal lines on top of the oxide and these could be used to link a number of active or passive devices as shown in Fig. 4.2 (planar technology conveniently allowed all connections to be made from above). He too recognised that resistors and capacitors could also be fashioned from silicon, thus allowing complete circuits to be designed on a single chip. Remarkably, Noyce was slow to pursue his idea, not even mentioning it to anyone other than his close confidant Gordon Moore, who was in charge of production at Fairchild. The reason appears to have been that Fairchild were just on the point of going into production with their first important product, a double-diffused silicon transistor and everyone's attention was concentrated on this (commercially speaking) life and death undertaking.

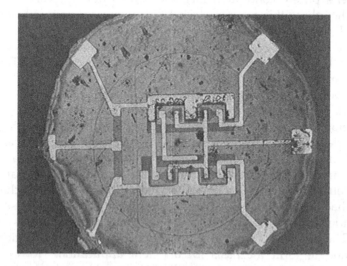

FIG. 4.2 The first planar integrated circuit. This was the brainchild of Robert Noyce of Fairchild Semiconductor and formed the basis for all future integrated circuit development. It represented a significant improvement over Kilby's approach in so far as it employed planar technology to effect interconnection. Courtesy of Fairchild Semiconductor.

Like TI however, they were pushed into taking action by a rumour (correct this time) that TI were about to disclose an important invention, and it was not difficult to guess something of its subject matter. So, in July 1959, Noyce's patent application was filed, clearly post-dating Kilby's but possessing one major advantage in that it made quite clear the means for interconnecting the various components.

The scene was set for a mighty patent war which rattled on for no less than 10 years and consumed well over a million dollars in legal fees. The final conclusion went in favour of Noyce but by that time hardly anyone was interested – the scientific community was agreed that he and Kilby should share equally the prestige associated with their invention and even the two companies had agreed to share royalties. The inventors were both awarded the National Medal of Science and both were inducted into the National Inventors Hall of Fame. Because of the rule that Nobel Prizes are never awarded posthumously, Kilby alone won the 2000 Nobel Prize in Physics but there can be little doubt that Noyce would have been standing alongside him, had he been alive to do so. Perhaps the more interesting question concerns the long delay (41 years) between invention and award, bearing in mind the mere matter of 9 years in the case of the invention of the transistor. There can be no doubt as to the global significance of the integrated circuit and its effect on 'Western' life, so was it a question of the physics? While the transistor triumvirate were clearly intent upon physics research, the invention of the integrated circuit can be seen as 'a mere matter of engineering' – but what magnificent engineering!

There can be no doubt that it was the combination of planar technology, photolithography and integration that produced a sea change in the electronics industry but the initiation was not altogether a smooth one. The initial reaction was far from positive, critics pointing out that integration did not make optimum use of materials – in other words, it was possible to make better resistors with nichrome, better capacitors with mylar and better inductors with a coil of copper wire. This was, of course, true but it was still possible to make quite adequate resistors and capacitors (inductors proved a little more difficult but the same argument eventually applied). Another complaint concerned the question of yield – if the yield of individual transistors was 90%, a circuit containing 10 transistors could expect an overall yield of $(0.9)^{10} = 0.35$ (or 35%) and considerably worse for larger circuits! On the face of it, this sounded like a serious criticism but it turned on the assumption that yield was uniform across a semiconductor slice and this was palpably not the case – areas of poor yield did occur but large areas were actually more or less defect free. The use of small chips succeeded in mitigating this particular problem. Finally, ICs were expensive. In reference to an early TI circuit, an unnamed director of Philips is quoted as saying: 'This thing only replaces two transistors and three resistors and costs $100. Aren't they crazy!' Hindsight suggests not; but the critic was not in a position to know just how steep the learning curves were going to be in this particular business. In any case, the integrated circuit would obviously come into its own when applied to much larger circuits – the computer, rather than the radio set was its ultimate target. The problem was that, whilst it was necessary to start with something modest in order to establish the methodology, this was not immediately viable as a commercial product. Fortunately, the military purchasing arm wasn't too worried about a little local difficulty like that!

Texas Instruments was first to the market place in 1960. Fairchild followed suit in 1961 with its Micrologic series and in the same year TI produced a computer for the Air Force based entirely on ICs. The following year TI was awarded a contract for an integrated control system for the Minuteman missile. In truth, quantities were small – only a few thousand units in 1962 – but it was a beginning, and then there was President Kennedy's 'Man on the Moon' project to keep things moving nicely. American rocket technology at that instant was lagging behind its cold war rival's and it soon became clear that weight and volume would be at a premium – guidance systems, vital to any hope of success, would have to be 100% reliable and, at the same time light and compact. Integration was the only way to go. The coincidence of this national prestige project with the needs of a nascent technology was truly remarkable – one could be forgiven for wondering whether Kennedy was in league with the electronics industry in much the same way that President Bush is today (2007) with the oil industry! But really it was all the fault of the Russians – had they not launched their Sputnik into space, progress in IC development might have been very different. Idle speculation, indeed – but events could scarcely have worked out more favourably for the Americans, particularly the fledglings such as Fairchild and Texas Instruments. It is also worth mentioning en passant another aspect of the space programme, to the effect that all electrical and electronic systems were to be powered by solar cells, yet another branch of the silicon fiefdom. Russell Ohl's wartime discovery of the photovoltaic effect in a silicon p–n junction suddenly became of major national importance, illustrating again the essential requirement for high quality single crystals, this time in much larger sizes. (We shall look in greater detail at solar cells in Chapter 10.)

Before delving into the long-term commercial development of integrated circuits, we should first consider the arrival (at last!) of the field effect transistor. For some years it had been clear that observation of the field effect was bedevilled by the presence of surface or interface states which trapped electrons and prevented them taking part in electrical conduction in the bulk semiconductor. It was this which had frustrated Shockley in his early attempts to make a semiconductor amplifier and the situation showed no sign of improving. In fact, the FET took something of a back seat as various versions of the so-called bipolar transistor (point contact, n–p–n or p–n–p) were wheeled out and gradually commercialised. Once it became apparent that such devices really worked, logic supported the practical approach of concentrating on perfecting them, particularly in regard of their reliability and speed of operation, this latter aspect being of special relevance to digital signal processing. The idea of a field effect device was still attractive but no-one could see a way round the interface problem (the interface, that is, between the semiconductor and the gate insulator). However, there was a solution and, as in so many other cases, it came from Bell Labs. The virtues of silicon oxide as an insulating layer on top of a silicon slice had been appreciated since its discovery by Carl Frosch in 1955, and, of course, it played an essential role in planar technology, but the possibility of preparing a silicon–silicon oxide interface with very low density of interface states was not appreciated until 1958, when 'John' Atalla stumbled upon it rather by chance. It was necessary to clean the silicon rather well and oxidise it under just the right conditions at a temperature of about 1000 °C in a stream of oxygen – then the density of interface states turned out to be so low (less than 10^{15} m^{-2}) that all the frustration of years was smoothed away. Thus, in 1960 the FET at last became a reality in the form of an MOS (metal-oxide-silicon) transistor (often referred to as a MOSFET) and, within the next few years, it was introduced into integrated circuit manufacture by RCA, followed by two more new companies General Microelectronics and General Instrument. It took some little time to become fully established but, once the teething troubles were ironed out, it gradually came to dominate the field. The majority of devices in integrated circuits are now MOS structures and the 'old-fashioned' bipolar transistors find application only in special cases. William Shockley was right all along – but, somehow, he always seemed to be doing something else when the vital breakthrough arrived!

Making a MOS device, in practice, was easier than making its bipolar equivalent. Typically, it involved a slice of p-type silicon into which were diffused a pair of n$^+$ contact regions, to act as source and drain, then evaporating a metal gate electrode on top of an oxide film, between them. In the absence of a gate voltage, source and drain were isolated by p–n junctions (one of which was inevitably under reverse bias and therefore acted as a barrier to current flow) but, when a suitable positive voltage was applied on the gate, the region between source and drain was inverted (converted to n-type conductivity) and current could flow between them. Because the conducting channel under the gate was n-type, this device was referred to as an NMOS. It was also possible to make a PMOS by starting with n-type silicon and forming p^+ contacts, the sign of the gate voltage then being reversed. Superposition of a signal voltage on the gate resulted in a corresponding fluctuation in source-drain current which produced an amplified signal voltage across a load resistor in series with the drain contact. The signal might take the form of an analogue voltage, such as an audio signal or, alternatively, a voltage pulse representing a digital 'one' in a computer circuit.

The advent of the MOS transistor had special significance for digital computers because, in addition to data processing, it also took over the role of storage element, a function previously supplied by an array of tiny magnetic cores which could be magnetised in either of two directions (one direction representing a 'one', the other a 'zero'). Bipolar devices had failed to replace magnetic cores because the flip-flop circuit made from them was too large and too expensive, whereas the smaller MOS transistor could, in the first instance, just about compete, then, as improved technology allowed smaller devices to be made, it could compete ever more effectively. It switched less rapidly than the bipolar transistor but, again, as dimensions became smaller, its speed increased and gave it an advantage in this regard, too. Finally, a new circuit element made from a complementary pair of MOSFETs, one NMOS and one PMOS, effected a considerable reduction in power dissipation because it consumed power only while actually switching – in the steady state it consumed no power at all. Thus, within a very few years, integrated circuits based on MOS transistors took over the computer functions of both data storage and data processing. They were easier to manufacture, they could be used as capacitors (indeed, they were capacitors) and they could also be used as resistors by connecting them so as the source-drain channel was always open. This was a tremendous advantage because it meant that all three functions could be incorporated using one simple device, thus simplifying manufacturing technology and achieving significant improvements in yield. Their low power dissipation also turned out to be vitally important as device size decreased and packing density increased.

The year 1968 may be taken to define a watershed in the development of the integrated circuit. At this point the market for discrete transistors was still about equal to that for ICs and, within the latter, digital circuits outsold analogue by a factor of five, $300–$60 million. However, MOS was still very much in its commercial infancy, even though working devices had been clearly demonstrated in 1960 – a measure of how difficult it was to introduce any new technology. The problem was that companies active in the solid state electronics business were fully stretched to bring their chosen technology to market and simply didn't possess the capacity to work on an alternative as well. RCA, having ventured into MOS, were faltering and none of the other large players had so much as dipped a toe into the water. The essential dilemma lay in the field of memory – a detailed analysis published in 1968 concluded that semiconductor memory would only put ferrite core memories out of business if an IC manufacturer could produce a memory chip which stored four thousand bits of information. In bipolar terms this meant 16,000 transistors! The general consensus was that this was simply beyond any possibility – data processing was all well and good but memory was best left to the science fiction writers. However, there were two important people who didn't agree. Robert Noyce and Gordon Moore at Fairchild were beginning to fret under the constraints of distant management (Fairchild Semiconductor was the junior offspring of a fairly conservative parent company on the American East Coast) and took the courageous decision to break away and set up their own company to make semiconductor memories. They called it Intel. It was a prodigious gamble and, as we now know, a prodigious success. By the end of 1968, they produced their first memory chip, holding 1024 bits – it used bipolar transistors but very soon they changed to MOS and scarcely looked back. In their first year their sales fell somewhat below $3000 – in 1973 they topped $60 million and in the year 2000 $30 billion! Fairchild, on the other hand, declined sadly and were finally taken over by the French company Schlumberger in 1979.

The key to developing future integrated circuits was, of course, the ability to reduce the size of individual transistors, measured, in the case of the MOSFET, by the length of the gate (the distance between its source and drain electrodes). In the early 1960s, this dimension was roughly 1 mm, in 1978 it was about a micron, while today it is approaching the 10 nm scale. Fig. 4.3 shows the example of an Intel 1103 microprocessor from 1971. While chip sizes have remained roughly constant at about 10 mm on a side, the number of transistors per chip has been rising dramatically. In 1965, Gordon Moore noticed that it appeared to be doubling each year and he tentatively suggested that this trend might well continue for another decade, a prediction which was seized on by the technical press and elevated in status to become 'Moore's Law'. In 1975, when considerably more data was available, Moore looked at the trend again and concluded that a better approximation would be a doubling every 2 years and a plot of the number of transistors in a long sequence of Intel processor chips does roughly follow this dependence. Another way to state it is to say that the gate length of MOSFETs

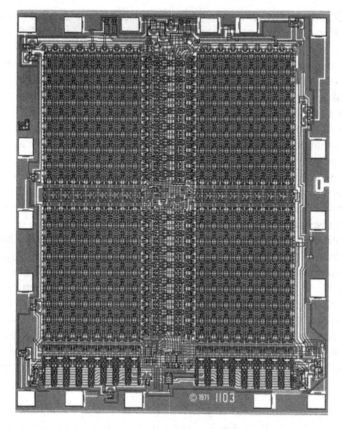

FIG. 4.3 An Intel integrated circuit of 1971, the 1103, showing how far the technology had developed during the thirteen years following Kilby and Noyce's pioneering work. Modern circuits contain so many transistors that their structure is difficult to resolve in an overview photograph. Courtesy of Intel.

decreases by an order of magnitude every 15 years. There has been some considerable confusion as to exactly what Moore said and what other people have said he said but the important point to grasp is that the trend has been quite remarkably steady over some four decades and still shows little sign of deviating from an exponential law. It has even been suggested that the law has become a self-fulfilling prophesy because the industry feels it just has to keep up with that rate of progress! Whatever the truth of this challenging proposition, it is certainly interesting to think whether the law must eventually run out of credence. For example, we might conclude that in about the year 2025 the gate length will have become equal to the distance between two silicon atoms in a silicon crystal and this appears to set some kind of fundamental limit. However,

20 years is quite a long time for scientists to evolve some completely new approach to making solid state amplifiers – it would be a rash prediction that defined the total demise of Moore's Law. Possibly we shall (or more realistically, I should say 'you will') see some significant change of slope!

In practical terms, integrated circuits have gone from containing a few transistors in 1962 to something like a billion transistors today. Small scale integration (SSI) in the 1960s gave way to medium scale integration (MSI) in 1970, to LSI in the 1970s, to VLSI in the 1980s and ran out of superlatives in the 1990s. As anyone who owns a personal computer knows, processing power, speed and memory size have increased dramatically and continue so to do. The demand placed on our ability to think of things to do with all this performance is considerable but man's inventiveness when it comes to filling his leisure time knows few bounds. There is probably a 'Moore's Law' about that too! But we should be careful not to rush ahead too rapidly – there are one or two details concerning this mind boggling race into the future which are worth our while to examine.

One of the most interesting is the birth of the pocket calculator. This was yet another product of the youthful TI organisation and was managed by none other than Jack Kilby himself. It was really the brain child of the TI CEO Patrick Haggerty who had previously been responsible for TI's push to develop the Regency portable radio. On seeing the first ICs being put together, he wanted a truly innovative product which would make use of the company's new technical skills, so in 1965 Kilby was given his boss's explicit order – create a brand new consumer product in the shape of a small, hand-held, battery-powered, pocket-size calculator, weighing no more than a typical paperback book! Mechanical calculating machines existed – so did electronic machines but they were heavy, bulky and frighteningly expensive. The new specimen had to be an order of magnitude smaller in all respects. It involved the development of a special 'shift register' – a means of storing digital information. The key here was to input the data serially, that is by sending all the bits along a single wire, one after the other, rather than in parallel, which would have involved many separate wires. The logic circuitry was something of a headache but they finally settled on a four-chip design. They designed a miniature keyboard, found a rechargeable battery and built a completely new display system based on thermal printing paper. It was all done in just over a year – but, because the circuitry was so advanced, it took the production people five more years to produce the prototype. Nevertheless, once ready, the calculator sold like hot cakes and started a trend.

Another thing it started was the microprocessor. In 1971, just as Kilby's calculator was coming to market, a Japanese company Busicom wanted to develop a new style desk-top calculator but, finding no-one able to design the necessary circuitry, they approached Robert Noyce at

Intel and arranged a contract with him. He, in turn, gave the problem to Ted Hoff who, as it happened, was keen to design his own computer. Hoff was appalled by what he saw as the crudity of the Japanese design and came up with a brand new approach which saved considerably in terms of complexity and was truly innovative in so far as it allowed the processor to be externally programmed to perform a range of different functions. It enabled him to put all the circuitry on a single chip which came to be seen as the first computer on a chip. The emergence of the pocket calculator caused such a stir that many calculator companies got into difficulties and Busicom were no exception. They found themselves unable to pay the agreed price for Intel's services and were obliged to allow Intel marketing rights, which, after some deliberation, they eventually took up and, in no time at all, there were intelligent traffic lights, intelligent lifts, intelligent tennis rackets, intelligent skis and so on ad infinitum. What Hoff had done was to reverse the whole trend of microcircuit design – instead of having to design a specific IC for each and every procedure, it was now possible to achieve flexibility by programming. It probably saved the industry from another 'tyranny of numbers'.

The subject of integrated circuit design is far too large and complex to be covered in any more detail here – readers wishing to pursue it will find much of interest in the bibliography. Our aim now must be to concentrate on some aspects, at least, of the essential technology. We have seen how vitally important was the introduction of improved methods of purifying semiconductor materials and growing highly perfect single crystals, also how necessary was the ability to control doping levels, once background purity was adequate. The other vital ingredient in semiconductor processing is, of course, the means for placing the dopant atoms in precisely the right places and, as the reader will well appreciate, as ICs became more complex and device sizes became smaller and smaller, this matter of accurate definition became paramount. I shall try to outline the main features of the lithography and chemical and physical processing which have been developed over the past five decades in face of the ever-increasing demands of the circuit designer. It constitutes a fascinating story in itself.

As a convenient way into the subject, it is helpful to begin with a brief account of the various stages involved in making an integrated circuit, starting with a single crystal boule of silicon and ending with a fully processed circuit. We have already looked briefly at the important question of semiconductor crystal quality and purity. It came to the fore during the Second World War in attempts to perfect microwave diode detectors for radar applications and was typified by the so-called high-back-voltage germanium detector. In order that the detector might withstand intense pulses of microwave energy, it was necessary for the diode to possess a reverse I–V (current–voltage) characteristic with a high breakdown value. When a reverse voltage was applied very little current flowed (this being an essential feature of the rectifying action) but at some value of voltage (known as the breakdown voltage) the current suddenly increased and, unless the applied voltage was removed, the diode would burn out. Pragmatic investigation of the effect showed that the breakdown voltage increased as the purity of the germanium improved. It was this motivation which first led semiconductor scientists to appreciate the virtue of material purification and gradually it became clear just how important this was for all manner of semiconductor applications – not least that of the bipolar transistor and, later, the MOSFET. As we have seen, success was first achieved with germanium because it was a much easier material to tame than silicon but, as silicon gradually became the semiconductor of choice, it was necessary to produce silicon crystals of comparable quality to form the basis of the integrated circuit. High purity in the starting material allowed optimum control of doping while a high degree of crystal perfection was essential for obtaining high electron and hole mobilities.

The first step in making an integrated circuit therefore involves growing a large cylindrical boule of single crystal silicon by the Czochralski technique, which we described earlier. 'Large' because it is economically expedient – the larger the size of the silicon wafer, the more chips can be obtained from it with the same amount of processing. In the early days of the integrated circuit the boule diameter would be about 2 in., whereas today it has expanded rather dramatically to 12 in. Most people's sense of semiconductor electronics would probably emphasise the minute size of an individual transistor, now measured in sub-micron units, but, by contrast, today's typical starting crystal is a veritable monster – it weighs in at something like 50 kg and, when polished, shines beautifully black (the band gap of silicon being smaller than the photon energies of visible light, it absorbs all visible wavelengths and therefore reflects no light to the eye). Each boule provides as many as 500 wafers and each wafer produces perhaps 500 chips – a quarter of a million chips in all! It represents a masterpiece of the crystal grower's art – not only has it to adhere to a standard of purity never so much as contemplated previously, it also has to be defect-free and almost unbelievably uniform. Just how all this is achieved would take us too long to discuss but let it suffice that it requires careful control of the rate of rotation of the crystal as it is pulled from the melt, the rate of pulling, the control of the temperature gradients both across and along the boule and the use of a large magnetic field applied along its axis. This latter interacts with electric currents flowing in the liquid silicon to improve dramatically the uniformity of impurity distribution within the boule. It may also be used to control the incorporation of oxygen in the finished boule, an important parameter, because oxygen may form a donor in silicon. Dozens of learned papers have been published describing numerical calculation of what happens in various practical cases – and thereby aiding its optimisation. We need make no attempt to understand the details but we can certainly marvel at its success.

The crystal is then ground into an accurately cylindrical shape and flats ground onto it, marking appropriate crystallographic directions. These flats are used during subsequent fabrication as references in what is a largely automatic process. The damage introduced by the grinding process must then be removed by chemical etching to dissolve a few tens of microns from the surface of the boule. Next, the boule is sawn into slices about 1 mm thick using an annular diamond saw and the slices are then heated to about 700 °C and rapidly cooled to eliminate oxygen donors – the dissolved oxygen being precipitated in molecular form, rather than being present as individual donor atoms. The wafers are then edge-rounded to facilitate the later spreading of photoresist, lapped flat and polished on a polishing machine, etched to remove damage, cleaned both chemically and mechanically and finally marked for future identification (see Fig. 4.4). They are now ready for processing into chips but, before we describe this part of the process, it would be well to emphasise one important point – the huge advantage gained by using such massive crystal boules is obtained at a price. It is reckoned that a modern Wafer Fab (as the wafer making part of the activity is known) costs something of the order of a billion dollars! Needless to say, only a

FIG. 4.4 A large processed silicon wafer. A wafer like this contains something like 500 individual 'chips'. Courtesy of iStockphoto, © Robert Hunt, image 985531.

few large enterprises can afford such an outlay. Most IC manufacturers simply buy-in ready-made wafers.

The process of forming electronic components on the silicon wafer and connecting them electrically together involves many different steps which vary according to the nature of the finished chip. Typical of these are oxidation, thermal evaporation, chemical vapour deposition, sputtering, diffusion, ion implantation, etching, all combined with various cleaning processes but we shall make no attempt to describe them all – it must suffice to look at one or two of them in some detail to give a feel for the complexities involved. In particular, we must deal with photolithography which is vital to all such processing as it defines the geometry of individual components and provides the key to the ever-decreasing size of each transistor, capacitor and resistor which make up the circuit. Photolithography was originally a process used in the printing industry and was borrowed by several of the companies involved in the early development of semiconductor devices, such as Bell Labs, Fairchild Semiconductor and Texas Instruments. (Actually, the first serious application to semiconductor technology occurred in a US Government laboratory the Diamond Ordnance Fuze Laboratory – DOFL – and was passed on to the commercial firms through their involvement in Government contracts.) We shall first examine the process itself then, in order to give some solidity to our discussion, we shall outline the steps involved in making an array of complementary MOSFETs – pairs of NMOS and PMOS transistors (usually referred to as CMOS pairs) which might well form the basis of a computer memory chip.

The name photolithography clearly implies that we are dealing with a process for defining shapes which depends on the use of light. We shall first outline the basic process as first developed in the 1960s, then discuss some of the modifications which became necessary as the size of individual features became smaller. Photolithography relies on rather special organic chemicals (known as photoresists) which are peculiarly sensitive to light. They come in two kinds – positive resists which become soluble and negative resists which become insoluble (in an appropriate solvent) under the influence of light. For simplicity, we shall discuss only the positive variety. To create a pattern on the surface of a silicon slice, it is necessary, first to spread a thin film of photoresist over the surface, then to illuminate certain areas to soften the resist and finally to dissolve away the illuminated regions with a suitable solvent (often referred to as a developer). To produce the initial film of resist, it is spun on. The silicon slice is mounted on a spinner which allows it to be rotated very rapidly, while a drop of resist solution is placed at its centre, the centrifugal force spreading the solution uniformly across the slice. (It all sounds simple enough but much research went into optimising the viscosity of the solution and the appropriate rate of rotation.) The resist solution is then dried, prior to illumination with UV light (from a mercury lamp) which is patterned by passing it through a photographic mask. It is this mask which contains the detailed information required to define appropriate features of the transistors, diodes, resistors, inter-connection patterns, etc. Finally, the illuminated regions of the resist are dissolved away in an appropriate developer and the slice cleaned and dried, ready for the next stage of processing. It is important that we recognise the need for several processing stages, each involving a different mask. One mask may, for example, define the size and shape of source and drain contacts (in a MOSFET), another mask the geometry of the gate, a third mask that of an interconnect pattern, etc, and it is crucial that all these masks are correctly aligned with respect to one another. For example, it is vital that the MOS gate lies accurately between source and drain contacts and that the metal lines which provide electrical connection make contact with the appropriate transistor features. This is achieved by means of alignment markers on each mask, the second and subsequent masks being adjusted by eye (under a microscope) until their markers coincide with those produced on the slice by the first mask. This, in a nutshell, is photolithography – it is only one aspect of the overall process – deposition, dissolution, diffusion, etc, are also involved in making any integrated circuit. (The length of this paragraph reflects the complexity of even this one aspect and perhaps emphasises the need for extreme care both in design and application of the mask set.)

Having said all this, it is clear that the mask set must be made to a high degree of accuracy, consistent with the dimensions of the devices involved. If, as in the early days, the MOS channel (the distance between source and drain) was some hundreds of microns in length, the masks had to be made to an accuracy of a few microns and this could be achieved relatively easily. The technique used was to cut out an enlarged mask (perhaps 100 times the finished size – about a metre square!) in a sheet of plastic and use this to produce a 10 times smaller replica on a glass slide with photographic emulsion (the 10 times reduction being obtained optically). This mask was then used in a step-and-repeat process to make a final mask defining the whole slice. Once again, a 10 times optical reduction was obtained while the mask was effectively scanned in steps over the area of the final mask to produce perhaps 20 chips. This mask could then be used to 'write' a whole silicon slice at once. A lot of work went into making the mask set but, once done, the set could be used again and again to manufacture large numbers of chips.

Now that the general principles are established, we can look in greater detail at the process of making an array of CMOS devices. First, we must remind ourselves that an NMOS transistor consists of two n^+ contact regions (source and drain) diffused into p-type silicon, with an oxide-insulated metal gate covering the space between them, while a PMOS device differs only in having p^+ contacts in n-type silicon. However, this difference raises the first challenge to the technologist – he must start with a silicon wafer which is either n-type or p-type – it cannot be both! Usually it will be n-type, which makes it straightforward to form the PMOS device but, in order to make the NMOS transistor, he must first diffuse acceptor atoms into a defined region of the silicon to make it p-type – then the NMOS device must be patterned within this special area. The processing sequence starts with the formation of this p-type region. Firstly, an oxide layer is grown on the silicon by passing wet oxygen over the slice at a temperature of about 1100 °C (in a quartz tube) and openings made in this oxide film at those positions where the NMOS devices are to be made. To do this requires a film of photoresist, window patterns defined by mask number one, holes etched through the oxide with hydrofluoric acid (HF) and boron diffused through the holes into the silicon. The next step involves regrowing the oxide and etching holes through it to make the p^+ contacts for the PMOS devices – this requires mask number two. A similar step defines the n^+ contacts for the NMOS devices, except that this time phosphorus is diffused in, rather than boron – mask number three. Then it is necessary to form the gates by very carefully growing a thin oxide layer which will act as the gate oxide. First the thick oxide is removed in the regions between the respective source and drain contacts – mask number four – and the gate oxide grown at 900 °C in pure oxygen. Then holes are made to access both sets of source and drain regions – mask number 5 – and aluminium evaporated all over the slice, forming electrical connection to all sources and drains and, simultaneously forming the gates, Finally, mask number 6 is used to define the individual aluminium connections to the various sources, drains and gates – effectively defining the wiring pattern. Some of these aluminium lines can be extended to the edge of the chip so as to make contact with the outside world.

The point of describing this particular process is not so much one of providing detail as one of creating an impression. In fact, though the above account obviously involves a fair amount of complexity, it actually represents a relatively simple example. Suffice it to say that manufacturing integrated circuits is a highly skilled and cerebral process and, as device sizes have diminished and device numbers have increased, the business has become ever more complex. An interesting corollary is the demise of the brilliant individual scientist or even the small group of scientists – gone (perhaps?) are the days of the brilliant Nobel Prize-winning individual turning the world on its head with an outlandish idea – the development of ever larger memory chips or ever more power-ful processors is a multidisciplinary activity, demanding the collaboration of large teams of experts and the investment of large sums of money. Though it becomes increasingly difficult to pick out individual contributions, this is certainly not to deny that tremendous strides have been and are being made. It would be counter-productive to attempt any detailed discussion of the vast amount of technological advance implicit in the development of today's spectacularly successful ICs but, at the same time, it would be irresponsible not to look briefly at one or two of the more important and, indeed, critical aspects. The techniques described in the previous paragraphs were entirely appropriate to the state of the art up to the end of the 1960s but would be totally inappropriate today – why? While not wishing to become bogged down in confusing detail, I am surely obliged to offer some, at least partial, answers.

Firstly, we should introduce one of many new technologies in the shape of ion implantation which has come to replace diffusion in many situations. Implantation makes use of a parallel beam of high energy ions carefully aimed at the silicon surface so as to penetrate a small distance into the silicon (typically 0.1–1 μm). To understand the point of this let us refer back to the structure of a simple NMOS device, consisting of a pair of n^+ contact regions in a p-type silicon sample, together with a metal gate covering the space between them. In a properly functioning transistor the gate must be carefully aligned so as to fill the space exactly but not to overlap either contact and, in the days when the gate length was about 100 μm (0.1 mm), it was relatively easy to achieve this alignment by eye, when positioning the appropriate photo-mask. However, as dimensions decreased, it became more and more difficult to achieve the necessary accuracy, until something drastic had to be done, in the form of a self-alignment process. Imagine the gate metal to be put down first and defined by conventional photolithography, then the gate used as a mask to define the edges of the source and drain regions. In principle, this automatically solved the alignment problem but, unfortunately, diffusion of donor atoms into these contacts to turn them n-type also resulted in donors diffusing sideways under the gate region, as well as downwards into the contact, this being the essential nature of the diffusion process – it is naturally isotropic. This was totally unacceptable and an alternative method had to be found. The answer lay with the use of ion implantation, rather than diffusion to place the donor atoms precisely where they were needed. It was a bold step because it involved a considerable investment in the necessary equipment (something like $3 million) – the required donor atoms had to be ionised (by having electrons knocked off them), then accelerated to energies of order 100 keV (10^5 eV) in a high voltage accelerator, then focussed onto the silicon slice so as to strike it normally (such high energies being required in order to obtain the necessary penetration depth within the silicon). The important point was that it worked!

Donor ions do tend to scatter very slightly to the side but to such a small extent that the resulting transistors still function more than adequately. Nothing comes cheaply in semiconductor micro- and nano-technology but without such investment the 'Moore's Law' charge into the foreseeable future would have ground to a halt many years ago.

So much for one expensive development – there are many others! Perhaps the chief of these lies at the very heart of the whole process and this is lithography. In the early days, and, indeed, for quite a number of years circuits were defined by photolithography based on the use of a standard mercury lamp which emits most of its energy at a wavelength of 254 nm (photon energy 4.88 eV) well out into the UV part of the electromagnetic spectrum (violet light having a wavelength of about 700 nm – energy 1.8 eV). The reasons for using such a short wavelength are two – firstly, photoresists tend to be more responsive to UV light and, secondly, and much more fundamentally, the resolution obtainable in defining circuit elements is limited by the wavelength of the light employed. A basic tenet of diffraction theory tells us that the smallest spot to which a light source can be focussed is approximately equal to the wavelength of the light. What this esoteric bit of optical physics means in down-to-earth practical terms is that the edge of a circuit feature must be blurred to a similar degree and, if we wish to define two lines alongside one another, they must be spaced apart by at least this amount. Consider again the case of a MOSFET. If the gate length (i.e. the channel length) is 100 μm, the source and drain will be this distance apart and there is absolutely no diffraction problem in defining them distinctly. However, if the gate length is 1 μm (i.e. 1000 nm), as it was in 1990, we can begin to see the beginning of a problem and if we move on to today (2005), when gate lengths are down to 0.1 μm (100 nm) we are in serious trouble. Two things have happened in response to this technological crisis. Firstly, the reliable and long serving mercury lamp has been gracefully retired, to be replaced by a relatively modern pretender, the so-called excimer laser. This was invented in the Lebedev Institute in Moscow in 1970 and takes various forms, depending on the nature of the gasses used. (It forms, as a matter of general interest, the basis of modern eye surgery.) Two versions have seen service in lithography, the krypton/fluorine laser emitting at 248 nm and the argon/fluorine laser emitting at 193 nm, the latter having demonstrated resolutions as low as 52 nm. The other thing to happen is the use of immersion lithography in which the space between the focussing lens and the silicon slice is filled with a liquid – this results in an improvement in resolution by a factor just equal to the refractive index of the liquid, about 1.5 for pure water and somewhat larger for one or two other more esoteric fluids. The mention of lenses also provides the opportunity to point out that these very short wavelengths demand special care in the materials used for their manufacture, the usual glass (which absorbs strongly at these wavelengths) being replaced by fused quartz.

Another aspect of change concerns the method of mask making. The early use of mechanically cut plastic sheet, followed by photographic reduction soon gave way to direct writing on the glass substrate and this, of course, ran into the same resolution difficulty that we have just discussed. However, in this instance there was a ready answer in the use of electron beam lithography, whereby a suitable resist was exposed to a scanned high energy electron beam in a high vacuum system. There is a fundamental reason for this choice in so far as the effective wavelength associated with electrons (which is related to their energy) is very much smaller than the optical wavelengths we referred to above, being typically about 10^{-2} nm. (In case any reader is puzzled by the idea of electrons having a wavelength, I should remind him or her that quantum theory associates wave properties to particles, just as it associates particle properties to waves.) Hence, the resolution is certainly not limited by diffraction effects – other problems such as electrostatic repulsion between electrons in the beam and electrons scattered in the resist and in the glass substrate tend to dominate. However, careful choice of beam energy and beam shape allow remarkably successful masks to be produced – albeit somewhat tardily because the focussed beam has to be scanned slowly over the area of the mask (electron resists are not very sensitive and require long exposures). It is this very slowness, by the way, which precludes EB lithography from being applied directly to silicon slice writing. It usually takes hours to scan a mask and scanning a 12 in. silicon slice would take all day, in contrast to the matter of minutes required for optical exposure. The time taken over mask making is of little consequence because, once made, the mask may be used hundreds of times. It is the time taken to expose the actual slice which is important commercially.

Once again, the use of electron beams involves esoteric and expensive equipment – for anyone familiar with the scanning electron microscope, this is roughly what the machine looks like. Little by little, the cost of lithography has risen rather frighteningly as feature dimensions have decreased. Roughly speaking, in the good old days of 1980 a typical set up might have cost about $300,000 ($0.3 million), in 1990 this had crept up to $1 million, while in 2000 it had reached $10 million and in 2005 it had reached $20 million. Predictions for 2010 suggest $40 million as the likely going rate, a sum large enough to discourage all but the bold (and rich!). The days when a small start-up company might be set up on the back of a $1 million bank loan appear to be well and truly over as far as integrated circuit manufacture is concerned. Only rather large and well established companies need apply – there is room for no more than a very few.

What of the future, one might ask? There still seems no let up in the race to reduce component sizes and the demand for even more sophisticated lithographic equipment intensifies. Many and various are the proposals for radical new approaches to meet these demands and it would be a rash prophet who claimed to know which stands the best hope of commercial success. I shall hide behind my avowed intention to write a history of the subject and make no attempt at predicting the future but it is worthwhile to look in some detail at one of the principal contenders, X-ray lithography, as an example of the issues involved. Research into this process has been under way already for several years and resolution of 15 nm features has actually been demonstrated, making use of X-rays with wavelengths close to 0.8 nm. Two essential aspects demand attention – those of finding a suitable X-ray source and of making suitable masks. Resist materials tend not to be very sensitive to X-rays so it is important to find a reasonably intense source and this has meant setting up purpose designed synchrotron sources. The synchrotron was invented shortly after the Second World War, by a Cuban physicist Luis Alvarez, for the study of high energy particle physics. It represented an improvement on the cyclotron accelerator for producing a beam of high energy electrons. Typically, the synchrotron consists of a roughly circular evacuated pipe, made from straight sections in which electrons are accelerated by an intense electric field, linked by curved sections in which the beam is bent by a powerful magnetic field (electric and magnetic fields must be synchronised in order to keep the beam within the pipe – hence the name). A basic result of electromagnetic theory tells us that whenever a beam of charged particles is accelerated it will generate electromagnetic waves – that is the beam will lose energy to the waves – and the curved sections of the trajectory imply acceleration because the beam is changing direction. Thus, the curved sections can be used to take off (through a suitable window) electromagnetic energy in the form of X-rays which can be piped into the semiconductor clean room to be used for lithography. The raw source contains a wide range of X-ray wavelengths but a degree of selection is possible using appropriate filters and reflecting mirrors. Needless to say, this large and complex apparatus represents yet another very expensive piece of equipment to add to the financial burden of the IC manufacturer.

The mask used to achieve feature sizes on the tens of nanometre scale is another wonder of modern technology and positioning a 3 cm × 3 cm mask to an accuracy of, say, 20 nm (one part in a million!) is yet another work of art. Consider, for example, the fact that typical solid materials expand by about 10 parts per million when their temperatures change by 1 °C. Consider also the effect of placing a very small weight on the mask – a mere 100 g (3 oz) is sufficient to distort it by a similar amount. To ensure the accurate alignment of a sequence of masks is something of a technological nightmare and to move such masks accurately in a step-and-repeat process is even worse. It demands the use of frictionless air bearings, for example. The mask is typically made from a thin film of a heavy metal such as gold, tantalum or tungsten, mounted on a membrane of silicon carbide or diamond about 2 μm thick (so as not to absorb significant amounts of radiation), stretched across an aperture in a slice of silicon and patterned by electron beam lithography. This one short sentence incorporates a wealth of advanced technology and one has little option but to wonder at the skill of those responsible for it.

Finally, in this brief survey of IC technology, we should not overlook the importance of the clean room (see Fig. 4.5). Dust in its many forms is a

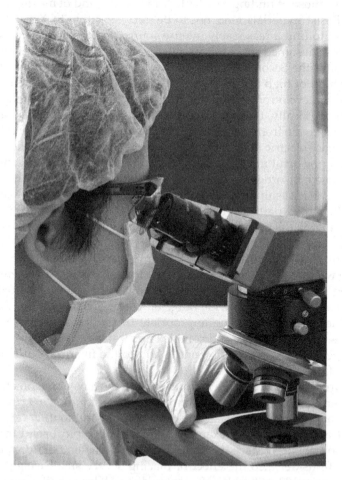

FIG. 4.5   An operator working in a semiconductor clean room. The minimisation of dust particle count is an essential feature of modern integrated circuit manufacture and all personnel are obliged to wear specially washed clothing designed to prevent the shedding of skin or hair. Human beings are far and away the dirtiest components of clean room equipment! Courtesy of iStockphoto. © Eliza Snow, image 2529737.

serious enemy. It takes only one dust particle to ruin a silicon chip and, what may be even worse to ruin a mask. Yield is an ever-present concern and the more complex the circuit, the greater the concern so dust must be eliminated at all costs. What is more, as circuit components become smaller, one has to worry over smaller dust particles. When resolution was measured in microns, dust particles of 10 nm size could be safely ignored but this is no longer the case – smaller and smaller particles must be filtered effectively. The standard approach has been to confine semi-conductor processing within a clean room which takes its air supply through appropriate filters, then perform lithography in special cabinets which maintain a slight overpressure so that a laminar flow of filtered air is blown past the work. This hopefully ensures that bits and pieces of human detritus (skin and hair) are kept out of range and clean room personnel wear complete suits of specially washed fabric as a further attempt to eliminate such deadly interference. It would be well to empha-sise that semiconductor processing involves many different processes and each one must be performed with a high degree of cleanliness. Dirt in a layer of photoresist, for example, is just as serious as dust in the atmosphere. Chemicals used for removing resist or for etching silicon oxide must be equally clean. Only when the chip has finally been mounted and packaged can the technologist relax and allow his vulnera-ble product out into the dust of day. Prior to this, standards have to be kept universally high.

It must be clear from this account that the technology of designing and making integrated circuits is an incredibly complex and challenging one and I make no pretence at having described it fully. I can only hope to have given the reader a feel for the kind of problems which have been and are being solved by a veritable army of semiconductor technologists. As I remarked earlier, the day of the single scientist or even the small group of scientists has long since departed in this business. Progress is made by large multidisciplinary teams, physics, chemistry, materials science, com-puter science and engineering all playing a role in what has developed into one of the largest commercial activities in the world today – only pharmaceuticals and automobiles compare with it – and it is truly remarkable to think that it all started only 60 years ago in a New Jersey science laboratory.

That is all I want to say about integrated circuits but it is certainly not the end of the silicon story - there is much more to be said about silicon. The title of this chapter includes an item called 'power electronics' and it is time now to turn attention to this. The ubiquitous appearance of the silicon chip in so much modern literature makes it all too easy to overlook the fact that silicon is seriously involved in many other fields of human endeavour. One physicist, working with silicon MOS structures, won the Nobel Prize for his discovery of the quantum Hall effect and we shall look at this towards the end of this chapter but of rather more everyday interest is silicon's dominance of those semiconductor device fields associated with the control of electric power, the generation of photo-voltaic solar energy and the newly developed flat panel television screens to which most of us are now addicted. These latter aspects will be discussed in Chapter 10 – for the present we shall concentrate on silicon power devices which crop up in all kinds of unexpected places and which deserve considerably greater publicity than they customarily receive. Current annual sales of semiconductor power devices are running at a level of $70 billion, several times smaller, perhaps, than the sales of integrated circuits but certainly of the same order of magnitude, and growing rapidly.

A complete list of applications for power devices would probably run to several pages, its very length illustrating their considerable importance. The scope of their usage is tremendously wide, running all the way from the kitchen sink to high speed trains and from domestic power tools to steel rolling mills so it would be impossible in this brief account to describe more than a small selection. Let us first look at a number of applications, with a view to obtaining a reasonable feel for their range, then learn something of the devices themselves before taking a closer look at one or two specific applications which illustrate their modus operandi. Battery chargers and power supplies represent an obvious example, where alternating current must be rectified to provide a necessary DC supply. Rechargeable batteries, of course, come in all shapes and sizes, from the modest specimens required to run a mobile phone or digital camera, through their somewhat larger cousins which may drive an electric drill to the hefty samples used in automobiles and fork lift trucks. They all have to be charged by way of a semiconductor rectifier and that rectifier is almost always made from silicon. While the rectifier is generally no more than a simple p-n diode, many applications call for something more sophisticated. A wide range of electric motors finds employment throughout our modern world and makes a correspondingly wide range of demands for electronic control. Examples cover the spectrum from kitchen gadgets such as the ubiquitous blender, through a range of power tools to heavier duty applications such as the fork-lift truck and to the huge power plants which drive our electric trains. In all cases there is a need for speed control based, typically on a silicon switching device known as a thyristor. Think, too, of the control necessary to bring a lift smoothly to rest at the desired floor of any tall building. This, again, makes use of thyristors, which, in turn, are tutored by silicon integrated circuits. Yet another sphere of application is concerned with automobile electronics, where engine control, battery charging, headlight switching and numerous other sophisticated functions depend on simi-larly sophisticated silicon power devices. Similar remarks apply to air-craft flight controls where the level of sophistication and the demand for reliability are raised to even higher levels. Many more examples exist but the time has come to take a look at the nature of the power devices, themselves.

As integrated circuits have grown rapidly in complexity, the size of a silicon chip has stayed roughly constant, in the interests of maintaining satisfactory yields. Too many circuit elements on a chip increase the risk of faults which render the chip unusable. On the other hand, power devices have been getting larger and larger (see Fig. 4.6) as this has become technologically more feasible. While chip sizes are restricted to square millimetres, power devices range all the way from square milli-metres to hundreds of square centimetres – that is, slices up to several

FIG. 4.6 An example of a silicon power device, mounted on heat sink. This is a silicon controlled rectifier (SCR), rated at about 100 amperes, 1200 volts and has a diameter of about half-an-inch. Note the thickness of the current carrying lead, which is appropriate to this large current capacity. Courtesy of WikiMedia.

inches in diameter for a single device. In terms of current and voltage handling, we are talking about currents as high as 10,000 A and voltages up to 10,000 V, in stark contrast to the microamps and volts appropriate to a typical IC. This is altogether a different world but, interestingly, it began life at about the same time as that of the junction transistor and in a sense it was William Shockley who can claim to have launched it. In 1950, Shockley proposed a four-layer p–n–p–n silicon structure, believing it to have a future as a switching device in microelectronics, whereas, in the event, it made its impact in the kind of macroelectronics to which we have just referred. It is ironic that, in the light of its future success, this device was largely responsible for the demise of Shockley Semiconductors because it proved too difficult to make with the technology then available (mid-1950s), a problem associated with Shockley's regarding it as a simple two-terminal device (i.e. having just two contacts, top and bottom). As we shall see, it was the introduction of a third contact to the upper n-layer which gave the device much needed stability and control. In this form it became known as a silicon controlled rectifier (SCR) and later as a thyristor.

Interestingly, the early development of practical SCRs actually happened at Bell Labs but in a different section from that ruled by Shockley. John Moll was in charge of the Group – Nick Holonyak was a young enthusiast for its experimental realisation. They were under pressure from Western Electric to develop a two-terminal switch for telephone routing to replace the current gas tube (which was a two-terminal device). In fact, they demonstrated a reproducible three-terminal switch but the telephone engineers were not prepared to compromise and the activity seems to have come to a premature end, with Holonyak being drafted into the US Army. Following this interlude, he joined the GE laboratory in Syracuse in 1957, rather than returning to Murray Hill and (coincidentally!) continued to work on the SCR under the leadership of Gordon Hall. GE were interested in developing a solid state power device to replace the well established thyratron, a gas-filled triode valve, developed during the 1920s, which was widely used for switching large amounts of power. The point of using this gas was that it could be ionised by an initial current of electrons and the ions were able to carry a very much larger current through the valve than would have been possible using only the supply of electrons emitted from the hot cathode. The initial electron current which triggered the switching action was controlled by a voltage applied to the grid electrode. Currents of up to 20,000 A and voltages up to 75,000 V could be switched in times a short as 200 ns. The thyratron had been very successful but there was a widespread tendency to look for solid state replacements for all kinds of tubes and this was not to be excepted.

The early GE SCRs were (from the top down) n–p–n–p devices and the upper p-type layer could be contacted in addition to the top n and bottom p-layers. This was achieved by alloying the top n-layer into the upper p-layer in a well defined circular region, thus leaving the p-layer exposed outside this circle. This p-type contact was referred to as a gate and could be used to switch the device on in much the same way as the grid electrode in the thyristor. Wafer sizes were about 5 mm in diameter and 250 μm thick, giving 300 V blocking (that means very small reverse currents at voltages up to 300 V) and 25 A current handling capacity in forward bias – very modest performance compared to that of today but a good start. At least these were clearly power devices and they were also three-terminal devices which could be switched by applying a current pulse to the gate electrode.

All this talk about switches rather begs the question as to just what the SCR is and does. To understand its behaviour requires a mathematical analysis of the interaction between two transistors. One can regard the device as a p–n–p transistor and an n–p–n transistor connected together by the two middle sections, n to n and p to p, and this allows one to analyse its current–voltage characteristic. The result shows that in reverse bias, it behaves like a semiconductor diode, the breakdown voltage being determined by the doping level in the lower n-type layer. The lower this doping level, the greater the breakdown voltage. In forward bias, the characteristic is more interesting. Initially, the forward current is again small but at a certain value of applied voltage (known as the breakover voltage) the device switches into a low resistance/high current state. It remains in this state as long as the forward current remains higher than a certain value (known as the holding current). The breakover voltage can be controlled by varying gate current, the larger the gate current, the lower the breakover voltage, so a gate current pulse can be used to switch from the high resistance to the low resistance state. Overall, these characteristics are very similar to those of a thyratron and this led to the SCR becoming known as a thyristor. It could clearly be used to perform similar functions.

Enough of theoretical principles, let us examine a few applications, the first of which serves as a simple example of how the SCR may be used in a domestic situation – as a dimmer switch. Here, we are concerned to control the brightness of, let us say, a 100 W filament lamp used on the standard 230 V AC supply. It would be possible to reduce the light level simply by putting a variable resistor (often known as a rheostat) in series with the bulb but this has the disadvantage of wasting power. By using a pair of SCRs in a simple circuit it is possible to achieve the same result without any waste. The principle is to reduce the time for which current flows through the lamp by including a gated SCR in series with it and controlling the time at which it switches on. Initially, the SCR presents a high resistance to current flow but, by applying a current pulse to the gate, it is arranged to switch to its low resistance mode after a certain fraction of the alternating voltage positive half cycle. If, for example, this switching happened mid-way through the half cycle, the average current would be only half that in the absence of the SCR. It simply requires a variable resistor and a capacitor to determine the precise moment of switching, which may be anywhere between the beginning and the end of the half cycle. A second SCR connected in parallel with the first, but in the opposite polarity effects a similar control on the negative half cycle of the AC supply. The current flowing through a 100 W lamp is rather less than half an ampere so this represents an example of a very low power application, the SCRs being quite tiny, no more than a millimetre in diameter and cheap enough to satisfy the demands of a simple domestic application such as this.

By contrast, we now consider a relatively high power application in the form of electric traction on the railways (see Fig. 4.7). Perhaps surprisingly, electric trains first came on the scene as long ago as 1879 though wide scale electrification had to wait until after World War II, partly on the ground of cost but also, perhaps because of a shortage of convenient control electronics. In France, Germany, England and Japan, for example, something like half the track has been electrified, while the remainder depends on diesel-electric power, largely in the interests of greater reli-

FIG. 4.7 Electric trains rely heavily on power electronics to supply electric current in the desired format and to control their speed. In this photograph we look further afield than the well known French TGV and take as our example one of the high speed (max 350km/hr) locomotives plying the Beijing-Tianjin line in China. Courtesy of WikiMedia.

ability and reduced need for maintenance – electric motors being extremely reliable and requiring far less maintenance than their more exotic steam engine predecessors. A modern electric power unit might be rated at about 5 MW and operate at a voltage of about a kilovolt which implies currents of up to 5000 A. Importantly, it must be based on the use of a DC motor in order to obtain the necessary high torque at low speeds – otherwise the train would take a painfully long time to get moving – and this means converting the AC supply to DC, having transformed it down from the line voltage of 25 kV to the 1 kV demanded by the motor. Note that it would be quite impossible to supply power at 1 kV because 5000 A of current flowing through long overhead lines would result in unacceptably large resistive losses. (Though, having said this, we should acknowledge that the London underground runs at a voltage of 600 V with currents of thousands of amperes – but this, of course, is supplied through a rail with a hefty cross-section which minimises its electrical resistance. Overhead cables have to be very much lighter and therefore thinner.) Looking at the demands on control circuitry, therefore, it is clear that we need some means of converting 1000 V AC to DC with a current handling capacity of 5000 A and including a method of controlling motor speed all the way from zero to an appropriate maximum value. Conveniently, the speed at which a DC motor runs is determined by the average current through its rotor so it is merely necessary to provide a circuit to vary this average, a requirement identical to that set by the dimmer switch – though at somewhat greater power level!

The control function demands the use of SCRs which can be switched from high to low resistance state in exactly the way we described for the dimmer switch but rectification of the AC supply to produce a DC output requires a circuit known as a bridge rectifier. This can be envisaged as a diamond shape with an SCR in each of the four arms, the AC input being applied across one pair of opposite points, while the DC output is taken from the other pair. It provides full wave rectification – that is it gives a positive DC output from both positive and negative half cycles of the AC. Each SCR is switched by a timing circuit so as to vary the temporal fraction of each half cycle for which current actually flows. The most obvious contrast with the dimmer switch lies in the magnitude of the current which demands a device area of about 50 cm^2 – a diameter of 8 cm (or 3 in.) – clearly a very different beast from that used in the dimmer. The other important difference concerns the necessary reverse voltage which the SCRs must handle without breaking down – the so-called blocking voltage. As we have commented earlier, there is a relationship between this blocking voltage and the doping level in the low-doped side of the appropriate p–n junction. To achieve a blocking voltage of 1000 V requires that the doping level be less than about 10^{20} m^{-3} (an impurity level of roughly two parts in 10^9 or two parts per billion). It also places a constraint on the thickness of the bit of silicon involved – this should be no less than 100 µm (or a tenth of a millimetre), not a difficult criterion to match, of course but an important design parameter for the device technologist. Another question that concerns him is the required speed of switching. At a supply frequency of 50 Hz a half cycle lasts for just one-hundredth of a second (10^{-2} s) and it is clearly necessary that the switching process should require much less time than this, say a thousandth of a second (10^{-3} s) or less. In fact, this is easily met – the switching time being roughly equal to the recombination time for minority carriers in the lightly doped silicon, a time typically about 10 µs (10^{-5} s).

Having considered some of the functional requirements, it might help to look at the actual structure of the SCR device and how it might be made. We shall consider a four layer structure of the form (from top downwards) n^+–p–n–p^+ with contacts to the top and bottom layers and a gate contact to the upper p-layer. This implies that the top layer is localised so as to leave access to the gate layer – note that the effective device area is determined by the area of this n^+ layer, not by the overall slice area. How is it made? The essential starting requirement is for a silicon slice with n-type doping level of 10^{20} m^{-3} into which are diffused the two p-type layers. Firstly, the bottom of the slice is masked while the upper p-layer is diffused in with an acceptor density of 10^{23} m^{-3}, then the top is masked while the bottom contact is diffused with acceptor density 10^{25} m^{-3}. Finally, a hole is etched in the top mask to allow the top n^+ layer to be diffused in locally at a donor density of 10^{25} m^{-3}. Metal contacts must then be made to the appropriate layers, two on top and one underneath – note that SCRs are not made in planar geometry, like their integrated circuit counterparts but are 'straight-through' devices. This represents the basic SCR but we should, perhaps, make one important comment – reverse blocking behaviour depends on the electric field within the n^--layer being uniform and this, in turn, requires the doping level to be highly uniform, too, not altogether easy to achieve at the required low densities. In fact, a particularly elegant technique for ensuring this was introduced in 1976, whereby the n-type doping was achieved using neutron irradiation of a highly pure silicon slice. Remarkably, this had the effect of transmuting Si^{30} atoms into P^{31} atoms which provided the doping at a very well controlled density. Doping levels were obtained with better than 1% uniformity using this revolutionary method, prompting the irreverent thought that someone up there really wanted to see electric trains with SCR speed control – such happy coincidences seem to occur extremely rarely in semiconductor (or, indeed, any other kind of) engineering.

While electric trains offer an excellent example of the use of large electric power units, electric motors abound in widely varying aspects of life, one of the characteristics of modern living (western style) being the plethora of such motors within the average home. Most domestic premises can probably boast something like twenty motors in appliances such as vacuum cleaners, food mixers, refrigerators, fans, hair dryers, small power tools, computers, CD players, tape recorders, etc. Most of these are likely to be induction motors, rather than DC motors on the grounds of the former's being significantly cheaper and (lacking any need for brushes and commutator) even more robust. AC current flowing through the motor stator induces currents in the rotor conductors which interact with the stator magnetic field to induce rotary motion. Because induction motors are essentially AC machines, they have no fundamental need for rectifiers but this is not to say that they have no need for electronic control. In fact, many applications require a circuit to provide gradual (or soft) starting and this takes the form of a pair of back-to-back SCRs, used to wind up the average current in much the same manner as employed in the dimmer switch. There are also important considerations when speed control is necessary because the induction motor runs at a speed which is determined essentially by the frequency of the AC supply. Basic machine design allows the motor to run at a convenient fixed speed when used with 50 Hz (or 60 Hz) mains but any variation can only be achieved by the introduction of a variable supply frequency and that involves some rather more subtle power electronics. It is done by first rectifying the mains supply, then using an inverter circuit to generate a variable frequency AC output. Without getting involved in too many electronic complexities, let us just note that this consists of a bridge circuit (similar to the bridge rectifier we discussed earlier) in which SCRs are switched on and off in pairs to produce a roughly square wave output waveform which approximates the desired sinusoidal shape. (Note the need to switch *off* the SCR, as well as on – we shall say more about this in a moment.) By changing the rate at which the devices are switched on and off, it is possible to vary the frequency of the AC output, as desired. Typically, this frequency may be varied between about 5 Hz and 100 Hz, providing a motor speed range of a factor of 20 times. However, in some applications there is a need for quieter running which can be achieved by increasing the supply frequency to values beyond those detectable by the human ear – that is greater than 20 kHz. This makes further demands on the switching devices by requiring switching times less than 10 μs – we shall come back to this too but, needless to say, the semiconductor technologists have risen to the challenge and high frequency inverter systems are in widespread use. One final brief comment – with regard to soft starting, it is worth noting that a similar requirement holds in the case of wind turbine generators (Fig. 4.8). Connecting a turbine abruptly to the electric grid would place huge stresses on the associated gear box and would cause undesirable perturbation to the grid so it is necessary to arrange for the connection to be made gradually, using a similar

FIG. 4.8 The UK rush to install renewable energy sources makes considerable demand on wind power but there are serious problems in matching wind turbine output to the national grid - problems which are efficiently solved through the exigencies of silicon power electronics. Courtesy of iStockphoto, © Terrance Emerson, image 3694914.

technique to that employed in starting an induction motor. Needless to say, the power levels involved with wind turbines tend to be rather large and the SCRs correspondingly large in cross-section.

Talking of wind turbines brings us to another major application for power electronics in a field which is growing rapidly in significance. There has been much discussion concerning the relative merits of fixed and variable speed turbines, and power devices play an important role in this. It is obvious that wind speeds vary considerably in most locations and some flexibility must be available in the design of a turbine/generator system to take account of it. One approach has been to arrange for turbine blades to change pitch in response to changes in wind speed, thus keeping turbine speed approximately constant. This has obvious advantages in so far as the frequency of the generator output is determined by its speed of rotation and if the intention is to feed power into the electricity grid it is obviously important to maintain a close match with the grid frequency of 50 (or 60) Hz. (There are also problems with small phase differences between the generator output and the grid voltage, often referred to as reactive power.) At the same time, it is also important that the turbine system produces as nearly as possible its optimum power and this can usually be achieved by allowing the blades to rotate at variable speed, while correcting for frequency mis-match with an electronic converter. The generator output is rectified to produce a DC supply which is then changed back into AC by an inverter circuit, operating at mains frequency. The trigger for switching the inverter SCRs can be taken from the mains, thus ensuring an accurate frequency match. In the case of off-shore wind farms, where the turbines are located at some distance from the grid, there are advantages in using a high voltage DC link between turbines and shore facilities, so the inverter can be located on land, remote from the rectifier. There is an obvious simplification in using DC to couple a large number of generator outputs together because it does away with phase matching problems – at the same time, it automatically takes care of reactive power problems. Well over half the world's wind power installations employ variable speed turbines and high power electronic devices appear to have bright long-term prospects in this particular market which promises to grow steadily into a rosy anti-global-warming future.

Yet another application of power semiconductor devices which has grown dramatically during recent decades is concerned with automobile electronics. For many years the car was a purely mechanical device which employed only the bare minimum number of electrical components required to provide lights, ignition and (later) direction indicators. Even the windscreen wipers were driven by the vacuum existing in the exhaust manifold. Suddenly (it seemed), the industry discovered electronics and the wiring loom grew rapidly in both volume and complexity – so much so, in fact, that a completely new concept has emerged recently in the form of a current carrying 'bus' (a fairly hefty copper conductor) running round the car which carries the current to each particular electrical component. This conveniently replaces the individual current leads but demands a semiconductor switch at every headlight, side light, flashing indicator, etc., to connect the bus to each one, as required. Operation of these switches is controlled by data signals carried by very much thinner control wires, thus reducing the size of the loom to more manageable proportions. However, the number of electrical components has increased to a remarkable degree. We might mention, for example, electronic ignition, fuel injection, engine management, rectification of the alternator output to provide a DC supply, voltage regulation and overvoltage protection, antilock breaking systems, traction control, electronic stability programmes, automatic headlight levelling, parking aids and engine immobiliser systems. Each and every one requires sensors and electronic control units based on a range of diodes, transistors and thyristors and results in a tremendous boost to semiconductor sales. Little wonder that the power semiconductor market is growing so rapidly. It also seems likely to boost sales of batteries, the trend now being in favour of separate batteries to provide power to the starter motor and to the plethora of other electrical functions.

Needless to say, these many demands on the semiconductor industry have posed significant problems and required corresponding efforts to meet a wide range of device specifications. Whereas the integrated circuit business has followed a steady linear course of miniaturisation, power has been obliged to innovate in all directions at once – current capacity, voltage blocking, switching speed, symmetrical switching, low loss, price reduction have all been subject to pressure from an ever growing range of users. It is fascinating to examine the various different devices which have emerged from this maelstrom of activity and we can list them roughly as follows: diodes, bipolar junction transistors (BJTs), thyristors (SCRs), gate-turn-off thyristors (GTOs), reverse conducting thyristors (RCTs), light-activated thyristors (LATs), diode AC switches (DIACs), triode AC switches (TRIACs), power MOSFETs, insulated gate bipolar transistors (IGBTs), static induction transistors (SITs), static induction thyristors and MOS-controlled thyristors. There may well be others! Though each and every one of them has its application, it would be a trifle boring to describe them all in detail, however, we should look at one or two important features.

Firstly, let us recall the need for a thyristor which can be switched off, as well as on. We noted an example of this in our discussion of the inverter circuit used to drive induction motors at variable speeds. It was satisfied in the early 1960s with the development of the GTO, a thyristor in which the gate and cathode (the top n-type layer in our n–p–n–p device) were arranged to be interdigitated – that is, they were in the form of interleaved narrow strips on top of the device. In this configuration a positive pulse on the grid switched the device on in the usual way but, in addition, a negative pulse was capable of sweeping free carriers out of the lightly doped n-region and thus switching the device off. Again, in considering high frequency drives for induction motors, we came across the need for very rapid switching and this implied the need for short (minority carrier) recombination times. A clever trick was used here. Certain impurities in silicon, such as gold or platinum introduce energy levels near the centre of the forbidden energy gap and these can be used to speed up electron-hole recombination by way of a two-step process – an electron from the conduction band is captured by the impurity level, then a hole from the valence band recombines with the captured electron. By introducing more of these impurities, the recombination time is progressively decreased, as required. An even neater method depends on the fact that irradiation with high energy electrons from an electron accelerator can introduce moderate amounts of damage into the silicon crystal, conveniently, producing similar 'recombination levels'. The recombination time decreases as the intensity of the electron irradiation is increased and the switching speed is correspondingly increased – another nice illustration of the way in which sophisticated control over materials can win battles on the commercial front.

The MOS transistor came to pass at Bell Labs, in its micro-form, in 1960 and its rapid takeover of a large slice of the integrated circuit market inevitably stimulated interest in the parallel field of macro-devices. However, it was not until the end of the 1970s that the power MOSFET came into being. Unlike its micro-counterpart, it is a vertical structure, in order to enhance its current handling capacity and this makes it more complicated to manufacture, one reason why it came rather late on the scene. It differs from the BJT in several important ways: having a high impedance, it can be used in different circuit configurations from the BJT (which, in its 'on' state, has a low impedance), it is switched by applying a voltage pulse to the gate (drawing only a very small current) whereas the BJT and the thyristor require current pulses to switch them, it is a majority carrier (rather than a minority carrier) device and can be switched very much faster (because it is not limited by the time taken for minority carriers to recombine). In practice, switching frequencies of up to 1 MHz are possible and MOS devices tend to dominate the market for high frequency switching. Against this, its current handling and its voltage blocking capabilities are significantly less than those of its bipolar rivals, emphasising the importance of horses for courses. The essential feature is, of course, that there are now power devices available to meet just about any challenge which might arise.

Two final comments are in order. Recent trends are moving in the direction of combining both power devices and data handling ICs on the same piece of silicon. Such smart power solutions have the obvious advantage of being less expensive than those based on separate components, while reducing the need for complex interconnections. Gradually, the dividing line between micro- and macro-electronics is being blurred and, in the foreseeable future, may well be totally eroded. The other aspect which is worth mentioning is the development of power devices based on materials other than silicon. In particular, silicon carbide is showing some promise as a material for devices which have to operate in high temperature environments. This ability arises from the considerably greater band gap of this semiconductor (3 eV), compared with silicon (1.12 eV), a factor which also results in potentially higher breakdown voltages. The down side is that it is a much more difficult material to tame and the resulting higher cost will probably ensure that its use is always confined to one or two specific applications.

This may feel to have been a rather long and involved chapter – silicon seems to be involved in so many different areas of application – but

I should like to bring it to an end with a brief reminder of one of the fundamental principles of the semiconductor story. The transistor was discovered as a result of some dedicated work on basic semiconductor physics and it developed its present position at the heart of our domestic and industrial life as a result of some dedicated work on material purification and structural perfection as well as some inspired device technology and it is the continuing interaction between semiconductor physics, technology and device design that has enabled semiconductors to satisfy a remarkable range of system demands, some of which we have just examined. As a kind of postscript to this account of silicon's domination of the semiconductor commercial scene, we shall now look at an example of the way in which silicon technology, developed specifically for device improvement, played an important role in the progress of pure science. I refer to the fact that Professor Klaus von Klitzing of the Max Planck Institute in Stuttgart was awarded the 1985 Nobel Prize in physics for his discovery of a novel physical effect known as the quantum Hall effect while studying the electrical properties of silicon MOS structures. The work was actually done in the High Magnetic Field Laboratory in Geneva (run jointly by the French CNRS (Centre National de Recherche Scientifique) and the German Max Planck Institute) for the essential reason that the experiments required very large magnetic fields. It was done in collaboration with another German scientist Gerhard Dorda from the Siemens Company in Munich and an Englishman Professor Michael Pepper from the Cavendish Laboratory in Cambridge, who supplied the MOS structures – a good example of the way in which much modern research is performed, across national and academic/governmental/industrial boundaries.

To understand the nature of their discovery we need to remind ourselves what the basic Hall effect is. We first discussed it in Chapter 2 as a technique for studying new semiconductor materials and measuring the density of free carriers contained therein. The experiment required a bar-shaped sample with electrical contacts at either end, allowing an electric current to be passed down the length of the bar, together with a second pair of contacts on opposite sides of the bar. A magnetic field applied along a direction normal to the upper surface of the bar caused electrons or holes to be deflected to the side of the bar, setting up a voltage between the side contacts. The sign of this Hall voltage provided information concerning the sign of the free carriers (i.e. whether they were electrons or holes) and its magnitude allowed the experimenter to measure their density. In the classical version of the experiment, the Hall voltage varied linearly with both current and magnetic field, an observation confirmed hundreds of times over on a wide range of semiconductors (and, indeed on metals, too). What von Klitzing observed were very significant departures from this simple behaviour when the magnetic field was increased from the usual value of about 0.3 T (T = Tesla – for comparison, the earth's magnetic field has a value of order 10^{-4} T) to much larger values, up to 18 T (which could be obtained using large superconducting magnets) and when the sample was immersed in liquid helium at the very low temperature of 1.5 K. However, it is important to appreciate one further aspect of the experiment, that the sample was in the form of a 'two-dimensional electron gas' – a jargon phrase which clearly demands some explanation! The MOS structures had a voltage applied to their gate electrodes so as to create a very thin inversion layer close to the interface between the bulk silicon and the gate oxide. The bulk silicon was lightly p-type so the inversion layer consisted of electrons, confined so close to the interface that they could move only in the plane of the interface – in other words, they could move in two dimensions but not in the third. In this respect they were very different from electrons in a normal bulk semiconductor and the QHE is essentially a property of this two-dimensional arrangement. The conceit of referring to these electrons as a gas simply reflects the fact that they are free to buzz about with random thermal velocities in much the same manner as do molecules in a normal gas. (It allows scientists to use the vulgar abbreviation 2DEG to describe a rather sophisticated sample configuration!)

So what exactly did the experiments show? At moderate values of magnetic field the usual linear dependence of Hall voltage on field was observed, providing a measure of the density of free electrons per square metre but as the field was increased, a series of abrupt steps appeared in the plot of Hall voltage versus field, steps which had never been seen before and which occurred in a remarkably precise sequence. If we define the ratio of Hall voltage divided by sample current to be the Hall *resistance*, then the steps occurred at values of Hall resistance given by a universal constant divided by an integer number, and this relationship was accurately reproducible to about one part in a billion. Such precision soon led to this phenomenon being used as a resistance standard and, what was more, the universal constant was found to be the ratio of the Planck radiation constant divided by the square of the electron charge, a remarkable and totally unexpected result. We shall discuss it further in Chapter 6 when we meet with yet another type of 2DEG and yet another type of quantum Hall effect.

CHAPTER 5

Laser Beams and Microwaves: Gallium Arsenide and Indium Phosphide

So far in our journey through the semiconductor landscape, we have been obliged to concentrate attention almost exclusively on germanium and silicon, for the simple reason that these two materials totally dominated the early development of semiconductor devices. The transistor was born of germanium and the first effective field-effect device of silicon (and its wonderful oxide). Then, as we saw in Chapter 4, silicon, with its larger band gap, gradually took over complete responsibility for both integrated circuits and power devices. Once the rather more recalcitrant technology of silicon had been tamed, there seemed to be no need for the involvement of any other material – silicon could satisfy all possible requirements! Or could it? There were those who disagreed with such an assessment and, in this chapter, we shall begin to examine some of their reasons. In fact, something like 20 other semiconductor materials have entered the commercial fray and the overall picture of the solid state electronics scene covers a very much wider canvas than the one we have painted so far. The fact that two thirds of this book is concerned with materials other than silicon surely bears out that statement but we should be clear that this certainly cannot be taken as denying the dominance of silicon in the market place. While the world market for silicon devices is close to $300 billion, that for its nearest rival gallium arsenide (GaAs) is less than $10 billion, something like 40 times lower. However, it would be an arrogant (or careless) business man who turned up his nose at a market of $10 billion – in its own right surely well worth some appreciable investment. We should be aware, none-the-less, that there exists a considerable discrepancy between the sales of silicon and those of all the other semiconductors that we are about to discuss.

Semiconductors and the Information Revolution: Magic Crystals that made IT happen   © 2009 Elsevier B.V.
DOI: 10.1016/B978-0-444-53240-4.00005-2

Not that this mere monetary superiority of silicon should be allowed to distort the relative importance of the various materials. Each new commercial semiconductor demands major investment in developing its material technology (we saw the truth of this in our earlier discussion of germanium and silicon) and such investment can only be justified by some significant return. Examples might include improvements in man's overall capability to communicate over greater distances, obtain more convenient entertainment, fight wars more effectively, light his way more brightly, generate energy more cheaply, reduce his carbon footprint more plausibly, provide medical care more efficiently, etc. Each new material has made its appearance on the commercial scene on the understanding that it could make a real contribution to one or more such activity. While making no pretence of being encyclopaedic, we might note, for example, that GaAs, indium phosphide (InP) and gallium nitride have been responsible for the development of microwave devices used in mobile phones and of semiconductor lasers used in compact disc and digital versatile disc players, several materials such as gallium phosphide, aluminium arsenide and gallium nitride have contributed to the development of light emitting diodes used in cars, traffic lights and special lighting systems, mercury cadmium telluride has provided superior night vision systems aiding military capability, while an alloy of InP and GaAs has been vital to the development of lasers for fibre-optic communication systems. Perhaps the principal point to emphasise here is that each of these several applications makes quite specific demands of the appropriate semiconductor – no one material can ever satisfy all these requirements and it explains why, on the one hand, there are so many materials of commercial interest and, on the other, why none of these materials can hope to rival silicon in sales volume.

As we have already seen, silicon has a number of properties which make it suitable for what we might call conventional semiconductor devices – transistors, integrated circuits, thyristors, etc. – and in many ways it seems to be the ideal semiconductor material. It can be extracted from sand, one of the world's most readily available sources, it has a suitable band gap, it has an almost ideal oxide and it is elemental (it contains only one type of atom). The technology involved in purifying it and producing high quality single crystals, while hardly trivial, is relatively straightforward and was mastered remarkably quickly, once its advantage over germanium was properly appreciated. One might therefore be forgiven for asking why silicon may be criticised at all – it sounds, at first hearing, to be almost the ideal semiconductor. The answer to such a question is implicit in the comment made above about each new application demanding new properties. The most significant of these is the band gap and, while silicon has a band gap which is convenient for its role in integrated circuits, etc., it is quite unsuitable for several of the applications referred to above. To illustrate this, we first note the importance of band gap in relation to light-emitting devices (discussed in Chapter 7). The key point here (as we shall see) is that the band gap must be equal to or greater than the photon energy of the light emitted and, given that the energies of visible photons lie between about 1.8 eV and 3.1 eV, silicon's band gap of 1.12 eV makes it quite unable to emit visible light. In fact, silicon is a very inefficient light emitter (because of its indirect band gap – see Chapter 3) but, in any case, it emits only in the infra-red part of the spectrum, at wavelengths close to 1.1 μm (1100 nm). Here we have one major area of application to which silicon is altogether inappropriate. Another is concerned with fibre optic communications (see Chapter 8), where the laser light used to carry data must be generated by semiconductors with direct band gaps – again silicon is effectively ruled out. A third example is that of the far-infra-red devices used to detect thermal radiation in night vision systems (Chapter 9). In this case the demand is for band gaps very much smaller than that of silicon – typical values being in the range 0.1–0.4 eV.

Clearly, there exists a number of applications to which silicon is totally inappropriate but there are also areas where, though silicon may compete, it has limitations which render it vulnerable to challenge by rival semi-conductors. Interestingly, it was one such application which provided justification for the early development of GaAs, namely the fact that electrons in GaAs are significantly more mobile than they are in silicon (they diffuse faster and acquire greater velocities when accelerated in an electric field). Another example concerns the use of a GaAs diode as a device for coupling signals between different parts of an electronic circuit using light as the carrier (rather than making a direct electrical connection). Because GaAs is a direct gap material, GaAs diodes can emit light (strictly speaking, infra-red radiation at a wavelength of 880 nm) with a relatively high efficiency, against which silicon diodes are unable to compete effectively. A third example where GaAs possesses an advantage over silicon is represented by a phenomenon known as the Gunn effect (after its discoverer J.B. Gunn) which has proved invaluable in the generation of microwave power in simple radar systems and microwave test gear.

Hopefully, this brief introduction will have persuaded the reader that there is life beyond silicon – the rest of the chapter provides an introduction to a group of compound semiconductors, the so-called III–V semiconductors which have made a strong challenge to silicon and established an important niche for themselves in the market place. But firstly, we should comment on the use of the word 'compound'. Silicon and germanium are referred to as elemental semiconductors because they are simple chemical elements – selenium and tellurium are two others – but there are very few elements which have semicon-ducting properties and, if we are to look for more semiconductors, this must inevitably involve chemical compounds. The simplest of these are the so-called binary compounds, containing just two species of atom – sodium chloride, NaCl (common salt) is probably the best known example but it is not, however, a semiconductor. The III–V semiconductors are also binary compounds consisting of one atom from group III in the periodic table and one atom from group V, so, in principle, we might select a group III element from any of: boron, aluminium, gallium and indium and combine it with any group V element: nitrogen, phosphorus, arsenic and antimony which gives us 16 possibilities (for readers familiar with the table, we should, perhaps, note that the known thallium and bismuth compounds are metals, rather than semiconductors). Most of these 16 compounds crystallise in a form very much like that of silicon and germanium, a tetrahedral arrangement in which each group III atom is surrounded by four group V atoms and each group V atom by four group III atoms, and in many ways they have similar properties. However, they provide a wide range of band gaps. The lighter atoms tend to give compounds with wide energy gaps – aluminium nitride, for example, has a gap of about 6 eV – while the heavier elements give narrow gaps – indium antimonide's (InSb) being 0.17 eV. Given that it is often possible to make alloys such as (aluminium/gallium) arsenide or gallium (arsenide/phosphide) which show gaps intermediate between those of the end members, it soon becomes clear that this group of compounds provides tremendous flexibility in semiconducting properties. Much agonising has gone into establishing the details but we need not concern ourselves with any of this at present.

The scientific study of the III–V materials began seriously in the early 1950s. It was only a very few years since the invention of the transistor and well before anyone realised just how important silicon was about to become. (Remember that it was not until 1954 that Gordon Teal at TI produced the first silicon junction transistors.) The pioneering work was done at the Erlangen laboratory of the German Siemens company by a converted academic Heinrich Welker who joined Siemens in 1951 and found there an environment peculiarly sympathetic to his interest in the systematics of this new complex of semiconducting compounds. He was particularly interested in the nature of their chemical bonding and the relationships between this and their semiconducting properties. Siemens, on the other hand, were interested in the possibility that one or two of these materials might reveal capabilities to rival the still somewhat insecure position of the elemental semiconductors in the world of commerce. It may have been a marriage of convenience but it certainly produced flourishing offspring in whom both parents could feel pride. Welker chose initially to concentrate on two compounds, aluminium antimonide (AlSb) and InSb which confirmed a number of theoretical ideas concerning the chemical bonding, band gap, free carrier mobility, etc. One characteristic was the relationship between these features and the atomic weight of the constituent atoms. For example, the heavy material InSb had a small direct gap, a low melting point (527 °C) and a very high electron mobility, while the lighter AlSb melted at a much higher temperature of 1050 °C, had an indirect band gap of 1.6 eV and quite modest electron mobility. This early success in establishing important basic properties owed much to Welker's recognition that it was vital to employ the purest possible starting materials and he devoted considerable effort to purifying aluminium, indium and antimony before growing small single crystals from appropriate melts. Following a visit by Shockley to his laboratory in 1952, he himself visited the Bell laboratory in Murray Hill and learned from Pfann the secret of zone refining. Such is the way in which scientific expertise spreads.

Welker continued his fundamental studies but Siemens were quick to capitalise on his work. Interestingly enough, in 1952, they took out a patent on the III–V semiconductors and any devices which might be made from them. It seems remarkable that they were able to do this at all and I cannot imagine they were able to defend it successfully – without doubt applications gradually appeared from numerous other establishments in Europe and on the other side of the Atlantic but whether Siemens were able to obtain royalties I know not! More specifically, Welker's most dramatic discovery had been the remarkably high electron mobility he measured in InSb (by way of the Hall effect) – in his purest material this reached values as much as 25 times those typical of germanium or silicon – and Siemens soon found practical ways of exploiting it. Previously, the Hall effect had been used as a tool to characterise the properties of semiconductors but it could, of course, be turned round and used to measure unknown magnetic fields (assuming the properties of the semiconductor were already known) and, the high electron mobility of InSb made it an ideal material for such an application. (The Hall voltage is directly proportional to mobility.) Fields could be measured to an unprecedented accuracy and Siemens engineers found several other applications based on this same effect. These included a non-contact method of measuring electric currents, not only in wires but also in the form of ion beams, and a method of measuring magnetic properties of certain materials.

Interesting though such applications might be, they were not exactly in the mainstream of solid state electronics and it was left to others to pursue the more conventional aspects. By the mid 1950s work on the III–V compounds was under way in Russia, France, England and the USA and it is probably fair to say that, in terms of scientific understanding, several of the III–V compounds were almost as far advanced as their elemental counterparts (though, as we shall see, their device application was proving somewhat more elusive!). By the end of the 1950s,

Cyril Hilsum and Christopher Rose-Innes, of the Services Electronic Research Laboratory at Baldock, were able to write an admirable book 'Semiconducting III–V Compounds', setting out the considerable amount of information which had already been accumulated. Hilsum became one of the foremost protagonists of the III–V cause and led a remarkably strong UK activity, which contrasted strongly with the uncertain and relatively unsuccessful national programme on silicon. British Government policy was apparently based on the principle of ensuring satisfaction of military needs but with relatively little insistence that such provision be made by UK firms. There was, however, a bias towards supporting technically advanced research and development, and this favoured work on III–V materials, where the scope was much wider than it was in the case of silicon. Whatever the inner reasons, there can be no doubt that Government laboratories such as Baldock and the Royal Radar Establishment (RRE) in Malvern (where all this work was eventually concentrated) made a major contribution to the rapid advancement of III–V materials. RRE, in particular was to play a leading role in the development of appropriate crystal growth methods and yet another as sponsor of wide ranging III–V research programmes in various industrial laboratories. (RRE has had something of a chequered history – it grew out of TRE, then, when SERL was incorporated, it became RSRE, later DERA and, finally, in 2001 it was made a private company under the name QinetiQ.)

There had, of course, to be valid motivation for the work and initially it was argued that effort should be concentrated on GaAs as a potential rival to silicon for making junction transistors. This was on the basis of its much greater electron mobility – roughly five times that in silicon – which promised transistor operation at significantly higher frequencies than silicon could hope to reach. The band gap of InSb was far too small, the electron mobility of AlSb was extremely modest and, in any case, it soon became apparent that AlSb crystals were unstable in moist air, slowly dissolving in their own juice, as it were! The band gap of GaAs was a convenient 1.4 eV and its melting temperature an achievable 1238 °C (slightly lower than that of silicon) and it was chemically stable. There was a minor drawback, however – the vapour pressure of arsenic at the melting point was extremely high, resulting in numerous explosions when crystals were grown from the melt in vulnerable quartz tubing. Hilsum describes the importance of surrounding the growth apparatus with a defensive brick wall and how small crystals could often be rescued from the ensuing catastrophe.

While (before the onset of Health and Safety at Work!) such dangerous living might be acceptable in the short term in a research laboratory, it was hardly a suitable basis for a commercial activity – something clearly had to be done to develop a viable crystal growth method. The initial breakthrough came from the Philips laboratory in Eindhoven, in the form of a two-temperature furnace, the GaAs melt being at 1238 °C but with a second region containing solid arsenic at a temperature of 600 °C. The point was that arsenic at 600 °C generated vapour at a pressure of about 1 atm which was just sufficient to prevent loss of arsenic from the molten GaAs at 1238 °C. There was no possibility of a build-up of the extremely high arsenic pressures existing in the early work. This clever trick, frustrating the natural tendency of GaAs to dissociate, allowed well controlled crystal growth to proceed and provided better samples for further measurements to be made. Unfortunately, measured electron mobilities reached no more than half the value predicted theoretically and it soon became clear that even this better material was still far from ideal. The next step, therefore, following the example set by germanium and silicon, was to experiment with the Czochralski technique. This was duly done in 1955 in Germany by R. Gremmelmaier who used a sealed tube in which to confine the arsenic vapour and magnetic coupling to transmit the pulling and rotational motion to the seed crystal, an admirable piece of technology which advanced the art of GaAs growth one important step further. It led, shortly afterwards, to the first GaAs p–n junction but attempts to make a junction transistor to challenge silicon's headlong progress were thwarted by the observation of undesirably short minority carrier recombination times. While such times in silicon would be measured in microseconds, in GaAs the appropriate units were nanoseconds so, in spite of the improved mobility, minority carriers recombined long before they could reach the collector junction, and gain was almost non-existent. Impasse!

This could easily have been the end of a III–V bubble had the protagonists not been made of sterner stuff. In fact, it was merely a second beginning. Within the next decade GaAs alone was to demonstrate no less than three unexpected virtues, all of which led to important commercial applications, and what was more, it also formed the basis for one of the most exciting scientific and technological developments since the first transistor, the development of nano-structures. This took place during the 1980s and will form the subject of Chapter 6. As we have already seen, other III–V compounds to meet specific challenges were gallium phosphide and gallium nitride in the world of visible light emission and the alloy InGaAsP in the realm of fibre-optic communications (Chapters 7 and 8). For the moment, however, we shall concentrate attention on GaAs and its three major device progeny, the Gunn diode, the semiconductor laser and the MESFET (metal–semiconductor field-effect transistor). The time scales of these three overlapped considerably but, in the interest of clarity, it makes sense to discuss them separately. We must also face complications resulting from numerous developments in material technology which occurred in response to challenges thrown up by these various devices. In particular, no less than five new crystal growth methods were introduced during the period between 1960 and 1980, each of which made a significant contribution to the commercial success of the devices in question. It was a complicated story and, in many cases, seriously confused by the inevitable rivalries between different techniques, different laboratories and different commercial interests, while, all along, there was an underlying struggle to justify the considerable investment involved. Silicon had taken the inside lane and GaAs, in the outer track, was permanently running against considerable odds. It became something of a cliché to comment that 'GaAs is the material of the future – and will always remain so'! It was a little unfair at the time and, with hindsight, it can be seen to have been quite untrue, but it nevertheless reflected a widely held view. Cyril Hilsum and his worthy band of supporters needed not only well developed powers of persuasion but pretty thick skins. I shall try to make all this clear in a moment but first, we should familiarise ourselves with what was still the 'material of the future'.

J.B. Gunn once described GaAs as 'only germanium with a displaced proton', on the basis that the gallium atom sits immediately to the left of germanium in the periodic table and has one proton less, while arsenic lies immediately to the right and has one proton more. This relationship implies that the two materials are closely similar in some of their physical properties such as density and lattice parameter (a measure of the distance between neighbouring atoms), though they differ in many electronic properties on account of differences in chemical bonding. Germanium crystals are held together by what the chemists call a covalent bond, while GaAs depends on both covalent and ionic bonds. The ionic bond is dominant in crystals such as sodium chloride (which is a I–VII compound), whereas in GaAs it plays only a secondary role, though sufficient to affect its electronic behaviour. In particular, the fine details of the conduction bands differ in important ways and, as a result, GaAs has a larger direct energy gap. It also has a small electron effective mass (free electrons in the conduction band behave as though they are very light and are therefore easily accelerated by an electric field) which results in a large electron mobility. The direct band gap, the structure of the conduction band and the small effective mass each played important roles in the material's device future, which we are about to explore. First came the laser, then the Gunn diode and, finally the MESFET. We shall look at each in turn and introduce the various crystal growth methods necessary for their development, as required.

The race to demonstrate laser action in GaAs and thus produce the world's first semiconductor laser was a remarkable example of the scientific competitive instinct. It all came to a head in the autumn (or, since this was another all-American race, the fall) of 1962 but we first need to understand a certain amount of background. The laser story has been well told by the Nobel Prize winner Charles Townes in his book 'How the Laser Happened'. It all began with his demonstration of a totally new type of microwave oscillator based on a phenomenon known as 'stimulated emission' which had been elucidated by no less a scientist than Albert Einstein in 1917. We can illustrate the concept by thinking of a p-type semiconductor and suppose that some minority electrons have been injected into its conduction band. There is a finite probability that these electrons will recombine with holes in the valence band, emitting light with photon energy approximately equal to the band gap energy, on a purely random basis. This is called 'spontaneous emission'. But there is another possibility. If photons of band gap energy are incident on the semiconductor, they may stimulate the recombination process and produce 'stimulated emission' of light. These two processes will proceed together but if the intensity of the incident light is increased, stimulated emission will gradually become dominant. Townes was working not with semiconductors but with ammonia gas and he was concerned with photon energies measured not in electron volts but in units of 10^{-4} eV. None-the-less, the principles were the same – he was concerned to manipulate his ammonia molecules so as to arrange for the stimulated emission to dominate. Success came in April 1954 – Townes and his team of graduate students at Columbia University had built the first microwave oscillator based on stimulated emission. They called it a MASER (an acronym for Microwave Amplification by Stimulated Emission of Radiation) and it soon came to be known as a maser. It operated at a frequency close to 24 GHz, determined by the properties of the ammonia molecule but people quickly started to think about other materials and other frequencies.

It was realised that this same principle might be used to make a highly sensitive microwave amplifier and work got under way at several laboratories, including Bell Labs and the Mullard Research Laboratory in Redhill, Surrey (where, at that time, I was a very youthful new employee). The aim was to make a maser to act as the first stage amplifier on the microwave dishes to be used for communication experiments via the Telstar satellite. In this case, the active material was to be an artificial ruby (an aluminium oxide crystal doped with chromium ions) and the whole amplifier was designed to work at a temperature of 4 K provided by liquid helium. It was a great success and the first trans-Atlantic TV signals were received at Goonhilly Down, Cornwall in July 1962. At much the same time, it was also realised that the maser principle might be used to generate light, thus making an optical maser – or laser. Once again, two very different materials were investigated – a gas, a mixture of helium and neon and a solid, again (coincidentally) ruby. Ted Maiman at Hughes Research was first to demonstrate laser action in a ruby crystal. This was in May 1960, followed shortly afterwards by Ali Javan's helium-neon laser at Bell Labs. The world of optics was literally over the moon– following the moon landing in July 1969, a laser beam was reflected back to earth from reflectors set up by Armstrong and Aldrin and the distance of nearly 3.9×10^8 m between moon and earth measured to an accuracy of about 1 cm.

Why should there be all this excitement? – beams of light were hardly a new phenomenon! Well, these beams really were new. Not only was a laser beam extremely narrow and remarkably parallel, the light was coherent – all the waves were in step with one another, a property of the stimulated emission process which generated them – and the spectral width of the emission was also extremely narrow – in other words, the light was very much closer to being at a single wavelength than could ever be achieved with conventional sources which relied on the spontaneous emission process. While it would be quite inappropriate for me to try and describe all the uses to which lasers have subsequently been put, we might note a very few, such as eye surgery, compact disc players, drilling fine holes in watch-making, measuring distances within and without the home, controlling lathes, optical communications, cutting the cloth for men's suits, prostate surgery, missile guidance, cleaning leaves from railway lines, holograms, artistic illumination and literally hundreds of scientific applications.

This is all very general and we need to home in on the semiconductor laser but, before doing so, it would be useful to look briefly at one important aspect of all lasers – how, for example, does anyone go about making one? The key feature is the use of an optical cavity which serves the purpose of concentrating the light into a specific direction, thereby allowing stimulated emission to build up and swamp the spontaneous process. This usually takes the form of a pair of accurately parallel mirrors which reflect most (though not quite all) of the light back into the laser medium. (This is known as a Fabry–Perot cavity, after its inventors Charles Fabry and Alfred Perot who made the first such instrument in 1897 at the University of Marseilles.) Suppose this medium is excited in some way to prod it into emitting light; the resulting photons will travel in random directions and most will be lost into the surrounding space, but a few will travel at right angles to the mirrors and will be reflected back along the same path. These few may stimulate further emission of photons which travel in the same direction and which are in phase (i.e. the waves are in step) with the photon that triggered them (this being the nature of the stimulated emission process). They, in turn, can stimulate further photons, leading to a rapid build-up of light in a direction normal to the mirror planes. While, spontaneously emitted photons get lost because they travel in random directions, the stimulated ones produce an intense beam reflected back and forth between the mirrors and this totally dominates the light within the cavity. To extract some of this laser light, it is only necessary for one of the mirrors to allow a fraction of the light falling on it to leak through. In other words, we need a partially reflecting mirror at one end to allow the beam to escape into the outside world where it can be useful.

Now, at last, we can follow the story of the semiconductor laser itself, a rather special example of the species. Remember that we need two essential features, some method of exciting light emission from the semiconductor material and some method of forming an optical cavity to encourage the dominance of stimulated emission. The first step in the process was the discovery that GaAs p–n diodes provided an excellent means of generating light (more accurately, infra-red radiation at a wavelength of about 880 nm). Implicit in this was the ability to dope GaAs both n- and p-type, a feat which proved relatively straightforward, though there was one small subtlety we should be aware of. First, consider n-type doping. This could be achieved by using a donor atom with one outer electron more than the host species (e.g. phosphorus in silicon) but here we run into a complication – there are two *different* atoms in GaAs! Which one should we use? The answer turned out to be: 'either' – but note that different donor atoms would be required according to the choice made. If the donor is to substitute for an arsenic atom, we need a donor from group VI in the periodic table, such as sulphur, selenium or tellurium but if it substitutes for gallium, we need a donor from group IV, such as carbon, silicon, germanium or tin. In fact, silicon has been widely used as a donor in GaAs, giving rise to the rather desperate witticism that 'silicon will always be important – as a donor for GaAs'. Notice, however, that these group IV atoms can also dope GaAs p-type if they substitute for arsenic – they are referred to as amphoteric dopants. Alternatively, it was possible to use an atom from group II, such as beryllium, magnesium, zinc or cadmium, as a substitute for gallium. Precisely what might happen in any particular circumstance involves things like the relative sizes of these various atoms and the orientation of the growing crystal as the dopant atoms are incorporated. I would prefer not to get bogged down in such details – suffice it to say that GaAs can be well and truly doped both n- and p-type. Excellent diodes were made by taking an n-type sample and diffusing in zinc to make the top section of the crystal p-type. The p–n junction is formed at a depth where the zinc stops – or, more strictly, where the zinc concentration has fallen to a level just below the original concentration of n-type dopant atoms. All this was understood in the 1950s and diodes were made with abandon. They showed excellent rectifying behaviour but, more importantly, as we pointed out above, they generated light.

The reason is clear. P–n junctions work well as injectors of minority carriers. We saw this in the case of the junction transistor, where the emitter was required to do just this – it was the diffusion of minority carriers across the base which produced the collector current and made the device work. In this case, the designer was at pains to avoid minority carrier recombination because that reduced the number reaching the collector and with it, the transistor gain. In the GaAs diode, minority carrier recombination was exactly what the device engineer was looking for – but vitally important was the fact that GaAs's direct gap encouraged the recombination process which generated light, rather than that which, in silicon or germanium, produced heat. Photons were good, phonons bad! To make a good laser, it was essential that this 'radiative recombination' process dominated its rival, the heat generation process (known, somewhat banally, as the 'non-radiative process'). However, one of the problems with GaAs in its rather primitive 1950s guise was the prevalence of accidental (and unwanted) impurity atoms which introduced electron states near the middle of the band gap and thereby encouraged non-radiative recombination – like the gold or platinum atoms in silicon which were deliberately introduced to speed-up recombination in certain types of silicon power device. This seemed something of an insuperable problem until it was discovered that cooling diodes in liquid nitrogen had the effect of minimising these non-radiative processes, while leaving the radiative ones unscathed. In the spring of 1962, there were reports on the American conference scene of GaAs diodes with radiative efficiencies close to 100%. At a temperature of 77 K nearly all the free carriers injected across the p–n junction in forward bias were recombining radiatively – that is each recombining electron-hole pair produced a photon with energy close to the band gap energy of 1.4 eV. A wave of excitement ran through those privileged laboratories which were in the know – this was exactly what was needed to make a semiconductor laser, and they were all too keen to try. The race was on. It was just 2 years after Maiman's dramatic demonstration of laser action in ruby.

It turned out to be a four-horse race. There were groups at IBM, Yorktown Heights, at the Lincoln Laboratory of MIT, at GE, Schenectady and at GE, Syracuse, led, respectively, by Marshall Nathan, Robert Rediker, Robert Hall and our old friend of silicon power device fame, Nick Holonyak. All four reached the finishing post, having successfully jumped several technological fences en route, and, remarkably, all four submitted their results for publication within the space of a single month. Hall won by a short head, his paper being received on 24 September, followed by Nathan's on 6 October, Holonyak's on 17 October and Redicker's on 23 October. Three of the papers reported laser action from GaAs diodes while Holonyak obtained his results from a semiconductor alloy gallium arsenide phosphide (GaAsP) which emitted in the red (i.e. visible) part of the spectrum. This, in itself, was of special interest from the viewpoint of future light-emitting diodes but also represented the first report of anything useful to emerge from a semiconductor alloy. There had been much scepticism concerning the possible usefulness of alloys, many doubting Thomases believing that the random nature of atomic arrangement in an alloy would compromise its semiconducting behaviour. Holonyak's undoubted success with his red-emitting alloy gave the lie to such speculation and opened the door to a vast range of alloys with widely varying properties which we shall meet in considerable numbers later in our journey through semiconductor space. The work was of interest, too, on account of the method he used to grow the crystals of GaAsP, which involved vapour phase deposition, rather than the melt processes used by others – that is, Ga, As and P were transported in gaseous form to the growing crystal. His success was all the more remarkable, given that his employers were pressurising him into dropping the alloy work in favour of 'more important' work on silicon. Fortunately, he happened to have an Air Force contract to pursue the GaAsP programme, otherwise it would almost certainly have been stopped. By such tenuous threads hangs scientific progress – more often than most people realise!

All four laboratories were familiar with the techniques for making diodes – what was missing was a method of forming an optical cavity to concentrate the radiation and encourage the stimulated emission process. This was a new realm for semiconductor technology and the different groups tried various different approaches. Hall moved first, forming a Fabry-Perot cavity by the simple artifice of polishing parallel facets on his GaAs crystals, thus avoiding the need for external mirrors. The reflectivity of these facets was about 30% which was considerably smaller than could be obtained with ordinary mirrors but it proved sufficient. The Lincoln Lab team used a similar approach, while Holonyak explored the idea of cleaving parallel facets on his small samples of GaAsP. This, as it happens, is the technique preferred in almost all subsequent work but Holonyak's crystals were too crude and he was obliged to learn from Hall's example (this appears to have been the only collaboration between the two GE groups – they were rivals in the race for publication). It appears that the IBM team made no deliberate attempt to form an optical cavity – their success must have depended on accidental feedback of the light. The important fact was that four different groups had demonstrated laser action in III–V compound crystals – a completely new type of laser had been born.

When all the excitement and euphoria had subsided, it was possible to evaluate their achievement. Whereas the gas laser and the ruby laser were large scale pieces of equipment, having dimensions of order of a foot and employing high voltage power supplies to excite them, the semiconductor counterpart was almost too small to be seen – it was no more than a millimetre in length – and it operated at only a few volts. Here was yet another example of semiconductor miniaturisation to rival that of the transistor in comparison with the thermionic valve. On the other hand, it was well to recognise that the new lasers operated only in short bursts (i.e. they worked under pulsed operation) because of their poor efficiencies (they generated too much heat to be run for more than a few microseconds at a time) and they would only work at all when immersed in liquid nitrogen. The scale of the Dewar flask required to hold the coolant completely negated their size advantage! The truly exciting achievement was, of course, one of establishing an existence theorem – semiconductor lasers were clearly possible – but, if the desired objective was to make a laser which operated continuously at room temperature, there was clearly a long way to go. In the event, the first commercial application of the semiconductor laser had to wait until 1978, when Philips and Sony launched the compact disc player, and there were good scientific reasons for the delay. We should take a few sentences to examine what they were.

Basically they were three. Firstly, and foremost, the material quality of GaAs available in 1962 was inadequate. Secondly, the simple diode structure used in the pioneering work suffered from the fact that injected minority carriers diffused away from the p–n junction and failed to recombine close enough to the junction region. Thirdly, there was no means of confining the light close to the junction where it was needed to stimulate further emission. In order to make an efficient laser, it was important to build up a very high intensity of light and this meant that light should be generated with high efficiency in a small volume of material and that this light should be confined as closely as possible to the junction region. Not only was the material inadequate, recombination at room temperature being dominated by non-radiative processes, but the device structure required serious modification, in order to concentrate both free carriers and light close to the junction. All three aspects demanded considerable development work. New crystal growth methods were imperative and device structures demanded the introduction of new materials. Neither of these improvements could be realised without a great deal of fundamental work and it was not until 1970 that the first semiconductor lasers would run continuously at room temperature. How were these various challenges met?

Crystal growth was key to all these problems so we must take time to introduce at least some of the important developments. Firstly, we should look at bulk growth in the shape of the Czochralski method. It was first applied to GaAs by Gremmelmaier in 1955 but the vital step was taken in

1965 at RSRE Malvern by Brian Mullin and his team when they introduced liquid encapsulation (see Fig. 5.1). As we have seen, one of the difficulties inherent in melt growth of GaAs was the tendency for arsenic to evaporate and generate a high vapour pressure in the growth apparatus – it also tended to corrode any motion seals. In the encapsulation technique this was countered by sealing the surface of the melt with a layer of molten boric oxide, thus preventing the arsenic from getting out into the surrounding space. This allowed the necessary background pressure to be supplied by an inert gas such as nitrogen or argon and the rotation and pulling motions to be transmitted through conventional seals in the pressure vessel. It was all a great deal more convenient, and high quality bulk samples became far more widely available. Nevertheless,

FIG. 5.1 A Czochralski crystal growth apparatus used at the Royal Radar Establishment to grow crystals of III-V compounds such as gallium arsenide and indium phosphide. The key to success lay in the use of boric oxide to encapsulate the growing crystal to prevent loss of volatile arsenic or phosphorus. Courtesy of Brian Mullin.

even the best bulk crystals proved to be unsatisfactory for making lasers (and, indeed, several other devices) because of their inherent background impurity levels. Something more radical was needed, and this took the form of epitaxy (from the Greek epi = upon and taxis = arrange). Epitaxial growth took various forms but basically involved growing a thin film of GaAs on a bulk GaAs substrate. In general, epitaxy occurred at much lower temperatures than melt growth and this very much improved material purity, so the idea became widely accepted that a device should be made from a thin epitaxial layer, while the physical support was provided by the much thicker substrate. Once the advantages were appreciated in the early 1960s, GaAs hardly looked back. Lasers, Gunn diodes, MESFETs and all other devices were based on one or other type of epitaxy.

There were numerous types of epitaxy but for our immediate interest we need mention only two: liquid phase and vapour phase epitaxy, usually abbreviated to LPE and VPE. LPE came first on the scene in 1961 when Nelson at IBM introduced the so-called 'tipping' method of growth. A saturated solution of GaAs in liquid gallium was tipped over a GaAs substrate at a temperature of about 700–800 °C and the temperature slowly reduced to cause GaAs to deposit on the substrate. After a while, the solution was tipped off again, leaving a film of crystalline GaAs on top of the substrate. Various modifications of this basic method were introduced during the 1960s but the most flexible was the so-called 'sliding' method in which several reservoirs of GaAs/gallium solution were arranged so that any one of them could be slid over a GaAs substrate, left for a time, then replaced by another, to grow a different layer (for instance a differently doped layer). This allowed, for example, p–n junctions to be grown in situ, rather than by the use of diffusion. It also led to a very significant development in multi-layer growth when aluminium was included in one or more of the reservoirs. Thus, it was possible to grow AlGaAs as well as GaAs films and this gave the device designer a degree of flexibility which paid special dividends in the development of the semiconductor laser. LPE represented a vital step in the struggle to obtain high quality GaAs – still not as pure as currently available silicon or germanium but certainly adequate for most applications. Not that LPE had things all its own way – VPE was able to mount a serious challenge and, in some cases, to provide even lower background doping levels. First to be introduced was the so-called chloride transport method in which Ga and As were carried along a quartz tube in the form of their respective chloride vapours by a stream of very pure hydrogen gas, and passed over a GaAs substrate where the chlorides dissociated to deposit a thin film of GaAs. This process was initiated at the Plessey Caswell laboratory by John Knight and co-workers in 1965 and was rapidly taken up in many other establishments. LPE and VPE were rivals for quite a number of years and were used more or less interchangeably in a variety of device programmes.

To return to our discussion of laser development, it is now necessary to follow the manner in which the various shortcomings referred to above were eventually repaired. Firstly, we consider the question of free carrier confinement, a problem first addressed in a paper by Herb Kroemer, who, at the time, 1963, was working with Varian Associates in Palo Alto. Kroemer had been interested in the applications of 'heterojunctions' since the mid 1950s, (when he had proposed a number of improvements to the bipolar transistor) and saw the semiconductor laser as an ideal example. (A heterojunction is a junction between two different materials and has a range of interesting properties, perhaps the most important being a difference in band gap which implies a step in the conduction and valence bands at the interface.) Kroemer's idea was to sandwich a slice of narrow gap semiconductor between two wider gap materials, forming a kind of well into which free carriers could fall and, assuming the sides of the well were deep enough, they would be unable to escape. Provided there existed suitable wells in both valence and conduction band, holes and electrons would be confined together in the narrow gap material and they would be obliged to recombine there – it would no longer be possible for them to diffuse away. The idea was excellent but probably before its time – the practical realisation of working heterostructures being extremely difficult with available technology. In any case, it appears to have been totally ignored by the laser community! It later transpired that Zhores Alferov at The Ioffe Institute in Leningrad (now St Petersburg) had applied for a patent for a similar idea in the same year but he too ran into difficulties when it came to practical realisation.

The important technological step came in 1967 when Jerry Woodall at IBM demonstrated that the alloy AlGaAs could be grown epitaxially on a GaAs substrate by liquid phase epitaxy. This discovery was greatly favoured by the fact that the lattice parameters of AlAs and GaAs are very nearly equal, which means that the two materials can grow together without introducing any mismatch – in other words, the crystal structure is continuous across the interface between GaAs and AlGaAs. Given that the band gap of AlGaAs is greater than that of pure GaAs, this wonderful piece of natural serendipity offered the opportunity for interested groups on both sides of the Iron Curtain to build laser structures with effective carrier confinement. The Russian approach naturally followed Alferov's patent (and Kroemer's suggestion) to make a double heterostructure though, rather surprisingly, the Americans, in the persons of Kressel and Woodall at IBM and Hayashi and Pannish at Bell Labs, chose a single heterojunction. Their idea was simply to bounce free carriers back to the p–n junction, rather than allowing them to diffuse away. Both methods achieved very significant reductions in the amount of electric current needed to achieve laser action and the Holy Grail of continuous operation at room temperature began to look like a realistic possibility. Alferov published his results in 1969 and stimulated the West into renewed effort, both groups achieving their ambitions in the following year, using the *double* heterostructure. The current density (amps per square metre) was now a 100 times lower than in the pioneering devices of 1962 and from this point no-one would think of making a semiconductor laser in any other way. Alferov and Kroemer shared half a Nobel prize in 2000 for their contributions (the other half going to Jack Kilby for the invention of the integrated circuit).

The combination of improved material quality, associated with the use of liquid epitaxy, and the incorporation of carrier confinement by heterostructure had, after 8 years, won the day for tenacity of purpose but even this was not the end of the story. It was still necessary to confine the light close to the recombining holes and electrons and this involved using an optical waveguide. To understand the concept, it is helpful to consider how light is confined inside the glass fibres used in optical communication systems (Chapter 8). The fibre consists of a fine core of glass with one value of refractive index, surrounded by a cladding of glass with a slightly smaller index, and light is confined to the core region, effectively bouncing from side to side as it propagates within this central region. A similar approach to confining light within the laser was proposed independently in 1973 in England, by Thompson and Kirkby at STC and in America, by Hayashi at Bell Labs. It was based on the fact that the refractive index of AlGaAs was lower than that of GaAs and gradually decreased as more aluminium was added. The idea was to add a second pair of heterojunctions (outside the pair used to confine free carriers) so as to create a step in refractive index and form a slab waveguide to keep the laser light close to the recombination region (the central well). Thus was born the 'separate confinement heterostructure laser' which brought the current density down yet further and provided a prototype for all future designs of semiconductor laser. To sum up this complex set of design rules, we can simply note that a typical structure would include five different layers of material, in the following sequence:

[n-AlGaAs50%][n-AlGaAs30%][GaAs][p-AlGaAs30%][p-AlGaAs50%]

The central GaAs layer was about 0.1 μm thick and was undoped. The optical waveguide formed by the two inner AlGaAs layers was about 1.0 μm thick. The overall shape of the structure was a stripe some

500 µm (half a millimetre) long by about 10 µm wide with carefully cleaved end facets which acted as Fabry–Perot mirrors. It represented a triumphant marriage of clever scientific thinking with inspired semiconductor technology – a combination of materials sophistication and structural wizardry – and was a very far cry from the crude devices employed by the early pioneers. What had begun life as a mere stroke of scientific inspiration had now become a viable commercial product. It had taken a long time but it had travelled a long way. It was a classic example of the need for a combination of brilliant scientific innovation and technological hard grind – neither could hope to succeed alone.

The semiconductor laser is now used in all manner of different applications from measuring systems to machining but the application that gave it a crucial start in life was the compact disc player cast onto an unsuspecting audio world by Philips and Sony in 1978 (see Fig. 5.2). At a stroke, the numerous deficiencies of the microgroove record and its delicately balanced pick-up arm were removed from the scene. The twelve inch disc was replaced by one slightly less than 5 in. in diameter and scratchy background hiss became a thing of the past. Philips had tried an earlier laser-based system known as the Video Disc in 1972 but it used a helium–neon gas laser and was altogether too large for convenience – it never caught the public imagination. The CD system was, as its name implies, very much more compact and took the audio world by storm – all

FIG. 5.2 The first major application for the newly developed gallium arsenide laser was as the light source used to read information from a compact disc in the CD player launched by Philips and Sony in 1978. The CD player shown here was manufactured by neither of those companies but is typical of the millions of players produced over subsequent years. Courtesy of the author.

because of the semiconductor laser and, let it be said, the availability of high speed integrated circuits to deal with the digital electronic requirements. However, the launch was not without its technical challenges. The amount of information that could be stored on the disc was limited by the size of the individual pits which formed the digital bits, and this, in turn, was limited by the wavelength of the laser beam – the shorter the wavelength, the more music on the disc. The wavelength of 880 nm characteristic of GaAs was reckoned to be too large and the compromise of 780 nm required that some 15% of aluminium be introduced into the well in order to increase its band gap appropriately. This, of course, implied a corresponding increase in the outer layers too. And this made life difficult for the production engineers because aluminium is blessed with a considerable appetite for oxygen and this had a bad influence on the quality of the resulting laser material. The struggle to keep oxygen out caused more than a few sleepless nights but eventually the 780 nm laser took its place and the CD system took off. With the introduction of the i-Pod, its long-term future looks now to be in doubt but there can be absolutely no doubt that it revolutionised the audio recording industry.

We shall leave the semiconductor laser for the time being, though it will surface in a number of other guises in subsequent chapters. It is time to move from the field of optics to that of microwaves, where GaAs has also played a major role, and our first call is on the Gunn diode, also invented during the early 1960s. It is an interesting, if somewhat tortuous tale. The development of microwave systems has a history dating from the Second World War when it became clear that radar equipment would give higher spatial resolution (i.e. pinpoint its targets more effectively) if it could utilise shorter wavelengths than were currently available. Conventional techniques were limited to frequencies of a few 100 MHz by the capabilities of thermionic valves and there was considerable urgency in efforts to develop microwave tubes which operated at frequencies of 3 GHz and above (wavelengths of 10 cm and below). These spawned the klystron, magnetron and travelling wave tube, devices which had major impact on the war but were all rather cumbersome and demanding of large, high-voltage power supplies. The advent of tiny, low-voltage solid state devices which were seriously challenging valves at lower frequencies stimulated a corresponding desire to replace their microwave counterparts with some form of semiconductor alternative.

This story, by contrast with the laser story, is an all-British tale and begins in the Mullard Research Laboratory, Redhill, round about 1960, when Brian Ridley and his colleague Ron Pratt were seeking experimental evidence for the existence of a 'negative resistance' in some samples of germanium. 'Whatever', you may reasonably ask 'is a negative resistance?' Firstly, we should remind ourselves as to the nature of a *positive* resistance. Imagine an electric current flowing through a wire or a uniform sample of a semiconductor and suppose we increase the voltage applied across it – the current will increase in direct proportion to the applied voltage. This is Ohm's Law, formulated by the German physicist Georg Ohm in 1827, and has been fundamental to electrical science ever since. (Interestingly, the law was first propounded by the English physicist Henry Cavendish in 1781 but, in common with much of his work, he failed to publish it. It only came to the notice of a wider audience when Clerk Maxwell reviewed Cavendish's unpublished work a century later – in 1878.) However, scientists are rarely completely happy with the status quo and will always be on the look-out for the unusual. In fact, there are many instances where the famous law does not apply – the forward current-voltage characteristic of a p–n diode is one example, where the current varies exponentially, rather than linearly. Nevertheless, it does increase as the applied voltage increases, whereas the concept of a negative resistance implies that the current will decrease when the voltage increases. The idea may sound crazy but Ridley and Pratt were perfectly sane in believing such a concept might be found in practice – its possible existence depended on the fine detail of the semiconductor conduction band and they had good reason to look for it in samples of germanium under high pressure. This theoretical concept was explained in a paper published by Ridley and a Mullard colleague Tom Watkins in 1961 in which they showed how a negative resistance might be used to stimulate an electrical circuit to oscillate (i.e. to generate spontaneous high frequency waves). They were aware that their choice of germanium was probably not ideal and their doubts were very soon confirmed by a paper from Cyril Hilsum of SERL Baldock, published just four months later. Hilsum had been thinking (quite independently) along similar lines and pointed out that GaAs was a far better material in which to seek the effect – its conduction band being far more appropriate.

The idea was a subtle one and involved a deep understanding of semiconductor band theory. We can formulate a simple model by acknowledging that the conduction band contains several different types of electron state. In GaAs, the states usually occupied by electrons have a small effective mass, which means that electrons are rapidly accelerated by an applied electric field. In being so accelerated, they gain energy, and this increase in energy allows them to jump into another kind of state which is characterised by a large effective mass. This change of mass implies that the electrons suddenly slow down. In other words, increasing the applied electric field results in a *reduction* in electron velocity, which implies a reduction in current flow. Over a certain range of applied voltage, therefore, the resistance of the GaAs is negative and we have the effect we were looking for, often referred to as the 'transferred electron effect'. The size of electric field required to produce this transfer was such that it would occur in a sample of GaAs about 10 μm thick with a voltage of about 3 V across it. Bearing in mind that epitaxial crystal growth methods were being developed in the early 1960s, capable of depositing a thin film such as this, we see just how exciting the transferred electron effect really was. The film could be grown on an n-type substrate which acted as one electrical contact, while a second contact could readily be made to the top of the film. Place a simple device like this in a microwave circuit and we have the simplest, cheapest, most convenient, low voltage microwave oscillator imaginable. It was wonderful concept – could it be realised?

The answer turned out to be 'Yes' but there was a twist in the tale. The effect has come to be known as the Gunn effect in recognition of its discoverer John Battiscombe Gunn (known to his friends as Ian). He was born in Egypt of British parents (his father Battiscombe Gunn was an Egyptologist) and joined IBM in 1959 with an interest in applying short pulse techniques to semiconductor research. It was during these studies that, in 1963, he observed microwave oscillations emanating from thin samples of n-type GaAs, though in the resulting publication Gunn was quite specific in saying that the mechanism for their production was not understood. He had convinced himself that he was definitely not seeing the transferred electron effect and, when Tom Watkins visited IBM shortly afterwards, Gunn also persuaded Watkins that this was the case. The Gunn diode spread rapidly round the microwave world, in spite of such doubts as to its parentage and soon proved to be the simple, cheap, convenient low voltage device everyone had hoped for, but it was not until 1965 that proof of its origins was obtained. The experiment was performed at Bell Labs by a group under A.R. Hutson and involved measuring Gunn oscillations as a function of applied pressure. Their results tallied perfectly with theoretical predictions based on the transferred electron effect and, from that time onward, it was recognised as the Ridley–Watkins–Hilsum–Gunn effect (Gunn effect, in short).

Ridley made a further important theoretical contribution to the story by analysing the detailed origin of the microwaves. When one thinks about a simple, uniform film of GaAs with a DC voltage across it, it seems altogether remarkable that it should produce high frequency oscillations. Wherever did they come from and why was the frequency (as seen by Gunn) dependent on the thickness of the film? Ridley demonstrated that the existence of a negative resistance, in such circumstances, must produce an instability in the electron density in the film such that electrons bunched together in what came to be called a 'domain' and this domain would travel through the film thickness with a velocity of about $10^5 \, ms^{-1}$, starting at the negative electrode and dying at the positive electrode. Only when the first domain died at the anode could another domain form at the cathode, so a regular sequence of domains would pass through the film. It would take just 10^{-10} s for a single domain to travel through a 10 μm film and the resulting frequency would therefore be 10^{10} Hz (10 GHz) which accorded perfectly with experimental data. If one needed an oscillator to operate at 30 GHz, all that was necessary was to grow a GaAs film just 3 μm thick, rather than ten, and again this agreed well with experiment. There were additional criteria for making successful oscillators, such as appropriate control over the n-type doping of the GaAs, and these proved challenging to the crystal growers – all was not quite as plain sailing as I have so far implied – but these challenges were duly met and the Gunn diode oscillator duly took its place in the microwave hall of fame.

In the period after its discovery, during the struggle to understand what was really going on and how to master the several crystal growth problems, the degree of interest in the subject rapidly reached fever pitch. Gunn himself recalled in a later review paper that:

> At one point, the field seemed to be so crowded that, if one did not perform an experiment on the day that one thought of it, it would have been done elsewhere and the opportunity would have passed. .... The key to progress became, not fast pulse electronics and new experimental techniques but materials technology and device engineering.

He was regretfully observing the changeover from basic physics to device-driven technology in which materials science becomes all-important. But life was like that when commercial interests began to sense the possibility of exploitation! Indeed, so exciting were the prospects that even Mills and Boone got in on the act and I was commissioned to write a monograph about it! Tempting though it was to choose a title such as: 'How I Came to Love the Gunn', it turned out to be the much less racy: 'Material for the Gunn Effect'. It was largely ignored in the general scramble to develop useful devices – and there can be no doubt at all that useful devices have been and are still being made. The Gunn diode has been widely used in innumerable microwave receivers and has acquired a degree of infamy from one of its applications in microwave radar systems designed to measure the velocities of automobiles speeding through restricted areas (see Fig. 5.3). Many an errant motorist owes his speeding fine to the efficacy of the Gunn diode as a microwave source. But many other applications of a less aggressive nature have also proliferated.

One specific area of interest is worth mentioning before we leave the topic and that concerns the highest frequency to which Gunn devices can be pressed. Most of the early work centred on frequencies in the range from about 3 GHz to 35 GHz but applications came into focus which

FIG. 5.3 A behind-the-scenes view of a Police radar speed camera in use on one of Britain's motorways. The device makes use of a Gunn diode to generate microwave radiation and measures the Doppler shift in microwave frequency to assess the speed of approaching cars. Courtesy of iStockphoto, image 3366444.

required frequencies up to 100 GHz and beyond – could the diminutive Gunn diode respond? The limiting factor was the speed with which electrons in the conduction band could jump back and forth between the different conduction band states. And this appeared to limit GaAs devices to frequencies up to about 100 GHz. It was possible to extract a certain amount of power at the second harmonic of the fundamental frequency and this allowed operation as far as 150 GHz but that seemed to be the upper limit. It was therefore of considerable interest when a new contender came on the scene in the shape of an alternative III–V compound InP.

Gunn himself had seen oscillations from InP samples as well as from GaAs but it was left to Cyril Hilsum (now at RSRE Malvern) to push the material firmly into the limelight. He and a colleague, David Rees, made a bold prediction in 1970 to the effect that InP should be a better material in which to exploit the transferred electron effect than GaAs on account of some subtle differences in its conduction band. They predicted that the negative resistance in InP should be greater and therefore it should generate microwave power with significantly greater efficiency. Experimental efforts to confirm their ideas were set in train in several UK laboratories in a flurry of exciting activity. The immediate problem was (as ever!) one of material technology. InP was known to be similar to GaAs in many ways – its band gap was direct and only slightly less than that of its rival (1.35 eV) and the effective mass of electrons just a little larger – but its technological development had lagged well behind. In the rush to catch up, Brian Mullin and his colleagues (also at RSRE) reported Czochralski growth of InP, using their liquid encapsulation method, which, by then, had come to be known as the LEC method. At the same time, both LPE and VPE techniques were under development and, within little more than twelve months material of adequate quality was available to allow meaningful experimental tests of the new theory. The results were certainly interesting and showed InP to have promise as a rival to GaAs but, alas, they did not, by any stretch of the imagination, support the detailed predictions made by Hilsum and Rees. However, there were signs that InP might be capable of generating higher frequencies than GaAs and this gradually came to pass – it is now the preferred material for Gunn oscillators operating in the 100–300 GHz range. This represents only a tiny corner of the microwave market and it could well be argued that the whole business was a waste of time and money. In fact, InP has become of major importance in the field of optical communications (we shall meet it again in Chapter 8) so this flurry of crystal growth proved of considerable value in giving the material a head start. Without the stimulus of the Hilsum-Rees theory it would certainly not have been available in the desired quality at the appropriate time. Such then is way of science – very rarely does it proceed in the well-ordered linear fashion in which we like to pretend.

Our final topic in this III–V chapter is concerned with the GaAs MESFET which was yet another product of the 1960s. As we saw in Chapter 3, the FET story began as early as 1930, in the shape of a patent application but it was only in the 1950s that the first experimental devices were made, first in germanium, later in silicon. They made use of a gate in the form of a p–n junction but the performance of these junction FETs (JFETs) compared unfavourably with their bipolar counterparts and the FET was effectively abandoned until the invention of the silicon MOSFET in 1960. This, of course, was a tremendous success, taking precedence over bipolar devices in many applications but the supporters of GaAs were clear that it suffered from one serious disadvantage – it could never perform at anything like the speed of a comparable GaAs device. This is because the speed of any FET is limited by the time taken for electrons (or holes) to be injected into or swept out of the channel region and this, in turn depends on the mobility of the carriers involved. Proponents of GaAs argued that, since the electron mobility in GaAs is six times that in silicon, the GaAs FET would operate at much higher frequencies than its silicon equivalent. In other words, GaAs was capable of working in the microwave region, where silicon could never hope to compete. It is no surprise, therefore, that the GaAs MESFET first saw the light of day in the period when the microwave spectrum was being opened up by the discovery of the Gunn effect. The Gunn device was essentially an oscillator and it soon became clear that full use of these frequencies required an amplifier – the ultimate goal being to develop microwave integrated circuits, both analogue and digital.

As one might expect, the first GaAs FETs were made as JFETs or MOSFETs, following the lead given by silicon but neither was very successful. To form the gate junction in a JFET involved diffusing zinc into an n-type channel and this process required temperatures of 800 °C which upset other parts of the structure. To form the gate oxide in a MOSFET involved deposition of a film of silicon oxide (GaAs has no native oxide of the requisite quality) and this proved much less satisfactory than the technique of oxidising silicon. There was a period of hesitation, topped by the invention of the metal Schottky barrier gate. It had been proposed by Mead in 1966 and was demonstrated practically by Hooper and Leherer at Fairchild just 1 year later. It involved nothing more complicated than the evaporation of a suitable metal film on the GaAs channel region, which could easily be done at a temperature of 150 °C, well below the damage threshold. The GaAs MEtal Schottky Field Effect Transistor was born. It depended on the fact that Schottky barriers on GaAs were relatively good barriers to current flow – one could apply the necessary gate voltage without any significant gate current flowing. This was a result of the larger band gap in GaAs, perhaps an unexpected bonus to go with the high electron mobility.

Let us take a moment to spell out the structure of a typical MESFET in order to make crystal clear its modus operandi. It consists of a thin (about 1 μm) n-type epitaxial film deposited on a non-conducting substrate, the upper surface of the film having a pair of electrical contacts, acting as source and drain electrodes, separated by the Schottky gate. The contacts have to be of low resistance to facilitate current flow along the epitaxial n-type channel, while the gate must have a very high resistance to minimise gate current. Application of a negative voltage to the gate repels electrons from the channel under the gate and tends to cut off the channel, thereby reducing the source-drain current. There is a compromise to be made with regard to the doping level in the epitaxial layer – it should be relatively high to encourage high conductivity but it must be relatively low to make possible effective control by the gate. A value in the region of 10^{23} donors per cubic metre works well and this is comfortably within the range of easy doping levels achievable with the VPE process used. The other aspect which needs mention concerns the non-conducting substrate. Clearly, it must have a very high resistance in order not to short-circuit the channel and this proved yet another good reason for choosing GaAs because it can be produced in a form known as 'semi-insulating' which satisfies this criterion extremely well. The existence of such material was recognised from a relatively early date, certainly well before 1960, but a detailed understanding had to wait a little longer. It soon became clear that it was related to the presence of high concentrations of impurities in the GaAs crystals grown from the melt and many different explanations were advanced, all based on the existence of deep impurity levels near the middle of the GaAs band gap. Such levels effectively trap free carriers and swamp the effect of shallow donors or acceptors. Two principal contenders for the crown of semi-insulating king-maker are chromium and oxygen – the former may be added deliberately to make reproducible semi-insulating material, while the latter is usually incorporated by accident. Both are used in practice for making high resistance substrates for MESFETs.

Given all these natural advantages, GaAs could be expected to march confidently into the microwave future, and it didn't altogether disappoint. Early devices employed relatively large gate lengths (i.e. channel lengths), typically of order 10 μm and this limited their frequency range to about 1 GHz. However, efforts to reduce the gate length led to the use of electron beam lithography to define the gate and in 1970 the Plessey group at Caswell were able to report MESFETS operating at frequencies close to 10 GHz. Doerbeck at TI used a self-aligned gate to reach slightly higher frequencies, then, in 1974, Nozaki from the Japanese telecommunications firm NEC introduced a further improvement in the form of an undoped epi-layer which isolated the channel from the substrate and achieved considerably higher frequency operation. Gate lengths crept down below the micron level and working frequencies approached 100 GHz. GaAs was, at last, fulfilling its early promise. But perhaps the most significant development came in mid 1970s when teams from Hewlet Packard in the USA and from Plessey Caswell in England reported (respectively) the first digital and the first analogue microwave integrated circuits. These were splendid achievements to crown two decades of strenuous effort to make an honest woman of GaAs – from this moment onward, she was certainly well able to earn her keep.

In conclusion, we might briefly refer to two further developments. The new girl, InP began to show promise in the MESFET field, promising even higher frequency operation than possible with GaAs. However, we shall leave the MESFET for the time being until it is time to take up the story again in our next chapter. The other point to make in defence of GaAs is that it provided the basis for several other microwave devices such as detector and mixer diodes. These benefited from the high electron mobility in much the same way as the MESFET had done and, being made from the same material, they could readily be incorporated into microwave ICs.

CHAPTER 6

Quantum Theory and Quantum Practice: The Nanostructure Revolution

One of the dominant characteristics of the solid-state revolution has been a dramatic reduction in the size of semiconductor devices, when compared to their predecessors, the vacuum tubes, klystrons, magnetrons, etc., of the early twentieth century. Here was an example where 'small' really was beautiful. Valve dimensions were measured in inches, transistors in microns – the contrasts were almost beyond belief. Suspend belief altogether, then because we are about to embark on an exciting journey into the ultra-small where dimensions are recorded in nanometres (10^{-9} m). Semiconductor films are now to be grown in thicknesses made up of literally a few atoms. The early years of the twenty-first century have made much of the nanostructure as a remarkable new development which has taken the world completely by surprise but this I find difficult to understand – the semiconductor technologist has been doing it ever since 1975!

How did it all begin? Like so many unusual developments, it began with a theory – a crazy idea that sounded almost too good to be true. In a paper published in the IBM Journal of Research and Development in 1970, Leo Esaki and Raphael Tsu put forward the idea of a semiconductor 'superlattice' and predicted several very unusual properties for it – including the possibility of a negative resistance. In much the same way that the Hilsum-Rees theory of the transferred electron effect in InP stimulated a flurry of crystal growth experiments, Esaki and Tsu's theory of superlattices set in train an activity of quite remarkable proportions that is still very much alive today. It is not stretching the point too far to say that, for the very first time, mankind was creating completely new materials based on physical, rather than chemical, design rules. Until now, semiconductor technologists were restricted to choosing suitable

Semiconductors and the Information Revolution: Magic Crystals that made IT happen   © 2009 Elsevier B.V.
DOI: 10.1016/B978-0-444-53240-4.00006-4

combinations of the chemical elements to build materials with specific properties but the science of nanostructures has added a new dimension. By growing semiconductor films with sizes in the nanometre range, it is now possible to create what are effectively new materials with properties never before realised. The fact that the idea has now spread far beyond its original sphere, into chemistry, metallurgy and medicine is a persuasive indicator of its versatility. Esaki and Tsu certainly started something – something probably very much bigger than they could possibly have imagined at the time. In a review talk Esaki gave many years later, he estimated that, at that time, roughly half the physicists working in the field of semiconductors were involved with superlattices and related topics, which certainly gives some idea of their significance.

After singing its praises so highly, it may now be appropriate to say what a superlattice actually consists of. It is made up of a regular sequence of layers, each very much thinner than any we have come across so far, typically of order 1–10 nm in thickness, and running to perhaps several hundred layers. An obvious example, given what we already know of them, would be layers of AlAs and GaAs and this particular combination has, indeed, been used in many cases (see Fig. 6.1). This follows naturally from their use in the double heterostructure laser which we met in Chapter 5. However, there is an important difference in philosophy between that example and the purpose of a superlattice. Whereas the laser structure was designed to confine free carriers to the plane of the p–n junction, the superlattice is designed so as to encourage electrons to propagate through the layers in somewhat the same manner as they propagate through the atomic structure in a simple semiconductor. In a very real sense, the superlattice represents an artificially constructed material in which the individual layers take on the role of rather large scale 'atoms'. Of course, there is an obvious difference in so far as the superlattice possesses its special properties only in one dimension – in the direction normal to the plane of the layers. We shall come across this question of dimensionality quite a lot in the present chapter – indeed, in its earlier incarnation, the subject was generally known as being concerned with 'low dimensional structures' (LDS), that is, structures which possessed special electronic properties in two, one or even zero dimensions, rather than the normal three.

To understand the nature of these remarkable structures, we need to advance our knowledge of quantum mechanics a little and, while this may sound forbidding, it really isn't difficult. We shall pick up the story in 1957 when Esaki was just completing his PhD thesis at the University of Tokyo. His thesis topic had developed out of the new science of semiconductor electronics and, in particular, the properties of germanium and silicon p–n junction diodes, which he continued to study during his first employment with the Sony Corporation. His work was of a remarkably

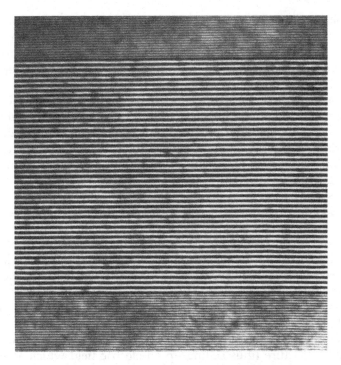

FIG. 6.1 An electron microscope picture of a semiconductor superlattice, grown by molecular beam epitaxy. The dark layers represent gallium arsenide, the light layers aluminium arsenide. Alternate layers are just eight molecules (about 2.25 nanometres) thick. Note the near-perfect regularity of the structure. Courtesy of Tom Foxon.

high standard and, even at this tender age, he invented a new device known as a tunnel diode, which showed yet another type of negative resistance, resulting from the phenomenon of electron tunnelling. He continued to explore tunnelling phenomena at IBM, where he moved in 1960, and, in 1973, he shared the Nobel Prize for his work. To appreciate its significance, we clearly need to know something about electron tunnelling. Imagine, therefore, a semiconductor structure made up of three layers: a thin layer of AlGaAs, sandwiched between two relatively thick layers of GaAs. Because the band gap of AlGaAs is greater than that of GaAs, an electron in the GaAs conduction band sees a step as it approaches the central AlGaAs layer and, in any classical picture, it would be quite unable to find its way past such a step unless it possessed sufficient energy to jump over the top. At room temperature, the electron has a small amount of thermal energy, of order 0.025 eV, but the step height might typically be 0.25 eV (for 30% aluminium in the AlGaAs), some 10 times greater, and the electron would therefore have no chance at all of climbing over. An electron in the left hand GaAs layer would have no chance of ever finding itself in the right hand layer. This, however, is far from true in quantum mechanics. If the AlGaAs barrier is thin enough, the electron has a significant chance of filtering *through* it. This apparently bizarre behaviour is known as electron tunnelling, a phenomenon which can occur in all manner of circumstances, *provided* the barrier is thin enough and the barrier height is finite (i.e., not infinite). The probability for tunnelling to happen depends strongly on both barrier height and thickness but we can put a figure on the kind of limiting thickness appropriate to typical semiconductor structures as follows: significant tunnelling will occur when the barrier is thinner than about 1 nm (10^{-9} m). Looked at slightly differently, we can express the limiting barrier as being something like 10 atoms thick.

Many strange results emerge from quantum mechanics – strange because we are accustomed to think in terms of the kind of mechanics which describe the behaviour of macroscopic bodies – that is, Newton's mechanics. They seem strange to us but they are quite natural to electrons. An electron thinks nothing of slipping through a barrier when the conditions are right, though we should not overlook the fact that this is only one option – it may also be reflected back into its GaAs host material. As with so much else in quantum mechanics, we have to deal with probabilities. However, the fact is that, given a thin barrier, if we apply a small voltage across our structure (thus setting up an electric field which accelerates free electrons) an electric current can flow through it. As each electron impinges on the barrier, there is a finite probability of its going through and contributing to the current. There is also a finite probability of its being reflected back, so the current is smaller than it would have been without the presence of the barrier. It is in this manner that we have to understand tunnelling currents. (Tunnelling has even been called up in support of the notion of ghosts apparently passing through solid masonry but this seems to me to be pushing analogy a little too far.)

This represents one leg on which the theory of superlattices stands – the other being known as quantum confinement. We have already seen how electrons and holes could be confined close to the p–n junction in a laser by building a double heterostructure. In this case, electrons fell into the bottom of the well formed by the central layer in an AlGaAs/GaAs/AlGaAs sandwich and were unable to escape because the well depth was very much greater than the electrons' thermal energy. The well width was set at about 100 nm in order to optimise the interaction between recombining particles and the light which stimulated the recombination process. But there is no a priori reason to limit the well width to 100 nm – it can be reduced well below this, if required. Two things happen, one rather obvious and fairly trivial, one far from obvious and rather important. In the first place, the volume of semiconductor material in the well is reduced, which means that the density of electrons and holes needed to reach the lasing condition (the so-called threshold) can be achieved with a lower drive current and this is useful in many cases (though it also implies a lower optical output). Secondly, if the well width is reduced below about 10 nm, a further strange quantum effect starts to occur, which is the quantum confinement referred to above. There is an interesting analogy between electrons bound to atomic nuclei (see Chapter 1) and those confined in a quantum well. In both cases, electrons are confined in a very small space and as a consequence their energies are defined by the confinement. In an atom we speak of 'binding energy' – the electron being bound to the atom, whereas in a quantum well we speak of confinement energy but these have similar origin. In an atom, the electron is bound in an orbit round the nucleus, while in a quantum well it is confined between the two barriers, so the geometries differ and this means that the mathematical descriptions also differ but the basic physics is similar. The atom has a series of energy levels from the innermost, strongly bound state to the weakly bound outer levels and the quantum well has a similar set of levels starting from the lowest, a little way above the bottom of the well to the highest close to the top of the well. The important point to grasp is that these ladders of electron states appear when electrons are confined in a small space, of the order of atomic dimensions.

Returning, now, to our discussion of superlattices, it is possible to see how electrons may travel right through such a structure. The barriers are thin enough for electrons to tunnel through them and the wells contain several energy levels from which and into which the electrons may pass. In a sense, electrons propagate through the superlattice in much the same way they propagate through a simple semiconductor but it must be clear from our discussion that the properties of the superlattice differ significantly from those of the semiconductors involved. We can reasonably think of it as a completely new semiconductor whose detailed behaviour depends on physical parameters such as the thickness and composition of the individual layers. The crucial question remaining is one of practicality. It is all very well to theorise about superlattice properties but it remains to be seen how such structures might be produced in practice. The answer came, rather remarkably, in the form of, not one, but two new methods of crystal growth. Esaki himself recognised the need for a radical change and, by 1970, had developed a form of epitaxy based on the thermal evaporation of atoms from heated cells containing the appropriate solid. This came to be known as molecular beam epitaxy (MBE) but coincidentally an alternative form of vapour phase epitaxy made its appearance at about the same time, also capable of growing superlattices. It was proposed by Manasevit of North American Rockwell Corporation in 1969 and made use of hydrides such as arsine (AsH_3) and metal-organic compounds such as gallium tri-ethyl ($Ga(C_2H_5)_3$), rather than the chlorides used previously. It came to be known as metal–organic vapour phase epitaxy (MOVPE) and these two techniques were to be arch rivals for many years in numerous different situations, both scientific and commercial.

Here, we need to pause a while to fill in a certain amount of background and to relate these new developments to the III–V semiconductor scene with which we are already familiar. Not all was idyllic in the AlAs/GaAs epitaxial landscape. Even though success with double heterostructure lasers had been demonstrated, difficulties still abounded. There was difficulty with both LPE and VPE in controlling the ratio between aluminium and gallium in the AlGaAs layers and another problem was that their growth rates depended rather sharply on the precise growth temperature. What was worse, a small error in temperature could result in the deposition process going into reverse and removing material, rather than depositing it. A whole layer, carefully deposited, might disappear as a result of a tiny error in adjusting the furnace temperature. Commercial growth of these complex structures would scarcely be viable under such circumstances – improved growth methods were clearly essential. Firstly, MOVPE and then MBE showed promise in providing solutions to these problems, while at the same time offering a capability of growing the very thin layers appropriate to superlattices and LDS. Let us examine them now in a little more detail.

Several features of MOVPE immediately came to the fore. In principle, the technique looked very similar to conventional VPE but differed in that the GaAs (or other) substrate was heated by mounting it on a graphite block and supplying heat from a radio frequency source which induced eddy currents in the block. This had the big advantage over conventional furnace heating in that the rest of the apparatus could be kept relatively cool, thus avoiding the transfer of unwanted impurities from quartz tubing and other parts of the equipment. It was readily possible to grow heterostructures by including a source of aluminium tri-methyl and good control over the ratio of aluminium to gallium achieved as a result of a third important factor. By growing epitaxial films under excess arsine conditions (i.e., supplying far more arsine than was actually needed), the growth rate was determined simply by the supply of gallium or aluminium and was independent of temperature in the convenient temperature range between 600 and 800 °C. This gave the method an important advantage when growing heterostructures and led to its being taken up in many laboratories and, ultimately, in many factories concerned with developing lasers and other heterostructure devices. A modern MOVPE machine is shown in Fig. 6.2. Though not part of its original motivation, it soon came to be recognised that growth could readily be controlled at rates very much lower than was possible with conventional VPE, slow enough, in fact, to allow the growth of nanoscale films. Within the space of a very few years

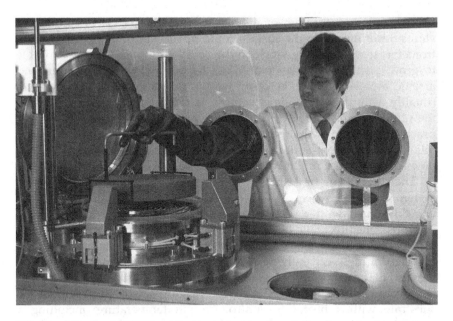

FIG. 6.2   A modern MOVPE (metal-organic vapour-phase epitaxy) machine used to grow thin, single crystal films of III-V semiconductors, such as gallium arsenide and indium phosphide. Courtesy of Aixtron.

MOVPE had changed the face of vapour phase epitaxy almost beyond recognition. But it had, as we have already hinted, a strong challenger.

More or less coincidentally with Esaki's work at IBM, there were important developments at Bell Labs. During 1968–1969, John Arthur was pursuing a programme of research designed to improve the understanding of GaAs surfaces and this involved some very sophisticated experimental techniques. In order to study clean surfaces, uncontaminated by unwanted impurity atoms, he was obliged to work in an extremely high vacuum (known as ultra-high vacuum or UHV) which required rather special vacuum pumps. It also demanded a vacuum container made from very clean stainless steel and several advanced methods of analysing the structure of the GaAs surface and recording the adsorption on and desorption from the surface of arsenic and gallium atoms. Surface structure could be measured using a technique known as reflection high energy electron diffraction (RHEED) in which an electron beam impinged at glancing incidence on the GaAs surface and a diffraction pattern detected on a fluorescent screen on the opposite side of the vacuum vessel. (Note that in this he was using the fact that electrons have an associated wavelength, related to their energy, just as light quanta, or photons have a wavelength associated with their energy). Gallium and arsenic atoms could be supplied to the surface by evaporating them from special containers through small holes which faced the GaAs sample and atoms desorbing from the surface could be measured with a mass spectrometer. It was exciting work, in the course of which Arthur was able to grow a single monolayer of GaAs by supplying both gallium and arsenic atoms together. In other words, he had grown a GaAs film roughly a third of a nanometre thick. It represented crystal growth at the ultimate limit and, while Arthur himself had no desire to take up practical crystal growth, his colleagues were quick to recognise the possibility. Al Cho, in particular, had an interest in improving lasers and saw how Arthur's work might be developed into a practical growth method.

The chief failing of Arthur's equipment, viewed as a realistic growth system, was its slowness. The special cells supplying gallium and arsenic were designed with a view to delivering an accurately known beam of atoms – the flux could be calculated simply by measuring the temperature of the cell. To build a practical growth system required a major redesign of these cells, giving them much larger apertures and shaping them so that they produced a well defined molecular beam which could be directed at the substrate. Other modifications involved providing the substrate with a heater to control growth temperature, including a vacuum interlock so that substrates could be manipulated into and out of the machine without letting it up to air pressure and adding shutters in front of each of the sources so that the beams could be switched on and off as required. For example, in growing a GaAs/AlGaAs structure it would be necessary to switch off the aluminium beam while growing the GaAs layer. It was also necessary to increase the number of sources. In addition to the gallium and aluminium cells, there was a need for doping cells, one n-type and one p-type. Not only was all this done, but it was done quickly and within 2 years films of GaAs, AlGaAs, GaP and GaAsP had been grown with good crystal quality, and doped both n- and p-type. Typical growth rates were 1 μm/h (about one monolayer – i.e., one molecular layer – per second) which, though very much slower than available from LPE and VPE were adequate for many structures – for example, the double heterostructure laser had a total thickness of perhaps 3 μm, so could be grown in 3 h. The important point was that it could be grown with excellent control and with sharp interfaces between the different layers. This latter point is particularly significant in the context of growing superlattices – the shutters in front of the sources could be moved in or out in a time less than a second so, with a growth rate of one monolayer per second, the interfaces were sharp on the atomic scale. Clearly, MBE was an exciting new growth method and it was not long before it had established roots in other laboratories round the world. Early in the 1970s there were facilities in England, where Bruce Joyce and Tom Foxon set up an activity in the Mullard laboratory, (see Fig. 6.3) in Germany (Klaus Ploog) and in Japan (Takahashi and Gonda). Together with MOVPE, it

FIG. 6.3 A typical MBE (molecular beam epitaxy) machine used for growing thin layers of single crystal material such as gallium arsenide or aluminium gallium arsenide. It relies on a stainless steel ultra-high vacuum system to keep out unwanted impurity atoms. Courtesy of Tom Foxon.

gradually came to dominate the III–V growth scene and, later that of other material systems, where appropriate.

It was several years, however, before one of MBE's greatest virtues was appreciated, an ability accurately to measure growth rate by actually counting the number of monolayers deposited. The story belongs to the Mullard (later Philips) group in Redhill who observed, one day in 1981, a strange variation in the intensity of the RHEED diffraction pattern during growth. This was puzzling in the extreme until someone pointed out that, since the laboratory was adjacent to the busy London-Brighton railway line, it was obviously an artefact of the regular vibrations produced by passing trains. But those who had made the observation were not altogether convinced – further careful study when trains were no longer plying their trade revealed that the period of the oscillation in intensity coincided exactly with the time taken to grow a single monolayer of material and thoughtful analysis of the RHEED phenomenon led to a theoretical understanding which confirmed such an interpretation. MBE had, at last, revealed one of its most endearing (and valuable) attributes. It was subsequently used in a variety of experiments to understand the subtleties of epitaxial growth but we shall not pursue this aspect further here – suffice it to say that its habitués were provided with a valuable technique for monitoring growth rate which was especially useful when growing the very thin layers appropriate to LDS.

But enough of this diversion into the wonders of crystal growth – we must return to those structures which were most to benefit from these recently acquired skills. (Though nor must we overlook the importance of material technology in making possible their realisation!) Superlattices were, perhaps, just a shade elaborate as an initial test of MBE's and MOVPE's capabilities and, though Esaki had some success in his efforts to grow them and Art Gossard at Bell succeeded in producing a mono-layer AlAs/GaAs superlattice in 1976, there were other new structures which attracted attention and which eventually led to more exciting device developments. In the first instance came studies of the optical properties of single and multiple quantum wells which led to the quantum well laser, followed soon afterwards by the two-dimensional electron gas (2DEG) which led to the high electron mobility transistor (HEMT). These, in turn, were followed by a wealth of other LDS which led both to sophisticated physics experiments and to other exotic devices. All in good time we shall touch on many of them but, for the moment, we must restrain our enthusiasm and concentrate on one development at a time.

One of the most important papers to excite the semiconductor community appeared in 1974 when Ray Dingle and colleagues at Bell reported on measurements of the optical behaviour of quantum wells. We have already made the point that confining electrons in a narrow well gives rise to a series of electron states within the well. In practical structures where the well depth is typically 200–300 meV (milli-electron volts), the number of such states is limited, and decreases as the well width decreases. Thus, assuming 30% aluminium in the barrier region and pure GaAs in the well, there will be just four states in a 15 nm well, three in a 10-nm well, two in a 5-nm well and just one in a 2-nm well. (There is a general theorem which says there will always be at least one state in any finite well.) In order to understand the optical absorption spectrum of a quantum well, we need to remember that there is a corresponding well in the valence band, too and this will give rise to a similar set of hole states. Optical absorption takes place when an electron in one of the hole states in the valence band is excited into one of the electron states in the conduction band. This is simply a modified version of the valence band to conduction band transition which gives rise to optical absorption in a simple semiconductor. The effect of the quantum confinement is to produce a series of absorption lines at energies greater than the band gap of the well material, GaAs. There are theoretical reasons why not all these possible transitions will be observable, in practice but, nevertheless, the spectrum is considerably more complex than is the case for the pure semiconductor. We need go no more deeply into the detail – suffice it to say that Dingle et al. observed spectra from a series of different samples with different well widths and were able to interpret them in terms of the ideas we have outlined. This was a tremendous step into a completely new field of interest and stimulated a wide range of activity, including, as we shall see, the development of the quantum well laser.

One such activity was a concentrated effort to measure the various parameters which were involved in the theoretical understanding of quantum well behaviour. For example, we talk glibly about a 5-nm well. How sure can we be that our well really is 5 nm wide? What if it were actually 6 nm? The theory used to calculate the energies of the various states assumes the well can be represented as being perfectly sharp (in terms of the transition from GaAs to AlGaAs). Can we be sure that the interface is not slightly blurred – that is that some aluminium atoms appear on the wrong side of the nominal interface? Finally, the theory has to make an assumption concerning the relative depths of the conduction band and valence band wells. This is a parameter very difficult to calculate a priori – it has to be extracted from the experimental data by fitting measured absorption lines to their energies calculated for a range of assumed ratios, the best fit giving the most likely value for the correct ratio. And this, in itself, proved almost equally difficult – it took something like 13 years before the value was known to an accuracy of 3%! The value determined in Dingle et al.'s original paper turned out to be well wide of the mark – they quoted a ratio of 85:15, whereas we now believe it to be about 67:33. (Science often moves like this in apparently uncoordinated jumps.) Application of sophisticated techniques such as X-ray diffraction and transmission electron microscopy showed, too, that the answers to our earlier questions were more than a little tricky – there was certainly some evidence for inter-penetration of the well and barrier materials and well widths were often not quite what the crystal grower had predicted. This was serious materials science at its best, urging a degree of caution in the way we should interpret what soon grew into a large volume of experimental data. The general picture was satisfactorily verified but details could be difficult to prove beyond reasonable doubt.

Needless to say, there followed a great deal of high quality scientific work in pinning down these details but we shall leave it as read and move on to consider some interesting developments in the device field. The first and, perhaps rather obvious, one was the quantum well laser. It was a simple matter, once the appropriate growth methods were in place, to shrink the active bit of the laser from its original thickness of 0.1 μm down to the nanoscale range in the interest of effecting several improvements. The first, and most obvious, we have already mentioned – the thinner the well, the smaller the volume of material to be pumped up to laser threshold, and this implies a lower threshold current. A theoretical argument concerning the nature of the optical transition involved – from the lowest confined electron state to the highest confined hole state – also led to predictions of much reduced threshold currents. The overall reduction compared with the original demonstration lasers of 1962 was dramatic. Even though these early devices operated at 77 K, their threshold current densities were in the region of $50\,\mathrm{kA/cm^2}$, while the first double heterostructure lasers operated at room temperature with values of $2\,\mathrm{kA/cm^2}$. Quantum well lasers eventually achieved thresholds as low as $50\,\mathrm{A/cm^2}$, a reduction of a 1000 times compared with their trail blazing predecessors and a reflection of some 20 years of dedicated development work. Another advantage of the QW laser is the fact that the threshold current depends less strongly on operating temperature (in general it increases as the laser runs hotter). At the same time it was clear that the laser emission wavelength should be significantly shorter than the original GaAs wavelength of 880 nm. Indeed, simply by varying the well width, it was possible to tune the laser output over a wide range of wavelengths, an effect clearly demonstrated by Karl Woodbridge and co-workers at what was now (1984) the Philips Research Laboratory in Redhill. A dramatic result of their work was the achievement of a laser emitting in the visible (red) part of the spectrum at 707 nm, a shift of $\sim$170 nm compared to the GaAs wavelength of 880 nm. The active material was still GaAs but by reducing its thickness to 1.3 nm the emission wavelength corresponded to that appropriate to an AlGaAs alloy containing 26% aluminium. There would, however, be a distinct disadvantage in including aluminium in the active layer as it has a strong affinity for oxygen and that is seriously detrimental to material quality. Thus the QW laser offered several important virtues, compared to its original double heterostructure cousin and it caught on in many different applications, including optical communications, CD and DVD players.

If the simple artifice of confining free carriers in one dimension (thus forming a two-dimensional structure – carriers being free to move in two dimensions), one might hope for further improvement by reducing the dimensionality even further. Instead of quantum wells, we might aim for quantum wires (with just one-dimensional motion allowed) or even quantum dots (with no free carrier motion in any direction). Theory certainly encourages such notions and many experimentalists have been so encouraged, but there are practical problems to be overcome and progress has been rather less rapid. We shall discuss quantum wires in relation to electrical measurements later in this chapter – for the moment we concentrate on quantum dots. It is a relatively straightforward matter to use MBE or MOVPE to grow ultra-thin sheets of material but it is far more difficult to produce dots in a controlled manner. In particular, it is extremely difficult to form dots with uniform size and shape, which is what is actually needed in order to obtain theoretically predicted advantages in laser performance. Surprisingly enough, Nature did try to help. Various attempts had been made during the late 1980s to fabricate dots by modifying quantum wells with electron beam lithography and some promising results obtained, such as wavelength reduction as the dots were made smaller. However, it was only possible to form a single layer of dots this way, which may have been useful for studying their properties but was not appropriate to making a practical quantum dot laser. The big surprise came at the beginning of the 1990s when it was discovered that dots could be formed quite naturally during MBE growth of InAs films on a GaAs substrate.

InAs has a band gap of 0.36 eV, very much smaller than that of GaAs, which means it will readily confine electrons from the GaAs. It also has a lattice parameter significantly greater – by some 7% – which means it is far from happy at the idea of growing on GaAs. Detailed study revealed that the first one to two monolayers of InAs grew with a lattice constant equal to that of the GaAs beneath it – that is, it was highly strained – but any further growth occurred in a quite different mode. Rather than forming a continuous film, the InAs grew in small islands which could then be covered in a GaAs film to isolate them. Once this cladding layer was complete, the process could be repeated with further InAs, more GaAs cladding, and so on until a relatively thick layer of isolated dots had been built up. The resulting structure can be thought of as a new material 'doped' with dots which act as radiative recombination centres – that is, they emit light when electrons and holes recombine within them. The process is similar to the one we have already met in the quantum well laser – electrons and holes are injected across a p–n junction, are captured by the wells (or, in this case, the dots) and recombine with emission of light. The wavelength of the light is determined by the size of the dots and is typically about 1.1 μm, further out into the infra-red from that appropriate to pure GaAs. Interestingly, a French group at France Telecom managed to measure the radiation from a single dot and found the emission line to be (as predicted theoretically) extremely sharp (close to being at a single wavelength). However, the spectral width of the overall emission was several hundred times greater, an indication of the large spread in dot sizes, each one emitting at its own wavelength but different from its neighbours. This represented a major difficulty in harnessing dots to make better lasers and remains something of a fundamental problem, there being an inevitable statistical spread in dot sizes which rely on this kind of random selection. It is important to recognise that dot formation depends on the degree of strain present and this suggests that there should be several other material systems which can grow dots – most combinations show some degree of mismatch in lattice constant, there being relatively few which match like the AlAs/GaAs pairing. It is also worth emphasising the fact that these dots occur without any deliberate activity on the part of the crystal grower – as a result they are known as 'self-organised' quantum dots (which all sounds a trifle unfair to the growth fraternity who actually have to work rather hard to produce a multi-dimensional layer of such dots!)

Indeed, the extent of the crystal growth problems is emphasised by the time it took to produce a viable QD laser. The first luminescence measurements on QD materials were reported in 1986 but it was not until 1994 that a QD laser was made to run continuously at room temperature. The threshold current was 1 kA/cm^2, much larger than the best QW lasers, but by the turn of the century things had improved to the point that threshold currents were competitive with those measured on QW devices. What was more, a clear advantage existed in so far as thresholds showed a very much smaller dependence on temperature, important for applications (as in fibre-optic communications) where devices could experience quite wide variations in ambient temperature. It is interesting, by the way, to note the make-up of the group that reported this first laser success. There were no less than 13 authors who came from the Ioffe Institute, St. Petersburg and Universities in Berlin and Weinberg, Germany, a collaboration set up in a deliberate attempt to help Russian scientists bereft of funds following the collapse of the Communist regime in the early 1980s. In fact, the QD laser has greatest relevance to the field of fibre-optic communications so we shall leave the topic for the time being and look instead at another exciting laser development, that of the vertical cavity surface emitting laser (VCSEL) which came to prominence during the 1980s.

Fully to understand this new development we need to recapitulate briefly. The laser devices discussed so far emit light in the plane of the p–n junction, while current flows across the junction, normal to the direction in which light emerges. This is a consequence of the use of mirrors formed by cleaving the sample in a vertical plane, and these mirrors can be placed any convenient distance apart. Typically their separation is about 500 μm, or half a millimetre, a distance which allows for a large degree of light amplification within the optical cavity. This makes up for the fact that the mirrors are only about 30% reflecting, so 70% of the light striking each mirror is lost. The VCSEL, on the other hand, emits light normal to the junction plane, and therefore normal to the surface of the sample, a geometry which has the advantage of allowing an array of lasers to be made over a single slice of material. The cross-sectional area of each device may be determined by defining the area of the top metal contact by photolithography. In contrast to the conventional laser, the VCSEL has a very short cavity length because its mirrors are separated by the thickness of the epitaxial layers used in making it. Thus, rather than being able to choose the cavity length more or less at random, it is limited to no more than a few microns and this implies that mirror reflectivity must be very much greater than the 30% available in the conventional structure. In fact, a simple calculation suggests it must be considerably greater than 90%, thereby suggesting the use of metal mirrors which is what the VCSEL pioneers actually did. However, even metal mirrors proved unsatisfactory. Evaporated gold films provide about 98% reflection but, when it was found necessary to reduce the cavity length in order to bring the threshold current down to acceptable levels (by reducing the volume of material that had to be pumped) the required reflectivity went up to something like 99.9%. The answer to this improbable demand came in the form of what are known as Bragg reflectors – multiple layers of materials having different refractive indices, each layer being just a quarter of a light wavelength thick. We saw earlier that AlAs and GaAs have different values of refractive index so these two materials were the first choice for these layers because they could easily be grown epitaxially on top of one another. To obtain an overall reflectivity of 99.9% required 30 pairs, a total thickness of about 5 μm for each mirror (for a laser emitting at about 1 μm wavelength), a modest challenge to the crystal grower but not impossible to fit into a rather long working day! (The need for so many layers arises because the change in refractive index at each interface is relatively small so each interface reflects only about 0.6% of the light.)

All this probably gives the impression that making a VCSEL to operate continuously at room temperature may well have been no easy matter – indeed, it certainly wasn't – so it may come as little surprise to learn that the first such success arrived only in 1990, some 12 years after the first tentative efforts. The 'winners' were a group at the University of California, Santa Barbara, under the leadership of Larry Coldren. They grew the structure by MBE – it involved a total number of distinct layers approaching 1000, all computer-controlled and pre-programmed – no human operator could hope to remember such a complicated sequence! They used an active region in the shape of a single GaAs/InGaAs/GaAs quantum well, which emitted at a wavelength of 963 nm. Threshold currents were of order 1 mA – densities of about 800 A/cm^2, comparable with values appropriate to conventional lasers. More recently, currents have come down to the order of 10 μA and current densities to about 100 A/cm^2. Again the VCSEL story takes on a major significance in the field of fibre-optic communications so we shall say no more until Chapter 8. Suffice it to say that the optoelectronic community now has available to it a wide range of light emitters, covering a wide range of wavelengths and output powers and able to satisfy a wide range of applications, from high power laser machining to high speed laser communications.

So much, then, for low dimensional optical devices – we must now turn attention to their electrical counterparts. Perhaps the first point to be made is that, while quantum wells confine electrons (or holes) and inhibit their motion in a direction normal to the plane of the well, these carriers are free to move in any direction in the plane. There are, indeed, two dimensions of freedom – electrons are just as mobile in the plane of the well as they would be in an uninhibited three-dimensional semiconductor. Thus, an electric field applied in the plane will generate an electric current in the usual manner, assuming only that there are free electrons (or holes) in the well, and we can satisfy that condition simply by doping the semiconductor in the usual way. All very well, you may say, so what? There doesn't seem much point in going to the trouble of making a quantum well just to obtain the same electric current as would pertain to a simple three-dimensional sample of the same material. But there is a point, of course – otherwise I would scarcely be taking all this trouble to discuss it! It was in 1978 that Ray Dingle and colleagues at Bell Labs explained exactly what the point was. We need to think a moment about electron mobility and what determines it. As we saw in Chapter 3, there are two important scattering mechanisms which limit mobility, thermal vibration of the crystal lattice and ionised impurity scattering. The former can be reduced by lowering the sample temperature but, like the poor, the latter is always with us and seriously reduces mobility at low temperatures. The problem is a very basic one – in order to produce free electrons, it is necessary to dope the semiconductor with donor atoms and when an electron separates from its parent donor, it leaves that donor with a positive electric charge. This sea of charged donor atoms then scatters free electrons by virtue of the electrostatic force between them, the more heavily doped the material, the stronger the scattering and the lower the electron mobility. Impasse! Well, not quite – this was the Bell Labs beautiful idea – suppose the donor atoms are arranged to be present *only* in the barrier material and not in the well, then free electrons which fall into the well will be physically separated from their parent donors and will not be scattered by them. Ionised impurity scattering is no longer of any significance – we can have lots of free electrons with much greater mobility than could ever be possible in a bulk, three-dimensional sample. Murphy's Law has been totally vanquished – not only do we have a high density 2DEG but also one characterised by high electron mobility. Pure genius!

I have slightly overstated the case – the electrostatic force acting on the electrons is reduced but does not disappear altogether. The trick, as demonstrated by the Bell workers, was to dope the barrier material so that there was a region close to the interface which remained undoped, thus separating the donors from the electrons even further. A compromise was necessary because, if the separation was too great, few electrons could reach the well and their density was too small. The combination of high density and high mobility needed to be carefully calculated. But, make no mistake, the idea worked. It was only necessary for the crystal grower to develop the skill to place donor atoms precisely where they were needed and nowhere else. Bearing in mind that the distance between donors and interface was typically about 50 nm, this was a skill which demanded a very well controlled growth system and MBE can be seen as ideal – to MBE, with its monolayer capability, 50 nm (about 18 monolayers) was massive. Art Gossard's MBE machine at Bell Labs was pressed into service to grow AlGaAs/GaAs structures and very soon produced some impressive mobility data. Electron mobilities at low temperatures were measured at values 10 times greater than those appropriate to three dimensional samples of GaAs. The advantage at room temperature was much smaller but still significant. The scientific world was really buzzing with excitement.

Once the existence theorem was proven, the MBE community took up the challenge to see just how far electron mobility could be pushed. It was a fascinating competition. The first step was the realisation that a 2DEG did not require a quantum well but could be formed close to the interface at a single heterojunction between an AlGaAs film and a GaAs film. The AlGaAs was doped with silicon donors, taking care to leave an undoped spacer layer near the interface, and the GaAs was undoped. Three criteria were vital – a high quality interface was essential, the spacer layer thickness had to be optimised and the GaAs had to be as pure as possible. The race began in 1979 and involved no less than 10 MBE groups from (in no particular order) the USA, Japan, England, Germany, France and Israel. It took 9 years to reach the ultimate plateau – a low temperature mobility some 300 times greater than the original value reported by Dingle et al. in 1978 – an international effort to rival anything seen outside total warfare. One bizarre aspect of the struggle was the use of superlattices as base structures on which to grow the crucial heterojunction. It turned out that each AlGaAs/GaAs interface tended to trap impurities, so growing a large number of them proved to be an ideal method of purifying the critical GaAs film in which the 2DEG was to be formed. (Leo Esaki may have seen this as being a far cry from his original application for the superlattice but it never-the-less proved an essential component in the MBE mix.) Fascinating stuff, but was it really worth it? (Remember this all refers to samples cooled in liquid helium – room temperature enhancement of mobility is far more modest.) Undoubtedly, the answer has to be yes – quite apart from the stimulus to MBE growth, international rivalry and scientific PR, it led to two major scientific advances, one of Nobel Prize significance, the other of considerable practical importance.

We have already come across the quantum Hall effect in Chapter 4. It was discovered in a 2DEG formed at the interface between silicon and silicon dioxide but the development of the AlGaAs/GaAs structure led to an enormous improvement in the quality of data that could be obtained. One reason for this was the fact that the III–V structure consisted of single crystal material throughout, whereas the silicon structure depended on an interface between a single crystal and an amorphous layer (the oxide). It was also considerably easier to observe the effect because it appeared at much lower values of the magnetic field, a consequence of the smaller electron effective mass in the GaAs, and at somewhat higher temperatures (4.2 K rather than 1.5 K). It was this which made it considerably more suited to the development of a fundamental standard of resistance. To recapitulate on our brief discussion in Chapter 4, we remember that the quantum Hall effect consisted of unexpected plateaux in a plot of Hall resistance (the ratio of Hall voltage divided by sample current) against magnetic field, and these plateaux occurred at values of Hall resistance related to the fundamental constant: (Planck's constant divided by the square of the electronic charge). From a practical viewpoint, the significant feature of these plateaux is that they occur with a high degree of reproducibility at values of resistance equal to:

$$25812.807/i \text{ Ohms}$$

Notice that 'i' is an integer, $i = 1, 2, 3, 4$, etc., which means the $i = 1$ plateau occurs at the highest value, $R = 25812.807$ Ohms, the $i = 2$ plateau at half this value, and so on downwards. In practice, only four or five of the plateaux are resolved, as they become gradually less well resolved as the integer 'i' increases.

It is significant that the above relationship is accurate to about one part in a billion and, on this basis, the quantum Hall effect was adopted as an international resistance standard, as from 1 January 1990. It was important to establish that the new effect was independent of individual Hall samples and that it could be measured in a wide range of laboratories and still give the same result. This was done on an international basis and the results showed agreement within a few parts in a billion, a very satisfactory outcome compared with similar measurements on standard wire-wound resistors which agreed to only a few parts in 10 million. What was even worse, in the case of wire-wound resistance standards, they had a marked tendency to drift in time, changing by about one part in ten million per annum. Their resistance also varied with temperature to such a degree that it was necessary to stabilise the temperature to an accuracy of 1 mK (a thousandth of a degree Centigrade). The standard resistor is usually a 1 Ohm resistor so comparison with a quantum Hall measurement involved several steps. Firstly, a resistor close in value to the $i = 2$ Hall resistance was chosen and compared, then this resistor was compared to a 100 Ohm resistor and, finally, the 100 Ohm resistor compared with the 1 Ohm standard. An overall accuracy of a few parts in a billion was achieved in one particular experiment. Using such esoteric equipment obviously has its drawbacks in standards work but the much improved accuracy and reliability of the method made it worthwhile and the new standard is widely established. It represents a quite remarkable application of what might otherwise be seen as nothing more than a scientific curiosity, yet another reminder that the usefulness (or otherwise) of 'blue skies' research should never be pre-judged.

The second feature brought out by the availability of 2DEG samples with extremely high mobility was yet another unexpected discovery made by the Bell group in 1982. Dan Tsui, Horst Stormer and Art Gossard reported observations of Hall plateaux at magnetic fields even higher than that corresponding to $i = 1$, that is corresponding to *fractional* values of 'i'. This was the 'fractional quantum Hall effect' and appeared to be a property of some completely new particle with a charge equal to one third of the electron charge, a remarkable observation after some 100 years of believing that the electron charge was the smallest charge in existence. Extreme puzzlement gradually gave way to qualified understanding, in the light of what was known as 'many body theory' – as a result of Bob Laughlin's theoretical intervention, groups of strongly interacting electrons were found capable of behaving like quasi-particles with fractional electronic charge. A whole new realm of theoretical and experimental physics had been exposed to view and, like the hunt, when granted such temptation, the hounds of science homed in for a kill. However, the fractional fox refused to give himself up without a fight. To change the metaphor slightly, it became clear that a whole nest of scientific hornets had been stirred into activity – attempts to tame them still continue and look likely to continue for some time yet. No matter what the final outcome, Tsui, Stormer and Laughlin were awarded the Nobel Prize for physics in 2000, while poor Art Gossard whose crystal growing skills made it all possible was ignored, yet another example of the undervaluing of materials science in relation to 'pure' science which should surely stir materials scientists into concerted revolt.

However, righteous indignation must never be allowed to deflect us from our course. The next topic to evolve from man's ability to make ultra-high mobility 2DEG samples acquired the designation 'mesoscopic systems', from the Greek meaning 'middle sized'. It was intended to cover those structures which were not quite small enough to be called 'microscopic', while being small enough not to qualify as 'macroscopic'. We shall make no attempt to deal with all aspects of the subject but merely mention one or two examples to give a flavour of its numerous delights. They will be concerned with electron transport – that is, electric current flow – under rather special circumstances. Let us not forget, however, that such ultra-high mobilities exist only at low temperatures so any devices which depend upon them can be expected to function similarly, only at low temperatures. That they qualify as mesoscopic follows from the use of lithography to define their structure in at least one dimension.

Having mastered the art of making well controlled two-dimensional structures by MBE or MOVPE, the next cast of the dice was aimed to control electron motion in the plane of the device. Imagine a pair of gate electrodes in the form of closely spaced parallel strips laid down on top of the 2DEG structure. By applying negative voltages to both, electrons can be repelled from the regions immediately beneath them, leaving an isolated stripe of conducting material between the two. This 'wire' has a cross-section defined in one direction by the thickness of the 2DEG and in the other by the lithographic definition of the two gate strips. It amounts to a one-dimensional conductor which has special properties of its own. Electrons travelling along it can be likened to cars travelling along a single lane motorway – the special feature resulting from their very high mobility being the fact that they can travel long distances without being scattered by an impurity or lattice vibration. In this respect, they have common experience with electrons in a vacuum chamber and a great many experiments have demonstrated behaviour which closely parallels such vacuum systems. Let us look at just one experiment, rejoicing in the name of the Aharonov-Bohm effect.

Extending our motorway analogy, imagine the case of a simple roundabout with a road entering at one side and a second road exiting from a point immediately opposite. An electron coming upon such a diversion has two options – it may choose the British, clockwise route or the American, anti-clockwise route and, in a macroscopic world, it would choose either one or the other. However, being an electron it actually goes *both* ways – another example of 'the special relationship'! This means that it meets itself at the opposite side and its electron waves interfere with one another. Because the two paths are equal in length, this interference is constructive and the result is an electron continuing along the exit road. However, if we can find a way to alter the effective length of one path, the interference may become destructive – one wave effectively cancelling the other – and the net effect is an electron reflected back along the incident road. This modification can actually be effected by placing a Schottky gate over one side of the roundabout and applying a suitable voltage to it, thus making a kind of quantum switch. Interestingly, an alternative method of inducing such switching behaviour makes use of a magnetic field applied normally to the plane of the roundabout. Quite tiny fields are sufficient.

The second mesoscopic device I should like to introduce you to is the single-electron transistor which was demonstrated during the 1990s. The philosophy behind it is concerned with reduction in power dissipation which is becoming an ever greater worry to the designer of integrated circuits in today's world of nanoscale transistors. A modern FET, small though it is, draws source-drain currents consisting of thousands of electrons – ideally it would be sufficient, if current could be limited to one electron at a time. It sounds almost too good to be true but it can certainly be done. The drawback, as we have already emphasised, is the necessity for low temperature operation, most of the experiments having been performed at temperatures below 1 K. The device consists of a minute quantum dot which is coupled to a pair of electrical contacts by way of so-called tunnel junctions, each being a thin barrier through which electrons can tunnel. This constitutes the channel. There is also a gate which is capacitively coupled to the dot so as to control the current flowing to and from the dot. There are several subtleties involved which need not trouble us but the demonstrable fact is that the gate voltage can be set so as to allow just one electron to tunnel from the source contact onto the dot, whence it is free to tunnel to the drain contact. Once this first electron has left the dot, a second electron may tunnel onto it and so on and on. However, a change of gate voltage may prevent even the first electron tunnelling, thus switching the transistor off. The tunnelling process is inherently very fast, which allows the transistor to operate at very high speed, just what the integrated circuit designer is looking for, but he is certainly not looking to build his circuits to operate at 1 K. The major question must be one of operating temperature and whether this can be raised to some acceptable level. Watch this space – but don't hold your breath!

I need hardly point out that the last few paragraphs have taken us into the world of scientific fantasy and, though it may be exciting to explore such exotica, sooner or later we must return to the world of reality. The art of growing heterostructures on a very small scale has made major contributions to more conventional transistor science and we shall end this chapter by looking in some detail at two important devices, the high electron mobility transistor (HEMT) and the heterojunction bipolar transistor (HBT). The HEMT represents a fairly obvious application of the 2DEG to making a field-effect transistor – it was first reported in Japan (at the Fujitsu laboratory) in 1980, just 1 year after the Bell Labs report of a single heterostructure 2DEG, but was rapidly taken up by a truly international phalanx of researchers, both pure and applied. By the end of the 1980s, one could list something like 20 different participants: Bell labs, IBM, Honeywell, Sandia, Fujitsu, Toshiba, Siemens, Thomson CSF, Hewlet Packard, Rockwell, General Electric, GEC, Philips, NEC, CornellUniversity, University of Illinois, Brown University and many others. The downside of this proliferation of contenders was a corresponding proliferation of device names – the French wanted to call it a TEGFET (two-dimensional electron gas field effect transistor), the Americans preferred MODFET (modulation doped field-effect transistor) while a more recent attempt to reach a consensus added a fourth version, HFET (heterojunction field-effect transistor). Being British, and neutral in this particular conflict, I shall stick to the original HEMT but the reader should be aware of these possible usurpers. Perhaps we can see this competition as demonstrating the intensity of interest generated by the device and we shall see, in a moment, just why there was such an interest. The HBT has a somewhat different pedigree – it was first proposed by William Shockley in 1951, taken up again in 1957 by Herb Kroemer and eventually put into serious practice also during the 1980s when the necessary technology had become available. Though both HEMT and HBT devices made important contributions to the upward extension of the transistor frequency range, they represent very different approaches to the problem. We shall examine them in turn – first the HEMT.

Unsurprisingly, the first HEMTs made use of the AlGaAs/GaAs materials system. The idea was to dope the AlGaAs barrier layer with silicon donors so as to produce a 2DEG in the GaAs, close to the interface and make this the channel of a conventional field-effect transistor. It involved diffusing n^+ source and drain regions through the barrier layer to make electrical contact with the 2DEG and placing a metal Schottky gate electrode between them in the manner adopted in the early GaAs MESFETs. The advantage accruing from the use of the 2DEG was, of course, the enhanced electron mobility which gave promise of higher gain and operation at higher frequencies or, in the case of digital applications, faster switching speed. (The response time of a FET is determined by the time taken for electrons to flow into or out of the channel, following a change of gate voltage, so the more mobile they are, the shorter this response time.) However, a word of caution is in order here because the mobility enhancement at room temperature (where all sensible devices function) is far, far smaller than that available when the device is immersed in liquid helium. Practical values are more like factors of three than the 300 times appropriate to 4 K – not, of course, that one should turn up one's nose at a factor three. No matter, progress was good and by 1984 (Big Brother permitting) HEMTs with 1 μm gate length had been demonstrated which operated at frequencies as high as 5 GHz, while devices with 0.35 μm gate length were operating at 35 GHz, a very encouraging result, indeed. However, there were problems with the use of AlGaAs as the barrier material.

It was during the 1980s that the III–V scientific world learned of a new hazard which, in true mystery story style, had been labelled the DX centre. It became clear that the silicon donors used to dope the AlGaAs n-type were also giving rise to a deep state within the band gap which could trap free electrons. What was worse, these DX centres became more and more of a hazard as the ratio of aluminium to gallium was increased and this imposed a practical limit on the fraction of aluminium at about 20–25%. In turn, this had a limiting effect on the supply of free electrons to the 2DEG, because this vital parameter depended on the conduction band step at the AlGaAs–GaAs interface. Further progress towards higher frequency operation depended on persuading more electrons to reside in the 2DEG (i.e., the channel) and this demanded a greater step, which meant more aluminium. It looked like an impasse. What was to be done? Remarkably enough, quite a lot of things were done, all of which provided solutions. An obvious method of avoiding the DX centre problem was to use a different material for the barrier. AlInAs was found to be free from DX centres but it could not be used in conjunction with GaAs – its band gap was too small and its lattice parameter was too large. This meant replacing the GaAs with GaInAs and growing the whole structure on an InP substrate in order to obtain lattice matching. This was materials science of a high order – AlInAs and GaInAs were found to be lattice-matched to InP when they contained 50% indium and, at this point the step in their conduction bands was large enough to produce nearly three times as many electrons in the channel. It was a brilliant solution to the DX problem but it suffered from two practical drawbacks – InP was not available in the same quality as was GaAs and it cost more money to buy. GaAs had been under development for so much longer and had reached a rather more sophisticated level of perfection. It was, for example, possible to buy 6 in. diameter semi-insulating GaAs boules, while InP was only available with 3 in. diameter. For pure scientific studies this would constitute no more than a minor inconvenience but in a seriously competitive industrial context it was closer to being a minor catastrophe. Nor should we overlook the additional difficulty associated with growing a new range of materials. Growth techniques for AlGaAs and GaAs were very well established by this time, whereas there was much less experience with GaInAs, and AlInAs was a complete new boy. All this could be accommodated if absolutely necessary but perhaps there were alternatives – materials scientists had met challenges before, and they were surely not to be beaten by this one.

   The DX problem lay with the association of aluminium and silicon atoms in the AlGaAs barrier layer, so why not just keep them apart? An ingenious method of achieving this was to replace the AlGaAs with a short-period AlAs/GaAs superlattice which was doped only in the GaAs layers. If the layers were thin enough, electrons could tunnel through the AlAs regions and reach the 2DEG, while, if the AlAs and GaAs layers were equal in thickness, the superlattice behaved rather like a layer of AlGaAs with 50% aluminium in it, just what was needed to increase the electron density in the 2DEG. Once again, the ability of MBE to grow ultra-thin films was utilised to the full in providing a solution to a very real practical problem. But this was far from being the end of the story – other clever solutions were in the pipeline. The first of these depended on a novel technique known in Europe as 'delta-doping' and in America as 'planar-doping'. The American term is probably more descriptive – the method involved stopping growth of AlGaAs while depositing a single monolayer of silicon atoms, then continuing the AlGaAs growth to completion. The result was a relatively heavy doping of the barrier in which the silicon atoms were arranged in a geometrically (and chemically) different way with respect to the aluminium – it succeeded in avoiding the formation of DX centres and allowed the aluminium content of the barrier to be much increased. Materials science seemed to be falling over itself to satisfy every whim of the microwave engineer but even this was not enough – there was yet another solution on offer. This again made use of GaInAs as the material in which the 2DEG was to reside – it had the advantage of even higher electron mobility than bulk GaAs – and it could be combined with an AlGaAs barrier. Because GaInAs has a smaller band gap than GaAs, the step in the conduction band was made larger, even though the aluminium fraction in the barrier layer was limited to about 20%. This minimised the DX centre problem, provided more electrons in the 2DEG, offered higher electron mobility and the structure could all be grown on a high quality GaAs substrate. What more could anyone ask? Well, there was a problem – because GaInAs has a larger lattice parameter than GaAs, it was difficult to grow it on GaAs. The answer was to grow only a very thin layer, a layer which adopted the same lattice parameter as GaAs and was therefore strained (it was referred to as a 'pseudomorphic layer'), but, provided it was thin enough, it remained as a coherent, well behaved film. Nevertheless, even a very thin layer was perfectly satisfactory in confining the desired 2DEG. This was an interesting new departure in crystal growth which was to have wide-ranging application and it seemed, finally to be the solution which gained favour with the engineers – it was called a 'pseudomorphic HEMT' (PHEMT). Semiconductor technology had come a very long way from its early life with bulk crystals of germanium or silicon – here was true materials sophistication harnessed to clever device design.

Competition between the AlInAs/GaInAs/InP HEMT and the AlGaAs/GaInAs/GaAs PHEMT continued for some time but the important point was that high frequency performance improved to a great extent. By the end of the 1980s, devices with gate lengths as short as 0.1 μm functioned at frequencies close to 200 GHz and were well able to satisfy an important military requirement for devices to operate at 94 GHz. Once again, we have an example of military funding providing a welcome boost to an expensive development programme but the longer term application was to be in mobile phones, (see Fig. 6.4) a market which made microwave technology commonplace and which called, in particular, for low-noise amplifiers. There were, and still are, concerns as to possible harmful effects of microwave radiation on the human body so it was in everyone's interest to keep radiation levels as low as possible and this implied the use of highly sensitive receivers. The HEMT proved the perfect answer to this requirement, and still does, but its application in such a competitive market meant that there was enormous commercial pressure to improve it even further. InP-based devices were themselves made into PHEMTs by the incorporation of larger amounts of indium but GaAs-based HEMTs fought back with the use of a graded buffer layer

FIG. 6.4 The ubiquitous mobile phone depends for its connections on microwave radiation which is processed by highly sensitive field-effect transistors made from carefully strained thin films of various III-V compound semiconductors. Courtesy of iStockphoto, © Jakub Semeniuk, image 5395519.

between the GaAs substrate and the GaInAs 2DEG layer, again in the interest of incorporating more indium. The cost of substrates is still fiercely debated – with annual sales of handsets approaching one billion, even small economies may have significant commercial implications. Technically speaking, cut-off frequencies climbed to 300 GHz and beyond and noise figures improved in consequence. It is ironic to see that III–V devices were now more than fulfilling the original hopes claimed for them by their early supporters – but doing so in ways not even dreamt of in 1960. At that time, the mobile telephone, itself, was beyond most people's imagination, let alone the PHEMT. But there is nothing new in this – such has always been the way of technological advance. Remember, for example, how the first transistor came into being and how it was so very rapidly replaced by Shockley's junction transistor. Remember, too, how germanium seemed to be the ideal semiconductor until displaced by the usurper silicon and how silicon itself was introduced because of its larger band gap at a time when no-one was aware of its advantageous oxide. If the future of silicon had been judged by early attempts to make field-effect transistors, the integrated circuit may never have happened. One should never be surprised by unexpected developments.

Low noise HEMTs may be ideal for receivers but we should at least take note of the need for high power devices at the transmitter. In the first instance, the InP-based PHEMT found preference but more recently it has been challenged by new materials with greater potential for power generation – silicon carbide and gallium nitride. Both of these came to prominence as light-emitters, and GaN, in particular, plays a major role in Chapter 7 but, during the 1990s, it became clear that SiC MESFETs and AlGaN/GaN HEMTs were capable of significantly greater power output at microwave frequencies. Both materials have relatively wide band gaps, which gives them an advantage in that they can operate at higher voltages and SiC has a further advantage in respect of its excellent thermal conductivity. High power devices inevitably generate a fair amount of heat and this must be conducted away to a suitable heat sink to avoid damage to the transistor. (We have also come across this problem in relation to integrated circuits.) In terms of operating frequency, SiC appears to be limited to the lower end of the microwave range, while GaN devices have greater potential towards higher frequencies.

It must now be clear that the MESFET and HEMT made possible a revolution in microwave usage (see Fig. 6.5), leaving the bipolar transistor lagging well behind in frequency capability, and this was, in part, at least, due to the fact that GaAs bipolar transistors had failed dramatically as a result of the very short minority carrier lifetimes measured on GaAs back in the 1950s. Given this difficulty, the design of a high frequency bipolar

FIG. 6.5 Microwave dish aerials such as those used for satellite communications and radio astronomy. Signal processing depends on sophisticated semiconductor devices and microwave integrated circuits. Courtesy of iStockphoto, © Richard J Gerstner, image 506802.

transistor demands a very thin base region in order that minority carriers can diffuse to the collector, before recombining. However, this implies an undesirably large base resistance unless the base is heavily doped, though this, in turn, leads to poor emitter efficiency. The emitter-base junction, under forward bias, must inject a high level of minority carriers into the base and this requires that the emitter doping level be very much greater than that of the base – increasing the base doping level obviously works counter to such a requirement. It sounds like another catch 22 situation but Shockley pointed to a very simple resolution in the form of a hetero-junction emitter. If the emitter were to be made from a semiconductor with a larger band gap than the base, the reverse injection of carriers from base to emitter would be blocked by the steps in the conduction and valence bands. Forward current would inevitably be carried in the form of minorities injected into the base, exactly as needed. The only problem was that no-one knew how to make such a junction in 1951. In 1957, Kroemer proposed using Si-Ge alloys but the first practical hetero-junction bipolar devices appear to have been made by combining a GaP emitter with a GaAs base. This approach was somewhat problematic on account of the lattice mismatch between these materials, though a small amount of power gain was recorded. What is remarkable is that this was achieved as early as 1958. However, the application to real microwave transistors had to wait until 1981 when III–V technology, in the form of our old friend the AlGaAs/GaAs combination, was suffi-ciently well established. In fact, the first successful HBTs were even cleverer than that, making use of a graded GaInAs base (starting from GaAs at the emitter and gradually increasing the amount of indium towards the collector side) which effectively introduced an electric field in the base to accelerate electrons to the collector junction (this, by the way, being yet another of Herb Kroemer's innovative ideas from the 1950s). It results in a quite important speed advantage compared with devices which rely only on minority carrier diffusion, cut-off frequencies as high as 300 GHz having been achieved with graded-base HBTs. Given that the base was rather narrow (50 nm), the GaInAs grew as a strained (ie pseudomorphic) layer, as in the PHEMT.

Needless to say, other material systems have been tried for HBTs, including the lattice-matched GaInAs/InP combination used in HEMTs, though the same arguments with regard to InP quality and price apply here too. One innovation which has found favour in many quarters is that of replacing the AlGaAs emitter with GaInP (lattice-matched to GaAs when the indium content is about 50%). It has the advantage of avoiding problems associated with aluminium's love for oxygen and offers better reproducibility and reliability. Another III–V combination involves the nitrides – AlGaN/GaN – which may be well suited to high power, high temperature applications. In general, HBTs offer better prospects than HEMTs as power devices, though they are generally less suitable as low-noise amplifiers.

Last, but not least is the material combination originally suggested by Kroemer in 1957, based on SiGe alloys. These alloys have band gaps smaller than that of silicon and the combination of a silicon emitter and SiGe alloy base has the advantage of providing a step which lies entirely in the valence band, ideal for an n–p–n transistor because it prevents hole injection into the emitter, while allowing easy electron injection into the base. For many years this idea remained nothing more than that but when new epitaxial growth methods such as MBE and LPCVD (low pressure chemical vapour deposition) became available, the situation was completely changed and Si/SiGe HBTs now form an important component of the microwave engineers' armoury. There are problems of lattice mismatch here too, the lattice parameter of germanium being about 4% larger than that of silicon, but, with care these are readily surmountable. Up to 10% germanium may be incorporated in a thin base layer while maintaining a stable pseudomorphic structure and as much as 30% may be included, provided the device is not subjected to high temperatures. This latter point implies that high germanium fractions are not compatible with typical silicon processing temperatures and there are therefore two distinct devices on the market – one giving higher performance but demanding careful thermal treatment, the other perfectly stable but with lower performance. Either way, this silicon-based technology is making a serious challenge to what had been thought of as strictly III–V territory.

No matter how hard compound semiconductors may forge ahead, silicon so often seems able to catch up – having successfully seen off germanium, can it now do the same to the III–Vs? The answer, of course, is 'no' – but one must never write it out of the script entirely.

I titled this chapter 'The Nanostructure Revolution' quite deliberately. The events chronicled here represent a completely new approach to semiconductor device design, the use of quantum wells, quantum dots, superlattices, tunnel structures, etc., having added a new armoury of techniques to the device engineer's tool box. We have seen how the GaAs quantum well laser has taken over from the earlier double hetero-structure laser, how the vertical cavity laser has improved on this yet again, how 2DEG structures have led to the discovery of amazing new concepts in fundamental physics, a revolutionary new resistance stan-dard and to field-effect transistors performing wonders at microwave frequencies. We have also seen how the extremely high mobilities of two-dimensional electrons have made possible a range of novel scientific experiments based on one-dimensional conduction. Perhaps the point which is still to be emphasised is that such new approaches have now become of universal application. When a new laser is developed to satisfy some demand from an optical communication system, it automatically takes the form of a quantum well device; when high brightness light-emitters are developed, they automatically use quantum wells, too; when a new transistor is needed to operate at hundreds of Giga-Hertz, it automatically makes use of the HEMT format; and numerous other devices similarly depend on nanoscale heterostructures for their success. Finally (and forgive me if I seem to be over-emphasising the point), all this has been made possible by the development of new epitaxial growth methods. Without MBE, MOVPE and LPCVD, none of it could have happened. In the introductory chapter, we saw how important was the control of material technology to the steel industry – in this chapter I hope I have finally convinced you that the same is also true of the semiconductor industry, but very much more so.

Finally, on a more sombre note, I should add an important rider to this account of semiconductor nanostructures. A measure of concern is, quite rightly, being expressed concerning the safety of nano-particles which are creeping into the manufacture of paints, cosmetics, nano-composites, self-cleaning windows, fabric treatments, catalysts, etc. (see the book "Nano-technology" by Toby Shelley); so it would be well to comment on this aspect of semiconductor technology. Though the benefits of the new technology certainly accrue from the nanoscale size of semiconductor films, wires and dots, it should be clear from the above account that these artifacts exist *only* as component parts of much larger scale structures and there is *no* possibility of their having a separate existence. While no power on earth can guarantee that such devices will never be used for undesirable purposes, it must be emphasised that they present absolutely no *intrinsic* danger to life, health or happiness. There is always a danger of one aspect of technology acquiring a bad name by association with another, particularly when they share similar titles, but I hope this brief assurance might avoid any unfortunate tendency to tar all aspects of nanotechnology with the same broad brush. While it is essential that each potential hazard be examined on its individual demerits, I would suggest that semiconductor nanostructures represent no such hazard.

CHAPTER 7

Light-Emitting Diodes: White Light for Green Consumers

One of the most exciting properties of semiconductors is their ability to convert electrical energy into light without the intermediary of heat. For thousands of years mankind had to be content with obtaining artificial light as a by-product of heat. Whether it was the warm glow of the camp fire, the deliberate burning of animal fat or vegetable oil, the nineteenth century sophistication of gas lighting or finally the amazing convenience of the electric filament lamp, light appeared as an indirect result of making something hot, and the colour of the light depended on the temperature achieved. (The reader may remember that in Chapter 1 we discussed briefly the concept of a spread of thermal emission wavelengths in connection with the Planck radiation hypothesis which represented an important aspect of quantum theory.) Direct sunlight has a colour temperature of 5400 K, meaning that it contains a distribution of wavelengths corresponding to a black body radiator at that temperature. (A black body is defined as one which absorbs all the radiation falling on it and which, in consequence, behaves as a perfect radiator.) For comparison, the carbon arc is characterised by a temperature of 3800 K, a 100 W incandescent bulb 2850 K and a candle flame 1900 K. It is important to appreciate that the hotter the body, the more the wavelength distribution is pushed towards the blue end of the visible spectrum and the colour temperature of a light source is therefore important in so far as it determines the human reaction, the 'feel' of the illumination. When we speak of 'cold' or 'warm' light we are reacting to the balance between the different wavelengths emitted by the source. Too much blue light means 'cold', an overall red balance means 'warm', though this is, of course, is an oversimplification, as too is the concept of colour temperature because real light sources are not strictly black bodies. Nevertheless, we all recognise the difference

Semiconductors and the Information Revolution: Magic Crystals that made IT happen © 2009 Elsevier B.V.
DOI: 10.1016/B978-0-444-53240-4.00007-6

between the 'colour' of a tungsten bulb and that of an arc lamp or fluorescent tube. Finally, an important point to note is that the process is intrinsically inefficient because a lot of the radiation produced lies in the infra-red part of the spectrum and a small amount in the UV – there is no way of avoiding this, except by using sources which create light using processes not dependent on temperature. It is the reason why we are now encouraged to throw away our tungsten lamps and replace them with compact fluorescent bulbs.

Fluorescent lighting made its commercial bow in 1938 when the American General Electric Company marketed a range of tubes producing a new kind of light, brighter but harsher than the customary incandescent lamp which had preceded it by some 50 years. These tubes functioned in an altogether different manner from their predecessors, depending on an electric discharge in a closed tube containing small amounts of mercury and an inert gas, such as argon. The gas molecules were excited by the discharge, and the mercury stimulated into emitting ultra-violet radiation which was absorbed by a phosphor coating inside the tube and converted into visible light. The colour of light produced depended entirely on the particular combination of phosphors employed and could, to a modest extent, be controlled by the skill of the blender. One important advantage of the new source was its improved efficiency, being roughly 5 times brighter than an incandescent lamp using the same amount of electrical energy. Disadvantages were the need for a starter circuit and a ballast load to prevent the discharge running away and, as many people would claim, a harshness which did not endear itself to the viewer. The problem was that people had become attached to the gentle yellowishness of the tungsten filament lamp and found the blue emphasis of the fluorescent tube somewhat jarring to their sensibilities. (Note that it makes no sense to speak of a colour temperature for a fluorescent tube as its wavelength distribution may differ markedly from that of a black body radiator.)

The fluorescent tube is just one version of a range of gas discharge lamps, dating back to the well-known neon tube used for outdoor advertising. The first neon tube was demonstrated by that genius of invention Nikola Tesla at a New York exhibition in 1893 but the first commercial tubes were sold in France in 1912. It is a cold cathode device in which a very high voltage is used to strike an arc through a long tube containing a low pressure of Neon or other rare gas. The characteristic red colour of neon may be modified by the introduction of an argon/mercury mixture, giving blue light, helium, pink or xenon, blue/violet. Other variants on the theme are represented by the mercury vapour lamp which is filled with mercury at higher pressure than used in fluorescent lighting and the now widely used sodium vapour lamp. The latter comes in two versions, the low pressure variant generating nearly monochromatic yellow light, characteristic of sodium vapour, while the high pressure version produces a much broader spectrum which can be coaxed into producing a plausibly 'white' colour, its colour temperature of 2700 K matching fairly closely that of the tungsten filament lamp. Similarly, the mercury lamp may be 'doped' with sodium iodide or scandium iodide, resulting in much improved light quality compared with the rather blue emission from the basic mercury lamp. Alternatively, a phosphor may be added to convert the UV component into red light. The range of options appears endless but the driving force behind these various modifications is one of improved efficiency, while obtaining some reasonable approximation to daylight or to conventional incandescent lighting.

So much for general lighting; but the story of light emitting devices does not end there. Displays designed to present information in its many different forms also demand light emitters and these come in a variety of guises. The neon tube is clearly one version which is well suited to large scale advertising but numerous smaller scale devices are even more widely employed, particularly in the modern world of digital information display. An early form which appeared during the 1950s was the so-called nixie tube (supposedly deriving from the phrase used by its originators the Buroughs Corporation: Numerical Indicator eXperiment number 1). It can be thought of as a miniature neon tube with a single wire mesh anode and a set of cold cathodes in the shape of the digits 0–9, the application of a voltage (typically 150–200 V) to a particular cathode determining which digit should glow with an orange-red discharge. The nixie tube saw service during the 1960s in various electronic instruments such as digital voltmeters, multimetres and frequency counters, also in the rather bulky desk-top calculators which were common at the time (soon to be replaced by the solid state pocket calculator during the 1970s). It was rather clumsy but functioned adequately. A rather more sophisticated device, the vacuum fluorescent display (VFD) was pioneered by the Japanese company Sharp in 1967. It worked on the same principal as a triode valve, employing a heated cathode, a control grid and a set of anodes in the shape of a seven segment format. The anodes were coated with a phosphor so that, when a small positive voltage was applied to the grid and a larger positive voltage to the appropriate anodes, electrons bombarded these anodes, causing the necessary segments to emit a bright green or blue light. Such displays are still widely used both in cars and in domestic audio and video equipment.

This, then, represents the appropriate context within which we should examine the phenomenon of semiconductor electroluminescence. To put the question bluntly, can semiconductors compete with these well established light sources, either to provide general illumination or to serve as display elements? This chapter will largely be concerned to provide an answer. We have already come across the GaAs LED in the form of a p–n junction diode which, under forward bias, emits 'light' at a wavelength of about 880 nm in the near infra-red part of the spectrum. As we discussed in Chapter 5, the device works because electrons are injected from the n-side into the p-side of the junction, where they recombine, generating both light and heat (photons and phonons), while holes are injected in the opposite direction, with similar result. We saw that the direct energy gap of GaAs favoured radiative recombination, though the presence of unwanted impurities or defects in the crystal might result in (undesirable) non-radiative recombination. The development of GaAs lasers which could operate continuously at room temperature involved both improvements to the structure (use of a double heterostructure) and considerable improvement in material quality (with regard to both purity and crystal perfection). Thus, the concept of a semiconductor device capable of converting electrical energy directly into light energy with relatively high efficiency was well appreciated in the 1960s and, what was more, Holonyak's success in making a red laser with the alloy material GaAsP pointed the way to a range of visible emitting devices. In fact, LEDs based on GaAsP first became commercially available in the early 1960s when the American General Electric Company marketed diodes at the outrageous price of $260 each. Needless to say, sales were modest. It was only when Monsanto launched the mass-produced GaAsP diode in 1968 that they started to become popular – sales shot skyward, prices plummeted and it is now possible to buy a (somewhat low-tech) GaAsP red-emitting diode for about a penny. But there is a great deal more to the romance of the LED than this.

The story began very much earlier, the first recorded evidence for a semiconductor light-emitting device being an almost throwaway report by an Englishman, Captain Henry Joseph Round, which appeared in 1907. Round was, at the time, working for the Marconi Company in the USA (in 1921 he became Marconi's Chief Engineer) and was studying the properties of cat's whisker detectors based on silicon carbide (carborundum) so, he addressed his communication to an American magazine, 'Electrical World'. It is seldom possible (or advisable!) to quote the total content of any scientific paper but in this instance we might reasonably make an exception. He wrote:

Sirs:- During an investigation of the unsymmetrical passage of current through a contact of carborundum and other substances a curious phenomenon was noted. On applying a potential of 10 V between two points on a crystal of carborundum, the crystal gave out a yellowish light. Only one or two specimens could be found which gave a bright yellow glow on such a low voltage, but with 110 V a large number could be found to glow. In some crystals only edges gave the light and others gave instead of a yellow light green, orange or blue. In all cases tested the glow appears to come from the negative pole, a bright blue-green spark appearing at the positive pole. In a single crystal, if contact is made near the centre with the negative pole, and the positive pole is put in contact at any other place, only one section of the crystal will glow and that the same section wherever the positive pole is placed.

There seems to be some connection between the above effect and the e.m.f. produced by a junction of carborundum and another conductor when heated by a direct or alternating current; but the connection may be only secondary as an obvious explanation of the e.m.f. effect is the thermoelectric one. The writer would be glad of references to any published account of an investigation of this or any allied phenomena.

New York, N.Y. H.J. Round

Round was a busy man, as his hundred-and-some patent filings indicate, and was more concerned with developing radio technology than with electroluminescence. He appears not to have taken the matter any further. He should, perhaps, be remembered for his work on the thermionic valve (he patented the indirectly heated cathode in 1913), the autodyne radio receiver and the development of direction-finding aerial systems. During the First World War his direction-finding aerials were use to detect the movement of the German battle fleet prior to the all-important Battle of Jutland which, in spite of the considerably greater British losses, led to the German ships being confined to harbour for the remainder of the war. During the Second World War he also made important contributions to the development of Sonar systems used in the anti-submarine campaign. He just possibly might also be remembered as the discoverer of semiconductor electroluminescence.

As so often happens when an out-of-the-way discovery such as this is made, there was no follow-up and the new phenomenon lay dormant. (It is a classic example of how new scientific developments only flourish in the presence of an understanding and sympathetic audience.) In fact, the next mention of the topic had to wait until the end of the 1920s, when it was re-discovered by a Russian scientist Oleg Vladimirovich Losev who was also working to further the development of radio (as mentioned briefly in Chapter 3, Losev may have come close to inventing a transistor-like amplifier). Born in 1903, he had little formal education but proved himself a remarkably effective research worker in his first job in the Nizhni Novgorod Radio Laboratory, moving to Leningrad in 1928 and being sufficiently well regarded that he was invited to spend time working in the Ioffe Physico-Technical Institute. He later joined the Leningrad First Medical Institute and was still doing exciting work on semiconductor devices right up to his death from starvation during the siege of Leningrad in 1942. Rather like Round, Losev worked on cat's whisker detectors and took time off to explore his own observation of electroluminescence in silicon carbide (being apparently totally oblivious of Round's earlier report).

He devoted rather more time to the matter than had Round, and was able to elucidate the effect in greater detail, showing that there were two distinct luminescence processes, one associated with what we now know as forward bias injection luminescence, the other with avalanche breakdown in a reverse-biased junction. The junction was an accidental artefact of crystal growth but Losev was able to show the presence of a thin (order of 10 μm thick) surface layer in his samples which was separated from the bulk by some kind of interface, across which there occurred a sudden change in conductivity. At that time there was no proper understanding of semiconductors or their band theory, let alone the concept of a p–n junction but that was almost certainly the explanation of Losev's observations. He can reasonably be credited with the discovery of the light-emitting diode – even though he was in no position to realise it! He also noted that the light output could be modulated at frequencies into the mega-Hertz range and applied for a patent to cover the use of the LED as a 'light relay' (ie a source for optical communications).

Losev may have provided a detailed description of the luminescence phenomenon in SiC but it was not until 1951 that it was explained in modern terminology, when Kurt Lehovec and colleagues at the New Jersey, Fort Monmouth Signal Corps Laboratory repeated much of Losev's experimental work, added some further detail and pointed out the likely nature of the different emission bands. They also measured the efficiency of the process as being of order 10^{-6} photons per electron. Comparing this with the hoped-for value of 100%, we can only wonder at the amazing sensitivity of the human eye in being able to detect the light at all! Silicon carbide could hardly be seen as the basis for a commercially viable LED – not only is it an indirect gap material but the quality of these rather primitive crystals was very far from perfect and the extremely high melting point (somewhere in the region of 2800°C?) made it an extremely difficult material to control. However, the important point was made that visible light emitting diodes, which operated at just a few volts DC, were, in principle, possible.

Stimulated by the recent invention of the transistor, the 1950s saw a wide-ranging search for new semiconductor materials, with their luminescent properties very much in evidence. In addition to GaAs and GaAsP, several other III–V materials showed initial promise, including GaP, AlP and GaN. Various II–VI materials such as ZnS, ZnSe and CdS also showed a capability for visible light emission. However, rather few of these materials could be doped both n- and p-type so they were not well suited to making LEDs. In fact, only GaP was pursued with any degree of seriousness, leading to modest success during the early 1960s, not only in the production of red light but also green. Efficiencies were modest, being of order 10^{-4} photons per electron, but the output was bright enough to be seen in normal daylight and more than adequate for use as indicator lights on car dashboards, for example, and, perhaps more importantly, as seven-segment numeric displays in the first pocket calculators. The early history of GaP is particularly interesting and worthy of a moment's diversion. During the early 1960s, three laboratories were seriously involved in developing GaP LEDs, each having its own special interest. The Philips laboratory in Aachen was concerned with possible long-term application in lighting, the English Government laboratory at Baldock, SERL was interested in military applications based on fast switching speed and Bell Telephone Laboratories were interested in using LEDs as indicator lamps in domestic telephones and telephone switching stations. While Philips regarded their initial success as justification for setting up a production activity in a lighting factory, they then had a sudden change of heart and shut down both production and research. SERL developed their technology to the point that it could be transferred to the Ferranti corporation who set up a manufacturing facility to market LEDs in 1966 but which left no lasting impression on the history of the subject. The Bell activity alone proved commercially viable – because of their low operating voltage, LED indicators could be powered from the telephone line, rather than requiring a separate 110 V supply and the greatly improved convenience won widespread approval.

Such applications were sufficient to support the fledgling industry for quite a while but it was clear that a wider range of colours was highly desirable in the interest of developing a full colour LED display technology and, if ever there was to be hope of white light sources to challenge the well-established tungsten filament and fluorescent light sources, a very considerable improvement in efficiency was imperative. A particular problem lay in the lack of an adequately bright blue LED. During the 1960s, a lot of hard work went into improving the quality of silicon carbide, both in the USA (at GE) and in England (at GEC) but the outcome was a blue emitter with efficiency in the region of 10^{-5} photons per electron and this remained a weak link in attempts to generate white light or to mount any kind of full colour LED display. Brighter red emitters were developed during the 1970s, based on AlGaAs alloys, followed by orange and yellow emitting AlGaInP devices but the blue end of the spectrum still presented a relative black-out. Hope was originally pinned on the direct gap material GaN but it stubbornly refused to be doped p-type and it was not until 1989 that the breakthrough eventually arrived. Within a very few years, not only was p-type doping established but highly efficient blue and green diode emitters were developed in what can only be described as a Japanese fairy story. All at once, bright, colourful LED displays, LED traffic lights (see Fig. 7.1), a blue DVD semiconductor laser and white light sources exploded onto the world

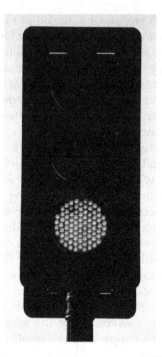

FIG. 7.1 LED traffic lights first appeared in Japan towards the end of the 1990s and are now widely used throughout the developed world. Their improved efficiency and longevity are saving local authorities large amounts of tax-payer's money. Courtesy of iStockphoto, © Harris Shiffman, image 964085.

stage in a surge of technical excitement not seen since the unveiling of the point contact transistor in 1947. It makes an amazing story so we should take a deep breath before plunging into the details.

Firstly, we need to establish a few basics. The word 'efficiency' has appeared in our discussion several times already, with units of 'photons per electron' but this requires elaboration. Electric energy is put into an LED in the form of an electric current of holes and electrons flowing across a p–n junction and light energy is emitted in the form of a stream of photons. The concept of efficiency concerns the number of photons generated for each electron-hole pair that recombines (a number always less than unity!). Expressed as 'photons per electron' it represents what is called a 'quantum efficiency' but there is another, perhaps more practical, definition in terms of 'energy efficiency' – that is, light energy out divided by electrical energy in. In practice these two quantities are usually very similar and I shall make no distinction between them in what follows. It is worth emphasising that this notion of efficiency can be applied perfectly well to any LED, whether it emits 'light' in the visible region or in the infra-red or ultra-violet regions. However, when we are specifically concerned with visible light, it is important to recognise one further aspect, that of brightness. In practical terms, we are always interested in how bright a light source appears to the human eye and this gives rise to a somewhat different expression of efficiency – that of 'luminous efficiency'. The units in this case are lumens per Watt – how many lumens of light are produced when the LED is driven with each Watt of electrical energy. And this brings us to the question of the eye sensitivity curve.

Because, we humans have evolved in an environment illuminated by sunlight, we have developed eyes which are sensitive to that range of wavelengths which peaks in the green part of the spectrum, near 550 nm, corresponding to the peak in the sun's spectrum (colour temperature 5400 K). We may not often be aware of it, but the detection efficiency of our eyes varies strongly with wavelength. Thus, 1 W of green light at 550 nm is equivalent to 683 lm, whereas the same 1 W of red light at 650 nm corresponds to only 73.1 lm (a factor of 9.3 times less) and for blue light at 440 nm only 15.7 lm (a factor of 43 times less). In terms of visibility, red or blue LEDs must work a great deal harder than green ones and this, of course, militated against the poor old SiC LED's efforts to provide an adequate source of blue light. Not only was it lacking in quantum efficiency, but it had to work against an unforgiving nature. There is another, rather more subtle aspect of this eye sensitivity curve which intrudes on attempts to improve the brightness of (for example) red LEDs. Roughly speaking, a downward shift of the wavelength by 10 nm (e.g. from 660 nm to 650 nm) increases the luminous efficiency by a factor of two (assuming the quantum efficiency remains unchanged) so, in practice, it may be easier to improve the luminous efficiency of a red LED by reducing its wavelength rather than by improving its quantum efficiency. We shall meet important examples in the case of the alloy materials GaAsP and AlGaAs. Similar considerations apply at the blue end of the visible range, except that here luminous efficiency is improved by increasing the wavelength.

Finally, we need to draw a distinction between 'internal' and 'external' efficiencies. Clearly, the object of the exercise is to produce light in the space external to the LED – it is only light which escapes from the device that is useful. This line of thought leads to the idea of an 'extraction efficiency' which refers to the fraction of internally generated light that actually escapes. It follows that the external efficiency is the product of the internal efficiency and the extraction efficiency. At first thought, the extraction efficiency may seem likely to be relatively large but, in practice, this is not always the case and external efficiencies can be very much smaller than internal efficiencies. Part of the problem lies in the fact that semiconductors tend to have fairly large refractive indices – values of 3.0–3.5 being typical – and one consequence of this is that only a small fraction of light generated beneath a flat semiconductor surface can escape through that surface. Most of the light striking the surface near normal incidence will be transmitted but any light making an angle with the normal greater than about 17° will be reflected back into the bulk of the semiconductor. In fact, only about 2% of the light escapes. Of course, some of this reflected light may be reflected again from the back surface of the LED and thereby have a second chance to escape but it may also be absorbed by the LED material or by the substrate on which the device was grown. It is by no means certain that it will eventually appear in the space above the LED surface where it can contribute to the desired output. We shall look at a number of specific examples when we consider particular materials below.

Now let us try to summarise all this in the context of attempts to design bright visible LEDs. Firstly, it is essential to use a material which emits in the visible part of the spectrum. Secondly, the material should have a short radiative lifetime (high probability of radiative recombination) which usually means a material with a direct band gap. Thirdly, we should aim to minimise non-radiative recombination by growing high quality and high purity crystals. Fourthly, it is helpful to use an active region in the form of a double heterostructure or (more usually) one or more quantum wells where electrons and holes are brought together in a small space so as to maximise the probability of radiative recombination. Fifthly, we should design the geometry of the device so as to maximise the fraction of the light which is extracted. This will depend on various factors such as the shape and texture of the emitting surface and on possible re-absorption of the light by the LED material and its substrate. Finally, we must always remember that the apparent brightness of any visible source depends (via the eye sensitivity curve) on its wavelength.

The history of light emitting diodes is more than a little involved so I shall attempt to simplify the presentation by dividing it into its individual colours. I shall first discuss the development of red-emitting devices, then move on to cover the shorter wavelength regions, orange, yellow and green and, finally, the epoch-making blue diodes. This will inevitably involve some jumping to and fro between the various decades (of which there were five) but it will, I hope, avoid, or, at least, minimise confusion. Broadly speaking, red diodes grew to maturity over the fairly long period 1955–1990, yellow appeared round about 1990–1995, while green, first available as early as 1960, achieved high efficiency in the 1990s. High brightness blue diodes emerged in the mid-1990s, followed fairly rapidly by brilliant white emitters which are gradually being perfected to challenge more conventional lighting sources during the 2000s. There seems to be an inevitability about semiconductor light sources eventually taking over the majority of lighting industry sales – only the time scale remaining uncertain.

One of the first 'proper' red LEDs was made by Hermann Grimmeiss in the Philips laboratory in Aachen, Germany round about 1960. It was based on polycrystalline gallium phosphide grown from a gallium melt containing phosphorus. The necessary p–n junction occurred somewhat by accident and the optical emission process involved two dopants, zinc and oxygen. Earlier work at the Fort Monmouth Signal Corps Engineering Laboratory in New Jersey had shown the need for zinc, a p-type dopant in GaP, but the role played by oxygen was not appreciated until somewhat later – research by John Allen at the Services Electronic Research Laboratory, Baldock made clear that red emission from GaP also depended on the presence of oxygen in the grown crystal. This was reported in 1963 but the full understanding of the emission process had to wait until 1970 when work at Bell Labs and at IBM showed the emission to be associated with a nearest neighbour pair of impurities, zinc on a gallium site, together with oxygen on an adjacent phosphorus site. This apparently unlikely occurrence was mediated by the fact that, during crystal growth at elevated temperatures, both zinc and oxygen were present as ions, Zn^+ and O^-, their opposite charges resulting in mutual attraction. The intensity of the emission depended on the density of such (Zn–O) centres and could often be increased by suitable thermal annealing of the crystals, following growth (the increased mobility of the ions at high temperature helping them to find one another). Initially, diode efficiencies were very low (being in the neighbourhood of 10^{-4}) but careful attention to crystal growth led to some quite remarkable improvement. The key development was the application of the Czochralski crystal pulling method to gallium phosphide. The RSRE group under Brian Mullin led the way in 1968, with their boric oxide liquid encapsulation technique (which we met in connection with indium phosphide in Chapter 5). This made available high quality GaP substrates on which epitaxial layers could be grown using either liquid epitaxy (LPE) or VPE and efficiencies reached the dizzy heights of 7% (Bell Labs) and 15% (Fairchild). LPE, in particular, produced material of considerably higher purity than anything previously available, thus reducing the density of non-radiative centres. Note, however, that this GaP red emission occurs at such a long wavelength (710 nm) that even 15% efficiency implies a luminous efficiency of only 0.2 lm/W – the poor old crystal grower having to work extremely hard to achieve very modest brightness!

There is some very interesting physics associated with this recombination process – which accounts for the length of time it took to pin it down! GaP is actually an indirect gap material, so it should not be expected to produce efficient luminescence at all. In addition, the band gap of 2.25 eV (corresponding to a wavelength of 550 nm) is much too large to explain the observed red emission at 710 nm. Much careful analysis went into understanding the answers to both these queries. What actually happens is that an electron from the conduction band and a hole from the valence band are attracted together by their opposite electrical charges (just as the Zn and O ions are) and are simultaneously trapped at the (Zn–O) centre. It is the nature of the (Zn–O) centre which accounts for both the strong radiative recombination and the wavelength of the emission. It turns out that GaP is rather special in possessing centres like the (Zn–O) pair which give rise to such efficient visible light production. We shall come across something very similar when we discuss GaP green emission, later.

A somewhat different approach to making red LEDs was the one adopted by Nick Holonyak, which coincided rather closely in time with Grimmeiss' early attack on gallium phosphide. Oddly enough, Holonyak was originally interested in studying the properties of the tunnel diode and hoped that the alloy GaAsP might give him better negative resistance characteristics than he could obtain with GaAs, his observation of red light emission being only incidental. However, as we saw in Chapter 5, this happened to coincide with the sudden burst of interest in the concept of a semiconductor laser so, showing admirable versatility, he rapidly changed course and threw in his lot with the race to demonstrate laser action in a semiconductor diode, which came to an exciting climax in the fall of 1962. But Holonyak can be credited not only with considerable versatility but with an even more remarkable degree of foresight. In an article in the February issue of the 1963 Reader's Digest, he is reported as forecasting the future likelihood that the LED would, one day, be developed as a practical white light source – a prediction coming rapidly to fruition today, some 45 years later. There can be few examples of such confident prescience in the whole dramatic history of semiconductor technology. Though much less well known, it certainly vies with Moore's famous Law concerning the growth of integrated circuit complexity.

As a light-emitting material, GaAsP promised two advantages – it offered a direct energy gap in the visible region of the spectrum, together with the possibility of varying the gap (and hence the colour) by changing the amount of phosphorus included in the alloy. In the event, both features materialised but with a rather frustrating limitation. Increasing the proportion of phosphorus certainly did result in an increase in band gap but only up to a point – at a ratio of 45% phosphorus the material flipped over from direct to indirect energy gap and remained indirect for all higher ratios. It was already well known that GaP was an indirect gap material, so the changeover was hardly surprising though one might have hoped for it to happen at slightly higher phosphorus content. The overall effect was to limit the shortest wavelength for efficient emission to about 650 nm, this being an interesting compromise between the fall-off in quantum efficiency as the amount of phosphorus was increased and the rapid increase in visual efficiency due to the increasing eye sensitivity at shorter wavelengths. The best quantum efficiencies were about 0.05%, corresponding to a luminous efficiency of 0.035 lm/W, and, at the time, this compared well with GaP red diodes, which glimmered rather miserably at values little better than 10^{-4} lm/W. Clearly, GaAsP had a significant advantage and this explains why GE (Holonyak's employer) and Monsanto selected it for their initial attempts to stimulate a commercial market for the new devices during the 1960s.

All went well for a while – GaAsP diodes were bright enough to serve as indicator lamps and in the seven segment numerical display elements used in early pocket calculators (see Fig. 7.2) but, as we saw above, GaP

FIG. 7.2   An early scientific calculator (c. 1980), with a display based on red LEDs. Many calculators made at this time employed GaAsP LEDs but these were gradually phased out in favour of liquid crystal displays which were far less demanding of battery power. Courtesy of the author.

made a serious challenge in the early 1970s and the alloy approach began to look distinctly dubious. However, the availability of GaP substrates offered a lifeline to GaAsP as well as to GaP itself. The early GaAsP diodes were grown epitaxially on GaAs substrates which had two disadvantages – there was a significant mismatch of lattice parameter which affected the structural quality of the epilayers, while GaAs strongly absorbed light reflected back towards the substrate in the manner we described earlier. Both effects led to poor external efficiencies, and, while the use of GaP substrates involved a similar amount of lattice mismatch (in the opposite sense) it neatly avoided the absorption problem because GaP has a much larger band gap than the GaAsP light emitting material. This innovation, introduced by the Hewlet Packard company, led to an improvement of a factor of ten in luminous efficiency to about 0.4 lm/W and saved the day for GaAsP – but at a price. GaP substrates were roughly 10 times more expensive than their GaAs forebears! In fact, the old technology suffered no early demise – there was room for all possibilities, a choice between cheap-and-cheerful and expensive-and-sophisticated provided the best of everything in the best of all worlds. It all depended on the application.

The problem of substrate-epilayer mismatch could be ameliorated slightly by growing a graded epilayer between the substrate and the actual LED material – that is by gradually increasing the amount of phosphorus in the layer from zero at the GaAs substrate up to the 45% needed for the diode but this was no more than an uneasy compromise. As we have seen on many previous occasions, high quality device material must be grown on a matching substrate and this leads us inevitably back to our old friends GaAs and AlGaAs which provided a solution to the problem of making a GaAs laser to operate at room temperature. Increasing the amount of aluminium in the alloy increases the band gap in a parallel fashion to the increase of phosphorus in GaAsP but, in this case, without compromising lattice match. Interestingly, the parallel between the two materials is even closer – AlGaAs also shows a changeover from direct to indirect gap at an aluminium content of about 40%, providing an optimum emission wavelength close to 650 nm. It was the use of liquid phase epitaxy, pioneered at IBM in the late 1960s, which resulted in really high quality material and correspondingly high LED efficiencies. Even so, it took several years before the full capability of this material system could be realised. Eventually, in 1983, a group at Matsuchita in Japan demonstrated a method of growing a thick AlGaAs layer before removing the diode from its GaAs substrate, thus achieving a structure with fully transparent supporting layer. The reward for such devotion to perfection was an external efficiency of 8% at a wavelength of 660 nm and luminous efficiency of 4.4 lm/W. Values as high as 10 lm/W became routinely available during the 1990s, representing a truly remarkable advance on earlier device performance. Of course, it also represented a significant advance in cost but, if such performance was not essential for your application, you could still obtain modest light output at an equally modest price. The important point was that luminous efficiencies were now approaching those appropriate to conventional light sources – tungsten bulbs at about fifteen and fluorescent tubes at about 70 lm/W.

But it turned out that this was far from being the ultimate method of making high brightness red LEDs – yet another material system was on the horizon and began to make an impact as early as 1972. It was first recognised that the compound GaInP was lattice-matched to GaAs at a composition having approximately 50% of indium and that it had a direct band gap of about 1.92 eV, allowing reasonably efficient light emission at 670 nm. This was interesting but frustratingly no better than AlGaAs – could anything be done to shorten the wavelength? Fortunately, the answer was 'yes' – it was only necessary to replace some of the gallium with aluminium and the band gap increased appropriately, while maintaining the desired lattice match. Here was a real breakthrough, because it opened up the possibility of not only red, but also orange, yellow and even green emission. However, we should concentrate first on the problems of making really efficient red LEDs. The first problem was a familiar one – AlGaInP was not at all an easy material to grow. The substrate was straightforward enough, GaAs being of long standing and sound, reproducible quality but there the straightforwardness ended. Attempts to use the, by now, reliable technique of LPE immediately ran into difficulty when it was discovered that it was necessary to include a large amount of aluminium in the melt in order to obtain a rather small amount in the grown crystal. It also made composition control extremely difficult. Fortunately, the alternatives of MOVPE (Metal organic vapour phase epitaxy) and MBE (Molecular beam epitaxy) became available during the 1970s and both, indeed, were harnessed to the task. Both, too, were successful, though MOVPE appears to be the preferred method for commercial production of LEDs. Then there was a problem to establish the true bandgap. It turned out that GaInP had a tendency to behave more like a short period superlattice than an honest alloy. Indium and gallium preferred to segregate so that the crystal contained regions rich in gallium and other regions rich in indium and this influenced the effective band gap of the material. Superlattices usually have smaller band gaps than the equivalent random alloy (that is the alloy in which the two metal species are randomly distributed) and, since the whole object of the exercise was to achieve as large a direct band gap as possible, this segregation effect was not at all to be desired. It took some time to discover that growing epitaxial layers on unusually orientated substrates could eliminate it. At long last, sometime during the early 1990s, it was possible to establish reliable criteria for making efficient red, yellow and almost green diodes.

Two important improvements were made. Firstly, the internal efficiency of the device was improved by using a double heterostructure. To make a red-emitting diode the active material was GaInP with confining layers of AlGaInP. Secondly, it was necessary to remove the strongly absorbing GaAs substrate. It was etched away and replaced with a wafer-bonded GaP substrate (basically, a slice of bulk GaP was glued on!). An epitaxial GaP film was also grown over the top of the structure to act as a window through which the red light could freely emerge. Finally, shaping the structure like an inverted pyramid improved the extraction efficiency and, by about 1995, external efficiencies as high as 60% had been contrived, a truly remarkable achievement. It meant that red diodes emitting at 650 nm were characterised by luminous efficiencies of over 40 lm/W, while orange (610 nm) diodes reached 100 lm/W. This was the breakthrough that lighting experts had been looking for, luminous efficiency even greater than that associated with fluorescent lighting. It was now even more important that someone should do something about the yellow, green and blue parts of the spectrum.

Green LEDs had actually been available from the early days in the form of nitrogen-doped GaP. This represented another surprising development – GaP, being an indirect gap material had no right to generate anything like the amount of green light that it did. Even in the early 1960s efficiencies of about 10^{-4} were available, yielding something approaching 0.1 lm/W, significantly better than the early red-emitting diodes. (Green had, of course, the advantage of being at the peak of the eye sensitivity curve.) It took some time before the significance of the nitrogen doping effect was fully appreciated – indeed, it was a quite fortuitous observation which allowed even the appreciation of its existence. It was in 1964, when Carl Frosch at Bell Labs was developing a vapour phase method for growing GaP which involved the use of water vapour that he noticed a significant difference between films grown in quartz or boron nitride tubes. In the latter case, the water vapour attacked the boron nitride to form ammonia which then doped the GaP with nitrogen. Thus, was it realised how important was nitrogen to the making of efficient green emitters and, subsequently, ammonia was introduced deliberately to achieve controlled doping. This aside, the interesting feature of nitrogen doping was the fact that it is neither a donor, nor an acceptor in GaP because it appears in the same column of the periodic table as phosphorus. It required some deep theoretical thinking to establish just how such a dopant might lead to a much enhanced probability for radiative recombination (it involves a neat application of the Heisenberg uncertainty principle – but that needn't worry us unduly – the fact is that it not only works but we now understand very well how it works!) In practical terms, GaP green LEDs with well controlled amounts of nitrogen achieved efficiencies as high as 1% (luminous efficiencies of 7 lm/W).

As to the yellow region, the first yellow diodes didn't appear until 1972, when George Craford at Monsanto heard of the Bell Labs work on nitrogen doped GaP. At the time he was working on red GaAsP devices (he had been Holonyak's student at the University of Illinois) and it struck him that nitrogen doping might be used to enhance their performance at shorter wavelengths. As we have already seen, GaAsP seemed to be limited to producing red light on account of the changeover from direct to indirect band gap as the phosphorus content was increased beyond about 50% but he argued that nitrogen doping of indirect gap material might allow yellow emission from an alloy containing about 75% phosphorus. (Another way to look at it is to think of replacing 25% of the phosphorus in nitrogen-doped GaP with arsenic.) The result was a roughly 0.5% efficient yellow diode with luminous efficiency of about 1 lm/W. Interestingly, it was also George Craford who was largely responsible for replacing the yellow GaAsP devices with much more efficient AlGaInP. In 1979, he had joined Hewlett-Packard, where he continued to play an important role in developing high brightness LEDs which led to many new applications such as the traffic warning sign shown in Fig. 7.3. We have already commented on the tremendous improvement that was obtained in red and orange diodes through the development of the AlGaInP alloy system. Yellow and green devices were also possible with somewhat lower efficiencies – typical luminous efficiencies achieved during the 1990s were 10 lm/W for green and about 50 lm/W for yellow. All this represented a mountain of achievement but it still leaves us to face the huge problem of obtaining comparable performance at the blue end of the spectrum – and thereby hangs an even more intriguing tale.

We have already referred to the efforts made to harness silicon carbide as a source of blue emission and to the disappointing results obtained. The fact of its being an indirect material was always bound to make it an unlikely candidate and its very high melting point made too many demands on the crystal growers' hopes of achieving adequately crystal quality. It was clearly necessary to find materials with both wide and direct energy gaps if efficient blue emission was to be obtained with any degree of reliability. Interestingly enough, there was no shortage of such materials. In the II–VI group of compounds possible candidates were cadmium sulphide CdS (2.5 eV), zinc selenide ZnSe (2.7 eV) and zinc

FIG. 7.3 A traffic warning sign using several differently coloured LEDs. Once the idea of using LED traffic lights caught on in the early years of the new millennium, local authorities vied with one another to introduce a wide range of active warning signs, also based on the new high-brightness LEDs. Courtesy of iStockphoto, © Nick Free, image 3967053.

sulphide ZnS (3.7 eV) and in the III–V group gallium nitride GaN (3.44 eV) and aluminium nitride AlN (6.2 eV), all of which were fairly well known as early as the mid 1950s. The problem was to obtain adequate crystal quality and, more fundamentally, to achieve both n-and p-type doping. While all these materials could be doped n-type, not one of them could be persuaded to show p-type conduction. Indeed, there was a strong body of opinion in favour of a theory that attributed this recalcitrant behaviour to fundamental properties of these particular compounds – accordingly, it would never be possible to tame them, no matter how hard anyone might struggle! And, gradually, there materialised an ever-increasing body of evidence that supported this pessimistic viewpoint. While it was possible to generate intense luminescence from each and every one of them, it remained frustratingly impossible to form the p–n junctions necessary for injection luminescence – after years of work, the hoped-for blue or green LEDs seemed to be as far away as ever. But remarkably, persistence paid off and, even more remarkably, it happened at almost the same time that ZnSe and GaN were successfully doped p-type, though the sequence of events in the two cases followed totally different paths. Success had had to wait until the end of the 1980s, nearly 30 years after these materials were first studied, but, as we shall see, it was well worth waiting for. GaN, in particular, gave rise to amazingly bright blue and green LEDs which changed the whole visual semiconductor scene. We shall look first at the progress of ZnSe, leaving the GaN success story till last.

Perhaps because of their luminescent properties, the II–VI compounds were among the first compound semiconductors to be investigated and quite a number of important discoveries emerged from these studies. For instance, the concept of radiative recombination at so-called 'isoelectronic' impurity centres was well understood from measurements on CdS as early as 1964 (the nitrogen centre in GaP, which we discussed earlier, is another example) and by 1967 the II–VI compounds were enjoying the luxury of their own annual conference. While it is certainly true that the early work was performed on rather crude bulk crystals containing high levels of impurities, there can be no doubt that considerable progress was made in understanding recombination mechanisms. And, let it be said, the observed luminescence was quite surprisingly bright, even though it had to be excited by shining UV light or an electron beam on the samples. It was apparently impossible to achieve p-type doping so p–n junction luminescence was not available and it was this frustrating limitation which led to a gradual loss of interest in these materials during the 1970s and 1980s. However, not everyone gave up – there was still some interest in finding the explanation for this recalcitrant behaviour. Was it fundamental to this class of materials or was it no more than another example of unwanted (and perhaps unknown) impurities. The way to find out lay in the development of new crystal growth methods – where have we met this idea before? – and it was the introduction of MBE to the growth of ZnSe that finally led to an answer. MBE had the virtues of allowing excellent control over the ratio of zinc to selenium so it was possible to grow stoichiometric films (films with equal amounts of the two constituent atoms) and, of course, very low background impurity levels. It was also found that nitrogen could be used as an acceptor. The only difficulty remaining was how to introduce atomic nitrogen into the films – nitrogen in its usual molecular form of

N_2 would simply not work – and this was solved by the use of a nitrogen plasma (molecular nitrogen was broken down by passing a radio frequency discharge through nitrogen gas). p-type doping of ZnSe was first achieved in this fashion in 1990, during a collaboration between the University of Florida and the Minnesota Mining and Manufacturing Company, thus ending 30 years of doubt and frustration. No scientific discovery is ever accepted unreservedly until verified by others but there was no shortage of those offering confirmation – it was also found possible to use ammonia as a source of nitrogen. The nature of the nitrogen acceptor meant that hole densities were fairly modest but p–n junction luminescence was soon demonstrated in several laboratories in both the USA and Japan.

The way was now clear to the development of bright blue LEDs but success in this direction was not to be achieved easily. There were difficulties in making good electrical contact to the p-type ZnSe, there were problems over the substrate and there were problems in making devices with adequate operating lives. The work described above was based on the use of GaAs substrates – GaAs and ZnSe being nearly lattice-matched – but there was a severe mismatch in thermal expansion coefficients which led to severe strain in the grown structures. Obviously, it would be better to use ZnSe substrates if at all possible and various attempts have been made to grow them – with no more than modest success. It also became apparent that the II–VI materials suffered from mechanical weakness in the sense that they are very sensitive to the presence of 'dislocations' (jogs in the crystal structure which can propagate through the crystal). Thus, though efficiencies of blue LEDs were as high as several percent and green LEDs could be made using the alloy ZnSeTe which reached luminous efficiencies of nearly 20 lm/W, there have been ongoing difficulties in achieving acceptable lifetimes. These problems have also been exacerbated by the more-or-less coincident success of LEDs based on the nitride alloy system AlGaInN which has proved considerably more reliable in use. We must now move on to consider the romance of the nitrides.

As we have mentioned previously, the Philips company took an early interest in light-emitting semiconductors based on its commitment to lighting technology in general and to phosphor materials in particular, and one of the III–V compounds of particular interest was GaN. Measurements on powder samples made by Kauer and Rabenau at the Aachen laboratory in 1958 showed that the band gap was somewhat greater than 3 eV, large enough to allow the generation of white light. Further work by Grimmeiss and Koelemans then showed that GaN doped with various 'activators' (as they were then called) produced a range of different emission lines within the visible spectrum and confirmed this initial promise. However, the considerable difficulty they encountered in growing single crystal material led to the abandonment of the work in favour of more tractable semiconductors such as GaP.

The next burst of interest in GaN appeared in 1968 at the RCA laboratory in Princeton, New Jersey, stimulated by a desire to make a flat panel TV display which could be conveniently hung on a wall (a kind of electronic holy grail long sought after in laboratories far and wide). The idea was to use a set of red, green and blue LEDs to make a full colour display, with GaN providing the blue component. The challenge was to find a method of growing single crystal material and a young researcher, Paul Maruska was given the task of meeting it. Based on his earlier experience of growing epitaxial GaAsP, he chose to deposit thin films of GaN on sapphire substrates by reacting ammonia with liquid gallium and was delighted to discover that a transparent film with the appropriate crystal structure and lattice parameter had formed on the sapphire surface. This was an important step forward but Maruska was soon to find that all his films were resolutely n-type, no matter which acceptor species he tried. However, the RCA team under Jacques Pankove were not yet beaten and succeeded in making the first blue LEDs from GaN, doped with first zinc and then magnesium. These probably depended on impact ionisation, rather than minority carrier injection (they never succeeded in making p-type material) and were very inefficient. The flat panel display had to wait for another 20 years – and then it took the form of a liquid crystal matrix, rather than one of LEDs (see Chapter 10).

Perhaps the greater step taken at RCA was the introduction of vapour phase epitaxy, which was also taken up at Bell Labs by Ray Dingle and Marc Ilegems and at the Lund Institute in Sweden by Bo Monemar. Gallium was transported as either the chloride or as the hydride while nitrogen was obtained from ammonia. Again the substrate was sapphire, there being no sign of bulk GaN crystals on anyone's horizon. This sparked a period of intense study of GaN's physical, electronic and optical properties which led to a tremendous increase in understanding but no further progress towards the production of p-type doping. It became clear exactly how minority carriers recombined in GaN but there was still no way to inject them across a p–n junction. Once again, frustration took over and people drifted away from this apparently intractable problem, the only significant development during the early 1980s being the production of bulk single crystals by the Polish group led by Professor Porowski at the High Pressure Research Centre in Warsaw. This represented a remarkable advance, given the high melting point of GaN (about 2800 °C) which made melt growth more or less impossible. The Warsaw method involved growth from solution in liquid gallium at temperatures of 1500 °C and under nitrogen pressures of 10 kbar (10^4 atm) and it took something like a week to grow a crystal whose dimensions were measured in millimetres. True, they have now perfected their techniques to the point of achieving crystals several centimetres in diameter but even this is too small and too expensive to be the basis of any commercial programme. Its principal benefit has been as an aid to better understanding of material properties – epitaxial layers grown on bulk GaN crystal substrates are relatively unstrained, whereas layers grown on other substrates such as sapphire and silicon carbide are inevitably strained and plagued by structural defects.

According to E Fred Schubert, 1982 saw the publication of only one paper on GaN, a startling indication of the loss of faith experienced by its previous devotees. However, as he also points out, Professor Isamu Akasaki and colleagues at Nagoya University on Japan's Pacific coast refused to give up the struggle. Akasaki experimented with the use of MOVPE for growing films of GaN and, in 1989, was able to demonstrate p-type doping. The important breakthrough had, at last, come to pass, though in a quite unexpected fashion. It depended on the use of magnesium as acceptor impurity but this alone proved to be insufficient – it was also necessary to irradiate the sample with a beam of electrons, an observation made largely by accident. Only after some of the doped samples had been studied in a scanning electron microscope was it noticed that they actually showed p-type conduction. In some mysterious way, the electron bombardment had activated the magnesium to function as an acceptor. Notwithstanding its doubtful provenance, the process very soon acquired a respectable name – LEEBI (low energy electron beam irradiation) – and was rapidly copied by other laboratories around the world. Suddenly, GaN was a suitable subject for investigation once more, and when Akasaki went on to demonstrate a blue LED with external efficiency of 1% in 1992 the compound semiconductor community really did sit up and take notice – within a matter of months nearly every electronics laboratory in the USA and Japan (but, strangely, not in Europe!) had started, or re-started a research programme on GaN and related compounds. The race was now on to settle once and for all the problem of making a high-brightness blue LED and, what was more, a blue laser to facilitate the new DVD players soon to be exploited by an eagerly waiting commercial world.

All this may sound improbably dramatic but it was at this point that the drama really took off through the person of yet another Japanese researcher Shuji Nakamura who was then working in a modest up-country organisation known as the Nichia Chemical Company. The company had been started in 1956 by an unusual Japanese entrepreneur Nobuo Ogawa who was motivated by a desire to provide employment in a rural area of the country well away from Japan's commercial Tokyo–Osaka axis. His factory was situated in the small town of Anan on the island of Shikoku, some 20 miles south of the city of Tokushima. It specialised in high brightness phosphors for use in fluorescent lighting and television tubes and Ogawa, an industrial chemist, had perfected techniques for producing top quality materials which he managed to sell against competition from much larger organisations. He started Nichia with some 22 employees but, by the time Nakamura joined the company in 1979, there were upwards of 200 – a clear measure of his success. Nakamura obtained a master's degree from the University of Tokushima where he learned the practical skills which were to prove invaluable in his later work, then persuaded Ogawa to give him a job with Nichia. He was put to work on the development of conventional infra-red and red LEDs – a new venture for Nichia – which probably represented a remarkable vote of confidence in his ability by Ogawa. Nevertheless, it proved extremely frustrating. Because he was competing with the large research teams employed by the likes of Bell Labs, Monsanto, Hewlett-Packard, IBM and RCA, he was unable to achieve any significant new results and spent his time simply chasing after the leaders. Eventually, fed up with his lot, he complained to Ogawa and asked to be allowed to follow his own inclinations and work on the nitride materials, which offered the possibility of a real breakthrough for the company. To his great credit, Ogawa said yes! It involved a huge gamble and cost a sum of money far greater than any which a small company might reasonably expect to invest in research but Ogawa was prepared to take the risk.

Thus, in 1988 Nakamura was dispatched to spend a year at the University of Florida in Gainesville where he would work with another Japanese researcher Shiro Sakai on the subject of MOVPE. He would learn the vital skills involved in growing high quality films of GaN and related semiconductors which surely held the greatest promise of making the long-desired blue LED (though remember that at that time GaN still obstinately refused to be doped p-type!). It was a bold move but with only marginal probability of success. The fact that it worked was, without doubt, down to Nakamura's persistence and to his well developed skill in adapting and modifying equipment to suit his immediate needs. On his return to Nichia, he installed his own MOVPE equipment, introducing an ingenious modification in the arrangement of gas flows which gave him an advantage over the many rivals who, as we noted above, flocked to the scene, following Akasaki's success (in 1989) with p-type doping.

Nakamura wasted no time in making a mark in his newly adopted field of research. In 1991, he made his first innovation by using an amorphous layer of GaN (rather than the AlN employed by Akasaki) as a buffer layer between the sapphire substrate and the single crystal film, which appeared to reduce problems associated with the large mismatch between sapphire and GaN. At the same time he discovered that p-type conduction could be obtained with magnesium doping, followed by a thermal anneal, rather than using the much less convenient LEEBI process. He also speculated (correctly) that the earlier difficulties with p-type doping arose as a result of hydrogen being incorporated in the GaN film – which was an inevitable consequence of the use of ammonia as a source of nitrogen. (An interesting consequence was the later observation at the Universities of Boston and North Carolina that when GaN was grown by MBE, using a plasma source for the nitrogen, p-doping could readily be achieved without resource to either LEEBI or thermal treatment.) But, from the viewpoint of his employer, these innovations in the growth process led, much more importantly, to startling improvements in the efficiency of blue LEDs. While Akasaki had used a simple homojunction to make his 1% efficient diodes, Nakamura chose to use the alloy GaInN to form, first of all a double heterostructure (like the first successful GaAs lasers) then secondly a number of quantum wells. GaInN has a smaller band gap than GaN but also differs in lattice parameter so a very thin quantum well is preferable to a relatively thick active region. It may be strained but it tends to contain far fewer defects. The result (in 1995) was a blue LED with an efficiency of 10%, comparable with that available from the best red LEDs at the time. Then, by using deeper wells, he was able to make a green LED with efficiency of 6% ($\sim$40 lm/W). The way was now clear for the development of full colour displays (which appeared surprisingly rapidly in various Japanese city centres) and the demonstration of bright white light sources. Nakamura achieved this particular breakthrough almost immediately by combining a blue LED with a phosphor which absorbed some of the blue light and emitted both red and green – the impact was stunning – dazzling white light from a tiny piece of semiconductor driven with a few tens of milliamps of direct current. The electric torch (see Fig. 7.4) would never be the same again! The city of Tokushima honoured its erstwhile student by introducing LED traffic lights in 1997 and the Nichia Chemical Company made billions of yen selling high brightness LEDs to a rapidly growing clientele.

As if this were not enough, Nakamura took his new-found expertise into yet another commercially vibrant area. The semiconductor laser had found its first serious market by helping to launch the CD player in 1978 but it had long been recognised that if the technology were to be extended to include video information, it would be necessary to develop a laser operating at much shorter wavelength than the original 780 nm device. The amount of data that can be stored on a disc depends on the size of the dot representing each bit of information and this, in turn, depends on the wavelength of the laser light employed in writing it and reading it. The original CD system used a dot size of about 1 µm the future DVD disc would require this to be reduced by at least a factor of two, implying a laser operating at a wavelength of about 400 nm, in the blue/violet part of

FIG. 7.4 An LED torch. The development of white LEDs in 1995, by Nakamura at the Nichia Chemical Company in Japan opened the way to a wide range of LED lighting. Initially, these devices were applied to a number of simple problems where the total amount of light required was relatively modest. One example is the internal lighting in a car, another the provision of illumination in remote native villages. The electric torch represents an ideal application, the increased efficiency of LEDs, compared to conventional torch bulbs, making for considerable saving in battery power. Courtesy of iStockphoto, © Murat Baysan, image 8428908.

the spectrum. We might recall that the GaAs laser required some 10 years of development to achieve continuous operation at room temperature with an operating life appropriate to its commercial use – Nakamura's blue laser took about 2 years to reach the same point and Nichia began selling sample devices in 1999. However, this certainly didn't happen by chance – there were two major problems to be overcome before a viable laser could be produced. First it was necessary to improve the structural quality of the material and secondly much better electrical contacts were needed to carry the much greater currents used in a laser, compared with those appropriate to an LED. Nakamura developed a quite remarkable growth technique known as ELOG (Epitaxial lateral over-growth) which produced material with very much lower dislocation density than was acceptable in an LED and he incorporated the use of AlGaN/GaN superlattices to replace the AlGaN cladding layers. This latter move improved his ability to dope the cladding by including dopant only in the GaN. Thanks to these efforts, Nichia had yet another winner on its order books – not only the technology but also the timing was immaculately suited to speeding the introduction of the new generation of DVD players and Nichia was once again in pole position to take advantage. Ogawa's gamble had paid off dramatically. There can surely be few better examples of an investment repaying its originator than this one.

This whole sequence of inventions is simply steeped in irony – that one man working almost alone in a small rural chemical works should succeed in beating a plethora of well-heeled rivals to both the blue LED and the blue laser is remarkable enough in itself but the final irony lies in the sadly deteriorating relationship between Nakamura and his employer which was to coincide with this almost unbelievable success. Unfortunately, Ogawa had effectively retired from the company in 1999 and his successor turned out to have very different ideas, particularly with regard to Nakamura's position. Rather than reward the company's benefactor he treated him with disdain and eventually, very much against his inclination, Shuji Nakamura was obliged to leave Nichia and take up a post at the University of California's Santa Barbara campus, where he was welcomed with open arms and where he is now picking up the threads of his scientific life. One of the most exciting and remarkable technical advances of the twentieth century (possibly of any century?) came to an end in recriminatory legal battles which may have provided a degree of financial satisfaction to Nakamura, but must surely have left him feeling bruised and deflated. It was an outcome that few could have envisaged and even fewer would have welcomed. (Anyone seeking a fuller account will find it in Bob Johnstone's book 'Brilliant!')

If the previous paragraph engenders a feeling of finality, I should hastily correct it – the development of the blue diode must surely be seen as a wonderful beginning, rather than an ending. Indeed, there can be little doubt that the semiconductor community as a whole saw it in such a light, and everywhere efforts were made to set up task forces to capitalise on Shuji Nakamura's inspired lead. By the early years of the twenty-first century there were at least ten companies attempting to rival Nichia as suppliers of blue lasers – the list including Toyoda Gosei (with whom Akasaki is associated), Cree, NEC, Xerox, Osram, Samsung, NTT, Matsushita, Sony and Fujitsu. Not all will survive to benefit from the blue laser market but their presence shows just how importantly it is regarded. Nor should we overlook the fact that the details of the light emission process are far from completely understood. Exactly what goes on in the quantum wells which serve as the active material is not at all clear. There is evidence for the indium in a GaInN well segregating to form something like quantum dots, which changes considerably the interpretation of the emission process – only time will lead to an adequate understanding. There is also great interest in the effect of strong internal electric fields in modifying the properties of the wells. Clearly, much fundamental work needs still to be done. More important, however, in the long run, is the market for domestic lighting and this must demand our attention in the remaining paragraphs of this chapter.

Lighting, we are told, represents something like twenty percent of all electricity usage in the United Kingdom (40% in Thailand) and most of this could be saved by switching to LED lighting with luminous efficiency roughly 10 times that of the tungsten bulb. It has been estimated that if half the lights in the USA were to be replaced with LEDs, the saving in power would amount to over 40 GW and would allow about forty power stations to be shut down. On the global scale, conventional lighting produces an amount of CO_2 equivalent to almost three-quarters of the emissions of the world's cars. A change to LED lighting would effect a tremendous improvement. LEDs have operating lives of perhaps a 100 times those of tungsten lamps which means that individual householders would make long term savings in capital costs, as well as in the more immediate running costs. Many city councils have already learned to appreciate the cost savings inherent in the use of LED traffic signals. Though LEDs have potential for little more than a factor two improvement in efficiency over the presently recommended compact fluorescent lamps, they certainly have the potential to avoid a serious pollution problem in the form of the mercury contained in CFLs, which looks likely to find its way into thousands of landfill sites at an ever-increasing rate. Yet another potential advantage lies in the possibility for controlling the effective colour temperature of the light, a facility not available to any other form of illumination. While the decision in the UK to replace tungsten bulbs with CFLs is understandable, the timing is unfortunate, when there appears every likelihood of LED lighting making inroads into the domestic and public spheres within the next 5 years. Why the delay? To answer this, we need to examine some of the problems in a little more detail.

As we noted earlier, white light may be produced from LEDs in a variety of ways. For example, an ultra-violet diode may be used to excite phosphors similar to those used in conventional fluorescent tubes, a blue LED may be used in conjunction with a phosphor giving yellow and red light, while an obvious alternative makes use of a set of three LEDs producing the three primary colours. It is not certain which of these approaches will finally bring commercial success but one general point is in order. The use of a phosphor to down-convert radiation to longer wavelengths is inherently an inefficient process. For example, using UV light with a wavelength of, say, 300 nm to generate red light at 600 nm wavelength implies a maximum efficiency of 50%, simply because of the difference in the energies of the respective photons. One photon of energy 4 eV (which implies an applied diode voltage of roughly 4 V) is effectively converted into one photon with energy 2 eV (which could, in principle, be generated with an applied voltage of 2 V). Maximum theoretical efficiency can only be obtained by using three different LEDs to provide three colours, but the present state of the art leaves something to be desired – there exists what is sometimes called the 'green deficit'. While very efficient red and blue LEDs have been demonstrated (close to 50% external efficiency), there is now a weakness in the yellow/green region,

FIG. 7.5 A green LED direction arrow. The inherent brightness of modern LEDs makes them particularly attractive for numerous different types of sign. They catch the eye much better than ordinary painted signs and they can, where necessary, be made to flash on and off to attract attention even more insistently. Courtesy of iStockphoto, © Ulrich Koch, image 117711.

where efficiency languishes at about 15%. (Only a few years ago this would have been regarded as rather good but times have certainly changed though the green arrow shown in Fig. 7.5 demonstrates that such efficiencies are more than adequate for many practical applications. In fact, LED efficiencies have increased quite dramatically over the years – from 1970 to 2000 at the rate of a factor of 10 per decade (sometimes referred to as Craford's Law) and are now approaching the end stops, represented by the ideal of 100%!

While some very interesting applications of variable colour architectural lighting have been initiated in different parts of the world, using individual coloured diodes, the main drive at present is towards finding a direct replacement for the humble domestic bulb and the list of contending firms is growing steadily. The approach more-or-less universally favoured involves the use of a blue LED, combined with a suitable phosphor and at the time of writing, the best performance, achieved by several companies, can be expressed in terms of a luminous efficiency of about 130 lm/W at a current drive of 350 mA – an output of about 130 lm from a semiconductor chip 1 mm × 1 mm in area. This has to be compared with the 1000 lm produced by a conventional 60 W lamp, which implies the need for an improvement of something like 8 times. It would, of course, be possible to combine the light from 8 chips but this would be prohibitively expensive – the cost of an LED being largely that of the material. As George Craford (now of the Philips company Lumileds) has pointed out, the manufacturer of tungsten bulbs can make a 100 W bulb for the same price as that of a 10 W bulb, while it costs the LED man 10 times as much! The way ahead is rather one of improving efficiency still further (say by 30%, to 170 lm/W) and, at the same time driving the diodes much harder. Currents of 2 amps have been achieved, which would be adequate to match the output from the 60 W bulb, though, unfortunately, efficiency drops significantly at such drive currents. One of the current tasks of basic research is to understand and rectify this annoying trend. Others are concerned with the long-standing problem of finding the most suitable substrate. While most companies depend on sapphire, which, being an excellent insulator, causes complications in making electrical contact to the diodes, the American Cree company relies on (home grown) silicon carbide, thus avoiding such difficulties. However, SiC costs significantly more than sapphire. A possible compromise may be the use of thick layers of GaN grown by vapour phase epitaxy on sapphire – large area slices are now available – but it is still too early to say which approach may win the day – if any! Notwithstanding all this, Craford looks confidently to the time when a 1 mm square chip will generate 1000 lm at a price of about two American dollars, very competitive with that of an equivalent CFL. Exactly when that may be is not clear but the omens appear good. Fifty odd years of LED research effort looks set to reach an impressive climax in the shape of the most efficient source of illumination known to man – just a small factor of two short of perfection.

CHAPTER 8

Information Highways and the Fibre Revolution

While the word 'revolution' tends to be applied to all manner of unworthy concepts, there can be little doubt that the introduction of glass fibres into the world's communication highways should be classed as one of the most remarkable and unexpected revolutions in modern technology. That it should happen at all was remarkable enough but that it should happen within the relatively tiny span of 20 years between 1970 and 1990 makes it all the more surprising. The communication industry has always prided itself on being well to the conservative side of the business spectrum, reliability generally being uppermost in its philosophical thinking, and forays into the unknown being encompassed, if at all, only with extreme caution. How, then, could it make this radical change at what, for it, was something approaching breakneck speed? It is a truly fascinating story, splendidly told by Jeff Hecht in his book 'City of Light' and, while I shall attempt to summarise it here, I can strongly recommend Hecht's book to anyone interested in following the story in its entirety.

As we saw from the introduction to Chapter 3, the development of communication technologies makes an exciting story in itself and it is interesting to recognise that two of the essential features of fibre-optic systems were incorporated at a very early stage. The use of light as an information carrier dates as far back as 1200 BC when, according to Aeschylus, the conquest of Troy was signalled some 600 km back to Argos by means of 'The Torchpost of Agamemnon', a sequence of fire signals. (It also signalled the return of Agamemnon to his wife Clytemnestra, who promptly murdered him!) In 200 BC, the Greek Polybios invented a visual code in the form of two walls above which a variable number (1–5) of torches could be displayed, allowing each of 24 letters to be represented and therefore permitting written messages to be transmitted to a distant

Semiconductors and the Information Revolution: Magic Crystals that made IT happen © 2009 Elsevier B.V.
DOI: 10.1016/B978-0-444-53240-4.00008-8

observer at a rate of some 8 letters per minute. Better known is the North American Indians' dependence on smoke signals to alert their compatriots to approaching danger and the Elizabethans' use of bonfire trains to warn of the Spanish Armada's imminent arrival in the English Channel. Nor let us forget the French-Revolutionary-inspired Chappe semaphore system (Fig. 8.1) which relied on ambient light to transmit intelligence between line-of-sight communication towers, spread across the French countryside, nor the British Admiralty version which connected London with the south coast. While these examples might just (at a stretch) be regarded as early forms of 'digital' technology, there can be no denying that Samuel Morse's nineteenth century electric telegraph transmissions were digital, in the sense in which we employ the term today. The use of short pulses of energy to represent written characters was clearly well established long before optical fibres came on the communications scene. Nevertheless, the generation of trillions of extremely short light pulses and their propagation along thousands of miles of tiny glass fibres to effect intelligent links between nations

FIG. 8.1 A nineteenth century painting of one of Claude Chappe's semaphore towers being demonstrated to an enthusiastic audience. The first line of these communication towers from Paris to Lille came into operation in 1793 and several similar systems were built across the French countryside. Napolean made much use of them in co-ordinating the many movements of his armies. Courtesy of WikiMedia.

constituted a unique advance in telecommunications technology which was very largely unforeseen even by those most closely involved in the communications business. Why did it happen? How did it happen? And what part did semiconductors play? Such are the questions that will concern us in this chapter.

To answer the first question, we need to understand one or two basic ideas concerning the nature of digital signals and the bandwidth required for processing them. But our starting point should, perhaps, coincide with the introduction of the telephone, an essentially analogue instrument, by Alexander Graham Bell in 1876. The human ear is sensitive to a range of frequencies between ~20 Hz and 20 kHz and any self-respecting Hi-Fi system will be designed to reproduce such a span. However, it is generally accepted that the human speaking voice can be adequately represented by a rather smaller range – from 200 Hz to 3 kHz and this was quickly adopted for telephone transmissions. Not only could the message be understood, but the characteristic of each individual voice could be adequately recognised. In technical parlance, we say that a frequency bandwidth of 3 kHz is adequate for telephone transmission and, conveniently, such a range is readily propagated along simple lines of copper wire. Thus, early telephone networks (Fig. 8.2) consisted of wire connections between each subscriber and suitable switching stations in town centres, where rows of (we like to imagine) glamorous telephone

FIG. 8.2 An 1896 Swedish telephone. Compared with the first Bell telephone (Fig 3.1), this shows more than a little evidence of designer influence, though it probably differed very little in its manner of functioning. Courtesy of WikiMedia.

girls plugged and unplugged telephone jacks in order to effect the necessary connections. Sadly, the girls were soon to be replaced by electro-mechanical switches which in turn were replaced by electronic switches – one overriding motive for the Bell programme to develop the transistor was, of course, that of replacing the clumsy mechanical relays with silent and much faster electronic devices. However, our imme-diate interest is in the transmission network. So long as the number of subscribers was relatively modest, such arrangements proved satisfac-tory, though problems soon became apparent in respect of connections between major centres. It was possible to route only one call along each line (otherwise, calls would be inextricably jumbled together) and this implied a need for a multiplicity of lines between centres, a multiplicity which grew to impossible proportions within relatively few years. Something radical was needed and this took the form of a 'carrier' onto which the audio signal could be modulated. Each call was to ride on its own carrier wave, the individual calls being kept separate by using a different carrier frequency (or wavelength) for each (a technique which came to be known as 'frequency division multiplexing' and later, when light waves became the carriers, 'wavelength division multiplexing', or WDM). Suppose, for example, that we chose a carrier frequency of 1 MHz and modulated it with the telephone audio range up to 3 kHz, the net result was a range of frequencies between 0.997 and 1.003 MHz (1 MHz ±3 kHz). To keep the next call separate, it would use a carrier frequency of 1.010 MHz (1 MHz plus 10 kHz), the next but one, 1.020 MHz and so on. If the total range of frequencies was limited to plus and minus 50% of the original carrier frequency, this allowed a 100 calls to be sent down each telephone line, a very worthwhile improvement over the earlier approach. (These numbers should not be taken too literally – but they serve to illustrate the principle.) Clearly, if the carrier frequency were to be increased to 10 MHz, this would allow a 1000 calls to be carried on each line so, as the pressure of ever-increasing subscriber numbers grew, this translated into a pressure towards ever-increasing carrier frequency. Imagine, for example, connections between large cities with several million subscribers – facilities should be provided capable of handling some 100,000 calls, which implied carrier frequencies of order 1 GHz (1000 MHz), a frequency which could no longer be transmitted along simple copper wires and demanded the use of co-axial cable (the type of cable used to carry TV signals). Note that the total bandwidth associated with such carriers is of order 1000 MHz – the increase in carrier frequency implies a corresponding increase in bandwidth and it is usually this bandwidth which is quoted when discussing communication capacity.

Two other considerations have a bearing on this question of required bandwidth. Firstly, with the development of commercial television, it became desirable to transmit TV signals along suitable cables and this implied a need for much greater bandwidth than appropriate to telephone signals because video signals contain very much greater amounts of information. The bandwidth for a TV signal is therefore measured in mega-Hertz, rather than kilo-Hertz. The problem was exacerbated considerably when it became necessary for computers to communicate with one another and this brings us to consider digital signals, rather than the old-fashioned analogue type. Computers, of course, deal in digital signals as a matter of course but we must also recognise that long distance communication links also led to even telephone calls being transmitted in digital form. The reason was simple – digital signals are much more robust in their resistance to noise and interference. To transmit a telephone message over long distances requires amplifiers at regular intervals because the signals inevitably become weaker as they travel along a transmission line. However, this is not the end of the story because, as the wanted signal becomes weaker, there is a tendency for the unwanted noise level to increase and, if this continues indefinitely, the signal disappears completely into the noise and all useful information is lost. Further amplification is of no value because both noise and signal are amplified equally. The advantage inherent in digital signals is that they can be regenerated at each repeater station with no loss of information. All that is necessary is for each pulse to be recognised – its precise amplitude, unlike that of an analogue signal, is of no consequence. Provided it can be detected, it can be regenerated and the information re-transmitted with perfect accuracy. The fact that we can now listen to telephone conversations over many thousands of miles, with crystal clarity results from the combination of digital signal methods (see Fig. 8.3) and the remarkable reliability of glass fibre transmission lines. However, there is a price to pay for digital techniques in terms of the increased bandwidth needed.

To convert an analogue signal into digital form requires that the amplitude of the signal be measured at frequent intervals of time and each reading expressed in binary form. If we suppose that each reading is expressed as eight binary bits, the resulting digital signal consists of eight pulses repeated again and again at the sampling frequency, which must be at least double the maximum frequency present in the analogue signal – 3 kHz in our case. This implies that we have roughly $6000 \times 8 = 48,000$ (say 50,000) pulses per second in our digital version, implying a bandwidth of order 100 kHz, some 30 times greater than was needed for the analogue signal. This may seem unfortunate but it is a price well worth paying in the interest of reliable long distance communication. There is, too, a further advantage to be gained from the use of digital methods – it is possible to send several conversations along a single line by means of 'time division multiplexing', a technique not available with analogue methods. This involves interleaving two or

FIG. 8.3 A modern digital telephone, which differs fundamentally from its earlier counterparts (see Fig 8.2), not only in looks but also in function. That the user can move freely throughout his home or garden whilst continuing a conversation represents one of the more significant changes to everyday life, though one that most of us take rather for-granted. Courtesy of the author.

more sets of digital pulses, the second (third, fourth, etc.) set fitting within the spaces between the first set. It implies that the individual pulses must be shorter than would otherwise be necessary and this doubles (or trebles or quadruples, etc.) the bandwidth required for their transmission. TDM offers flexibility and may be valuable in some instances but, as we see, the penalty is always one of increased demand for bandwidth. There is no way of avoiding this.

The obvious conclusion to be drawn from this very brief introduction to the principles of communication theory is one of rapidly increasing demand for transmission systems with ever-increasing bandwidth. Thus, when the telephone first appeared in the 1870s, analogue signals were sent along simple copper wires employing bandwidths of 3 kHz while the use of RF (radio frequency – typically 100 kHz) carriers and frequency division multiplexing was well established by the time of the First World War (aided and abetted by the thermionic valve). The development of wireless transmission, following the War, threatened serious competition to land-lines, though, in practice, the quality of the resulting communication suffered from fading and general lack of reliability. The first coaxial cables were introduced in 1940 with a carrier frequency of 3 MHz and a capability for transmitting 300 voice channels. Microwave links at frequencies in the region of 10 GHz came into prominence after the Second World War and offered correspondingly greater bandwidth but these were 'line of sight' links and therefore confined to overland connections. They also depended on exotic vacuum tubes such as klystrons and travelling wave tubes which were fragile and worked at inconveniently high voltages. International links saw competition between the first transatlantic telephone cable TAT-1 (using co-axial cable) which was laid in 1956 and the exciting new satellite microwave systems which began with Telstar (employing a frequency of 4 GHz) in 1963 and brought with them characteristic (and often annoying) time delays. However, the demand for more and more bandwidth demonstrated that there was still a need for even higher frequencies, for which both Bell and the British Post Office had a mind to use millimetre waves (frequencies of 60 GHz), piped down suitable metal waveguides. Several trials were instigated during the 1950s with some degree of promise, though there were worrying difficulties as a result of the inevitable waveguide bends which upset the desired smooth propagation of the waves. Perhaps the most serious of these was the observation that settling of earth around buried guides was sufficient to cause perturbations in transmission characteristics. Apart from these somewhat inconclusive trials, no practical systems emerged. Meanwhile, a number of visionaries saw the future in terms of light, rather than millimetre waves – a particularly strong protagonist being Alec Reeves, the inspiration behind optical fibre research at the STC (Standard Telephone and Cables) laboratory at Harlow in Essex. Optical frequencies lie typically in the region of 10^{14} Hz and the huge step up in bandwidth available to optical systems seemed altogether unnecessary in 1937, when Reeves first made his proposal (though modern systems are already struggling to satisfy the many demands being made upon this huge capacity!) and the idea of using optical carriers began to be taken seriously only during the 1960s once laser sources became available. The problem at that time was one of finding a suitable means of transmission and it was only when low-loss optical fibres were developed during the 1970s and 1980s that effective light-based communications became possible.

This, then, was the maelstrom of activity within which the fledgling fibres made their bow in the early 1970s but a long history of experiment had preceded it. The first demonstration of light guiding within an optical waveguide dates back to 1841 when a Frenchman, Daniel Colladon observed that light could be confined within a jet of water. As the water jet curved downwards under the influence of gravity, so the light followed it, staying obediently within the jet until it disintegrated in a luminous cloud of spray. The effect provided a basis for numerous fanciful fountain displays at grand events such as The International Health Exhibition in South Kensington in 1884, The Royal Jubilee Exhibition in Manchester in 1887 and the 1889 Universal Exhibition in Paris. At much the same time another Frenchman, Jacques Babinet noted that light could also be guided within bent glass rods. Both effects were based on the phenomenon of 'total internal reflection' for light striking an optical interface from a material of high refractive index (such as glass) towards a medium of low index (such as air) at a near-glancing angle. In effect, the light bounces to and fro within the jet (or rod), being reflected several times on its journey along the guide. Because each reflection is 'total', there is no associated energy loss, no matter how many reflections occur, and (unless the optically dense medium serves to absorb it!) all the light emerges from the end of the guide. This is an interesting and remarkable result and forms the basis for the use of glass fibres as optical transmitters. Without it, we might still be using much clumsier microwave systems with relatively minute bandwidths – and struggling to book long distance telephone calls in competition with hundreds of neighbours!

The year 1880 saw an arresting and far-sighted approach to the use of light as an information carrier. Once again, it was the inventive Alexander Graham Bell who led the way into optical communications with his so-called 'photophone', a device which contrived to modulate a light beam with a voice signal by reflecting it from a vibrating mirror. The modulated beam was then collected by a convex mirror some distance from the source and detected by a selenium photocell, connected to a telephone receiver. Not having the luxury of a laser source, Bell made do with a shaft of sunlight and became almost poetical in his enthusiasm for the resulting sound transmissions: 'I have heard articulate speech produced by sunlight! I have heard a ray of the sun laugh and cough and sing – I have been able to hear a shadow, and I have even perceived by ear the passage of a cloud across the sun's disc.' It was an exciting innovation but, alas, proved impractical with the technology available at the time. The maximum distance over which these marvels could be experienced was no more than a matter of a few hundred feet and the sun could not always be relied upon to co-operate at the required moment. As we now know, it was a prophetic demonstration but a century before its time. Bell may have been unaware of the possibility of guiding light along glass rods but, in any case, the idea of making miles of flexible glass 'wires' was even more remote in

1880 than it seems to have been to many communications engineers in the post-Second World War period.

The next development of note had nothing to do with light guiding, though it did involve the making of long glass fibres. Charles Vernon Boys was a physicist with an interest in the accurate measurement of the gravitational constant G using a torsion balance, and in 1887 he recognised that thin quartz fibres possessed the most suitable elastic properties for his experiments. He had difficulty in acquiring anything suitable, so set about making his own by attaching an arrow to a quartz rod which he then melted with an oxy-hydrogen torch before shooting the arrow down a long corridor! The resulting quartz fibres were so thin that he believed them to be 'beyond the power of any possible microscope'. They were, in fact, on the order of 10 μm in diameter (1 μm $= 10^{-6}$ m) and were ideal for a number of ultra-sensitive measurements, as well as for that of G. He left no reference to light guiding, possibly feeling this to be altogether too trivial.

Others, however, were very much concerned to utilise the optical properties of glass fibres or rods. John Logie Baird in England and C Francis Jenkins in America both experimented with the use of bundles of fibres as an aid to mechanical scanning of images for their incipient television systems, while C.W. Hansell at RCA and a German medical student Heinrich Lamm used similar bundles in attempts to make flexible fibre scopes for remote viewing of images. Lamm's attempt in 1930 to make a gastroscope which would allow doctors to look inside their patients' stomachs was certainly well conceived and did, indeed, allow an image to be viewed but its brightness and resolution left much to be desired. There were two difficulties to be overcome: the first was concerned with finding a reliable method of packing a large number of fibres together, the second, which apparently escaped Lamm's cognizance, with the fact that light could leak between adjacent fibres, where they were in close contact, resulting in considerable muddying of his images. The answer to the first of these was provided by a professor at Imperial College, London, Harold Horace Hopkins who developed a technique for winding long lengths of fibre round an annular spool, clamping the resulting stack at several points round the periphery, then cutting the stack into appropriate lengths. His paper appeared in Nature in 1954, alongside a brief letter from a Dutchman, Professor Abraham van Heel, describing a concept which would solve the second difficulty. This involved the use of clad fibres – fibres having a core of one glass, surrounded by a cladding of a second glass with a slightly smaller refractive index. Light would be totally reflected from the interface between the two glasses and would never reach the point of contact between fibres. This crucial idea appears to have been the brainchild of an American professor at the University of Rochester and President of the

Optical Society of America, Brian O'Brien. He is on record as discussing such an approach with van Heel in 1951. Van Heel was interested in using an array of fibres in making a periscope for the Dutch navy and was probably the first person to publish the idea of using clad fibres in 1953, much to O'Brien's annoyance – there seems to have been an unfortunate breakdown in communication between the two. Various others experimented with the use of plastic coatings or with coatings of oil but the real breakthrough in making reliable cladding for glass fibres occurred at the University of Michigan. In 1956, a student named Larry Curtiss rejected the advice of his professorial betters and succeeded in pulling a composite fibre from a blank made from a glass rod inside a piece of glass tubing. The following year the Michigan group applied Curtiss's fibres to making a working gastroscope and, by the end of the 1960s, this model was being widely used by surgeons throughout America. It represented a huge improvement over earlier versions based on lenses, which were completely rigid and very difficult for the unfortunate patient to swallow! Very shortly after Curtiss's success, an employee of American Optical, John Wilbur Hicks also managed to draw clad fibres by using a pair of concentric crucibles, one containing the core glass, the other the cladding, and used them to solve a problem for the Pentagon. There was a military interest in coupling a pair of image intensifier tubes together so as to achieve higher brightness in the final image but attempts to use lenses between the two tubes were unsuccessful because the output widow of the first tube was excessively curved. Hicks was able to couple the image by way of a fibre bundle, one end of which was machined to the same curvature as the tube faceplate. Military faces lit up in instant gratitude – American Optical had an exciting vision of bundles of fibres turning into bundles of dollars – everyone was happy.

Glass fibres were beginning to make a name for themselves and, during the early 1960s, communications engineers began thinking the unthinkable – could these clever optical guides be used to transmit messages? After all, there now existed a wonderful light source in the shape of the laser? The 1960s may have been the heyday of technological optimism, when almost anything might be possible – witness Prime Minister Harold Wilson's 'white heat of technical revolution' speech to the British Labour Party Conference in 1963 – but there was a serious problem attached to this particular idea. All applications hitherto had involved extremely short distances – even the gastroscope required only a couple of feet of fibre – while telephone lines tended to be measured in miles (or kilometres, if you happened to live in mainland Europe). It soon became clear that available fibres suffered from losses due to optical absorption in the core glass which appeared to rule them out from carrying messages much further than an adjacent room – typical losses were quoted as 1 db/m and 3 db corresponded to a factor of two, so 3 m of fibre was sufficient to reduce the signal to half its starting value. A 100 m would be enough to reduce it by a factor of 10^{10} – to all intents and purposes, zero! A kilometre didn't even bear thinking about! Even digital signals were unable to live in an environment such as this. Clearly, if light was to be useful as a carrier of information something better than this lossy fibre would be needed for a transmission medium.

One obvious answer lay in the use of air. Claude Chappe's semaphore system had worked well enough over distances of kilometres so there was no doubting the transparency of air, and the advent of the laser, with its ultra-narrow beam and high light intensity, could be relied upon to improve considerably on Chappe's example. This was confirmed in dramatic fashion in 1969. Following the first moon landing, when astronauts Armstrong and Aldrin had set up reflectors on the surface of the moon, a laser beam fired from earth produced a reflection which was detected slightly more than a second later back on earth and provided a measure of the distance to the moon, accurate to within an inch. The nearly 500,000 miles round-trip was a trifle more than anyone had in mind for terrestrial communications, and most of the journey was through relatively empty space, but the principle was clear – laser beams could be transmitted over long distances without being attenuated to the point of undetectability. Alas, this particular experiment was undertaken in near-ideal weather conditions – in practice, free-air transmission proved unreliable as a means of carrying telecommunications (not surprisingly perhaps this was appreciated much more rapidly in Britain than on the other side of the Atlantic!). No matter – there were ideas a-plenty for confining the laser beam within a narrow pipe, an environment which would never be subject to cloud, rain, fog or temperature gradients such as confounded the free-air experiments. First attempts relied on multiple reflections from the silvered inner surface of a metal pipe – not unlike the reflections within a glass fibre except that, in this case, they were not 'total' and the, admittedly small, losses at each reflection proved to be too large. In any useful length of pipe there were just too many reflections. Next in line came the use of 'confocal lenses', arranged so that the image formed by one lens lay at the focal point of the next. In theory, this did away with the need for light reflections from the pipe walls altogether but, once again, the small reflection losses at each lens added up to worryingly large collective loss. Even with lenses employing anti-reflection coatings performance was marginal. However, the real difficulty lay with the inevitable bends which any practical guide must make and, as with the similar

FIG. 8.4   "A bunch of glowing fibres". Fibre optics certainly lends itself to the production of interesting artistic effects, as well as to extremely high density telecommunication traffic. This aspect dates back to the nineteenth century, when light guiding by water jets was all the rage at Victorian trade exhibitions. Courtesy of iStockphoto, © Rob Friedman, image 2524046.

case of millimetre wave guides, small bends associated with settling of earth around buried guides caused the beam to destabilise – to walk off its pre-programmed line. Unhappily, this also proved fatal to hopes of using gas lenses – an extremely clever idea which relied on thermal gradients inside a heated tube to control the refractive index of the gas within it in such a manner as to focus light travelling along the tube axis. Long before the end of the 1960s, it was widely, if reluctantly, accepted that 'the only thing left is optical fibres' (see Hecht, p. 104 and Fig. 8.4).

Initially, Bell were still hopeful of being able to rely on millimetre waveguides so the early running in the direction of fibres was taken up at STL (the STC laboratory in Harlow), where some very advanced thinking took place. They concluded that clad fibres represented the best way to go and gave considerable thought to methods of making them, but perhaps their most immediate conclusion was a precise specification for the minimum absorption loss in the core glass – if fibre was to be of even marginal use for communications, the loss must be no greater than 20 db/km, demanding an improvement in quality of some 50 times, compared to the glass used to make fibre bundles for imaging. This implied, firstly, a careful choice of the optimum glass to be used and, secondly, some serious material study to ascertain the origin of the losses observed. The really exciting development made at STL, however, was the result of some very careful measurements of absorption losses in pure samples of fused silica glass which came up with the shock result that the loss was close to 5 db/km, considerably better than their estimate of the limiting value required. It was just the stimulus that Charles Kao, then leading the STL work, needed to push the case for optical fibres around the world. However, the STL team was still uncertain as to how to grow fibres of this wonderful material and it was at this point that the trail skipped back across the Atlantic to the Corning Glass Works in New York State. Bob Maurer had also been thinking about glass fibres for communications and his experience with many types of glass told him that, if purity was an important criterion, then fused silica was likely to be the best hope. Staking his reputation on such a hunch, he plunged into the far-from-trivial task of growing fibres (pulling silica fibres requires temperatures of 1600 °C or more) and, after months of work succeeded in producing clad silica fibres with losses of 20 db/km, just equal to the figure quoted by Kao. By 1970, this figure had been reduced to 16 db/km and telecom companies around the world sat up and took notice. Corning redoubled its effort and, 2 years later reported a fibre with loss of only 4 db/km. Equipment for drawing silica fibre appeared miraculously at Bell Labs, at STL, at the British Post Office laboratory at Martlesham Heath, at the Japanese Telegraph and Telephone Company NTT, at Fujitsu, at Southampton University and, doubtless, at many other locations. The race was now on to develop practical systems. Everyone was convinced in principle – all that was needed was to solve a mass of detailed problems, such as how to launch light waves into tiny fibres with diameters of only a few microns and how to couple two fibres together and how to generate ultra-short pulses of light at appropriate wavelengths and how to minimise the spreading of these pulses as they travel down their guiding glass highways.

Then, quite suddenly, the centre of gravity shifted to Japan, where in 1976 Mashahara Horiguchi of NTT and Horoshi Osanai of the Fujikura Cable Company reported fibre loss of 0.5 db/km, measured at a wavelength of 1.2 μm, then in 1978 0.2 db/km, measured at 1.55 μm. Theoretical calculations suggested that these values represented practical limits and that the absolute minimum loss did, in fact, occur at a wavelength of 1.55 μm, results that changed the whole complexion of the activity. It had been tacitly assumed that practical communications networks would make use of the now well-established GaAs laser diode operating at a wavelength of 880 nm but it now became clear that long-haul systems must utilise the longer wavelength region in order to capitalise on the much lower loss available (the minimum loss at 880 nm being about 1.5 db/km). This conclusion was confirmed when it was also realised that minimum fibre dispersion occurred at a wavelength close to 1.3 μm. (We shall say more about the importance of dispersion in a moment.) It was all very well – but it raised one rather significant problem – there was no suitable light source which operated in this spectral region, neither laser, LED nor any other! Nor were there any light detectors designed for these wavelengths! And that brings us (at long last!) to the subject of semiconductors. The best light guides in the world are of little value unless sources, modulators and detectors can be found to send and receive the appropriate signals. Having dwelt at some length on the fascinating story of fibres, we shall devote the rest of this chapter to the even more fascinating story of the requisite semiconductor devices.

The saga begins, inevitably, with gallium arsenide. We saw in Chapter 5 that the first CW (continuous wave) GaAs/AlGaAs double heterostructure laser was reported from the Ioffe Institute in Leningrad, and shortly afterwards from Bell Labs, in 1970. It had been 10 years in the making and was, even then, far from the finished article. Any device capable of making an impact on the world of consumer or professional applications must first demonstrate its long-term reliability and the DH laser was no exception. It took another 8 years before Philips and Sony were confident enough to use it as the basis for the CD player, while telecoms companies were likely to submit it to even closer scrutiny. As we noted earlier, reliability was a sine qua non in the communications business. But the laser was up to the challenge – in 1976 Bell Labs were able to demonstrate (extrapolated) operating lifetimes of 10 years for GaAs lasers and in May of the following year Bell engineers incorporated them in their first real fibre link, sending live telephone traffic over a 1.6 mile circuit in down-town Chicago. It was supposed to be a 'first' but Bell were embarrassed to find that one of their smaller rivals, GTE had beaten them to it by about a couple of weeks! And it was little consolation to know that GTE had used a GaAs LED source, rather than a laser. The British Post Office was not far behind, routing live traffic through 5 miles of fibre in the region of Martlesham Heath in Suffolk in June. Thus, began the first trickling photon flow which was destined to reach flood proportions within a decade.

But already we have an example of one of the many technical dilemmas facing telephone engineers – why use a laser if a simpler (and cheaper) LED might perform as well? The simple answer is that the LED does not perform as well. There are three important factors. Firstly, it is much easier to focus the light from a laser into the end of a tiny optical fibre, secondly it is possible to modulate a laser at a much higher rate than an LED and thirdly, laser light suffers much less from the problems of fibre dispersion. Let's look at each point in turn. A simple calculation shows that for rays of light to enter a fibre they must approach within a rather small cone angle and it follows that the narrower beam from a laser allows much more light to enter the fibre than is the case for the LED – the laser beam is, in any case, considerably brighter. Clearly, the laser has the advantage of launching a greater intensity of light and therefore allowing it to be transmitted over a greater distance before the need for regeneration. Also, in the interest of generating short pulses for pulse-code-modulation, it is important that the light source can be chopped at high speed (by chopping the drive current) and this depends on the effective recombination lifetime for electrons and holes. The LED is characterised by the spontaneous lifetime which may typically be about 10^{-8} s, whereas the laser depends on stimulated emission, having significantly shorter lifetime. Thus, very roughly, the LED is limited to producing pulses of length 10^{-7} s, while the laser is capable of about 10^{-9} s, some two orders of magnitude shorter. In other words, laser systems can work at gigabits per second, while LEDs are capable of only 10 Mb/s. Finally, we come to the important question of fibre dispersion and this is so important that it needs a paragraph to itself.

We should first recognise that dispersion has to do with different packets of light travelling along fibres at different rates and, secondly, that there are two kinds of dispersion in a fibre waveguide. These are known as 'modal dispersion' and 'material dispersion'. In general, electromagnetic waves travelling along any appropriate waveguide may exist in several different configurations – that is with different patterns of electric and magnetic fields. This is as true of microwave radiation in metal waveguides as it is for light in glass fibre waveguides and in both cases these different modes travel at different speeds. But, in addition to this modal dispersion, material dispersion refers to the well known fact that different wavelengths of light also travel at different speeds through glass. It is this effect which produces the spectrum of different colours when white light is passed through a prism. Why is all this so important for optical communications? Because it causes nice, sharp, clearly-identifiable light pulses to spread out as they progress along the fibre. Take 'modal dispersion; when we launch a light pulse into a fibre, the energy in the pulse is distributed between the various waveguide modes so different parts of the pulse travel at different velocities and it therefore spreads out progressively as it travels. Similarly, each pulse contains a spectrum of wavelengths (which depends on the nature of the light source) and material dispersion causes the different wavelengths to travel at different speeds. The effect is the same – pulses spread out as they travel. The problem for communications is that, eventually, the various pulses in a pulse-coded signal start to overlap and merge together so that it is no longer possible to distinguish them – in other words, the information they carry is lost. This limits the length of fibre between regeneration stations, which, in turn, demands more such stations and makes the system both inconvenient and expensive. And it is clear to see that the narrower the pulses and the closer they are spaced (i.e., the greater the 'bit-rate') the more serious is the effect – the shorter must be the distance between repeaters.

Could anything be done about it? Fortunately, yes – three things could be done. Firstly, fibres were designed so as to minimise modal dispersion. For example, many early systems made use of a subtle idea propounded by a Japanese scientist at Tohoku University, Shojiro Kawakami, whereby the refractive index of the core glass was graded from high at the centre to low at the periphery and these so-called 'graded index' fibres showed much less modal dispersion than their simple fixed-index predecessors. Even more radically, fibres were designed capable of supporting only one mode, though this had the disadvantage that the fibre diameter must be no more than a certain minimum size. Eventually, 'single mode' fibres came to be widely used but only when it was discovered how to join (splice) fibres with core diameters of less than 5 μm and how to launch light into such minis-cule fibres. Secondly, it was possible to make use of the fact that modal dispersion and material dispersion may work in opposite senses and thus tend to cancel one another, a feature of the long wavelength spectral region. Thirdly, it was desirable to choose a light source with a narrow 'linewidth' – that is one containing a narrow spectrum of wavelengths. In general, lasers show much narrower widths than LEDs but it is also possible to work with 'single mode' lasers which have the narrowest possible linewidths. We shall see how these vari-ous aspects came into play in the development of fibre systems and the demands made on the necessary semiconductor devices.

We might begin by returning to the question which first led us into this discussion – what are the relative advantages and disadvantages of laser sources compared with LEDs? From the viewpoint of fibre dispersion, we are concerned with the spread of wavelengths emitted by the two sources – that is, their respective linewidths – but the comparison is compli-cated by the different nature of their outputs. A GaAs LED emits a single broad line with a width of about 30 nm, whereas the laser emits a number of sharp lines (longitudinal modes – see Chapter 5), spread over about 3 nm. While the individual lines may be extremely sharp, with widths as small as 10^{-3} nm or less, in a multi-mode laser, it is the spread of modes which determines the pulse broadening effect. Thus, the laser is about an order of magnitude better than the LED in this respect. Summarising the various aspects of the comparison, we can see that LEDs may provide acceptable performance over relatively short distances and at relatively small bit rates, which makes them suitable for local area networks – around an office, between buildings, within a small town – but for anything more demanding it is essential to use a laser. And, for the absolute minimum dispersion, it is clear that a laser which emits in just a single longitudinal mode is far and away a better bet. We shall come back to this in a moment.

The first commercial fibre systems built in 1977 were designed to transmit only 8 Mb/s – over distances of order 10 km. This provided a (bit rate) × (length) product $BL = 100$ Mb/s km but the bit rate was imme-diately raised to 140 Mb/s, giving $BL = 1$ Gb/s km. This performance was closely in line with that then available from co-axial cable but coax was rapidly left behind as fibre technology developed, the BL product increas-ing by a factor of two each year (shades of Moore's Law for integrated circuit complexity?) – in 2000 it had reached the staggering figure of 10^6 Gb/s km, something like 10^4 Gb/s through 100 km of cable. And results reported in research were generally at least a factor ten greater! How was it done? It involved many steps, a move from the GaAs wavelength of 880 nm, first to 1.3 μm where dispersion was minimised, then to 1.55 μm where fibre loss was minimised, then to dispersion-shifted fibres where both loss and dispersion were minimised at 1.55 μm wavelength, then to wavelength-division-multiplexing, then to time-division-multiplexing, then to the use of fibre amplifiers (rather than electronic repeaters), then to all-optical processing of signals and, somewhere in and amongst all the other developments, optical pulses were generated by optical modulators, rather than the original switching of laser current. To describe all this in its amazing detail would require far more space than is available here – I shall try to pick out the most important factors, as seen from the semiconductor device angle.

By far the most important development concerned the shift to the longer wavelengths of 1.3 μm and 1.55 μm. As we saw earlier, when the advantages of using these wavelengths were first appreciated, there were neither sources nor detectors available and it was necessary to invent a completely new material system to meet these demands. It began with an innovation in a completely different field. In 1972, George Antypas at Varian Associates was struggling to develop a photocathode for an infra-red imaging tube which might extend the sensitivity to longer wavelengths than could then be detected and he came up with a brilliant solution in the form of a new alloy material, indium gallium arsenide phosphide (InGaAsP) which could be grown on an InP substrate. Growing this alloy by liquid phase epitaxy, he had demonstrated a cathode sensitive at 1.1 μm wavelength (a band gap of roughly 1.1 eV) which encouraged others to think in terms of even longer wavelengths for possible communication lasers. This material was of special interest because of its wonderful flexibility. Starting from InP, it was possible to replace some of the phosphorus with arsenic, which had the effect of increasing the lattice constant, and some of the indium with gallium, which had the effect of reducing the lattice constant. So, by selecting appropriate ratios of arsenic and gallium (roughly twice as much arsenic as gallium), it was possible to maintain the lattice constant at the InP value while varying the resulting band gap from a minimum value of 0.75 eV (for $In_{0.55}Ga_{0.45}As$) up to a maximum value of 1.34 eV for InP. The alloys which correspond to wavelengths of 1.3 μm and 1.55 μm are, respectively:

1.3 μm In 0.72 Ga 0.28 As 0.61 P 0.39
1.55 μm In 0.59 Ga 0.41 As 0.90 P 0.10

Both of which could be grown on InP substrates with considerable success (provided considerable care was taken to control the alloy composition – no trivial matter!). It was also possible to grow lattice-matched alloys with somewhat larger band gaps to serve as carrier-confining layers and optical waveguides – ideally suited to DH laser making. The initial work followed the Varian example and employed LPE and, by 1980, CW lasers were operating at 1.3 μm (and, in some instances, at 1.55 μm) in many laboratories around the world – NTT and KDD in Japan, RCA, Bell and MIT in America and STL in England, for example. However, it soon became clear that this was not the best way to grow DH lasers and the MOVPE (metal organic vapour phase epitaxy) method gradually took over. Much of this MOVPE pioneering work was done in France at the Thomson CSF laboratory and, by 1983, they had achieved CW operation of laser diodes with threshold current densities as low as those measured on comparable GaAs lasers. After 1990, no other growth method gets so much as a mention in the scientific literature (MBE, which might have mounted a serious challenge, has generally proved unsuitable for the growth of phosphorus-containing compounds).

As an aside, I should emphasise that we have, here, yet another wonderful example of the vital contribution made by materials scientists to the development of new technologies and it would be reprehensible not to comment on it. Without such new materials, progress in many different spheres would be impossible and one can only admire the ingenuity and innovative skill that goes into satisfying a wide range of demands from the device and systems people. Developing low loss glass fibre material was admirable in itself but the development of long wavelength lasers was even more so. Imagine the incredible skill required to control the composition of a four-component alloy so as to deposit five different layers (p-type light confining layer/p-type carrier confining layer/active layer/n-type carrier confining layer/n-type light confining layer), while achieving accurate wavelength control and then doing it again and again to provide reproducible laser performance! I feel we should all take our hats off to those responsible!

In the mid 1980s, the move to long wavelengths was proceeding with some urgency as long-haul communications assumed increasing importance and the demand on bit rate grew and grew (people wished to exchange more and more information) so there was inevitably some concern for the reliability of the new laser devices. As we know, GaAs DH lasers took something like a decade to achieve acceptably long operating lifetimes and there was a worry that any similar problems with their long wave cousins might hold back the communications flood for a commercially unacceptable time. Fortunately – and here we have an example of fortune smiling benevolently on the hard-worked technologist – it didn't happen. Progress with InGaAsP lasers was surprisingly rapid. They were found not to be susceptible to the various failure mechanisms which dogged GaAs lasers (possibly because the energy of the emitted photons was so much smaller than in the GaAs case). The one negative aspect of these long wavelength devices was a significantly greater dependence of threshold current on ambient temperature. It was a well-known feature of GaAs lasers but their temperature coefficients were relatively small and posed little in the way of practical problems but long wavelengths brought with them much more serious temperature-dependences. This turned out to be a direct consequence of the longer wavelength and simply had to be accepted – it meant that some care was necessary to keep the ambient temperature within reasonable limits, a minor, but irritating restriction for devices which inevitably were to be used in widely different circumstances. But, if the device technologist thought he could relax with a job well done, he should have known better. The history of human progress suggests that advanced technology never stays put for long – no sooner is one 'impossible' challenge met, than the next one appears over the horizon, and optical communication was never likely to prove the exception. Once the challenge of low fibre loss was overcome by the move to longer wavelengths, the need to minimise dispersion became paramount. It was not possible to work precisely at the wavelength where fibre dispersion was actually zero, so the device designer was faced with a demand for a single mode laser which, as we mentioned earlier, was characterised by a very much narrower spectral linewidth than that appropriate to its multi-mode parent. To reiterate, the multi-mode laser emitted in a number of longitudinal modes covering a spectral width of order 3 nm, whereas each individual mode could be at least 10,000 times narrower (the precise value depending on how hard the laser was driven – the harder the drive, the narrower the line).

The problem of multi-mode emission stems from the nature of the Fabry-Perot cavity formed by the two reflecting mirrors on the ends of the laser structure. These mirrors reflected more or less equally well at all wavelengths – in order to make a laser which emitted in only one mode it was necessary to include some wavelength-selective feature and this took the form of a Bragg mirror. We discussed the concept of the Bragg reflector in Chapter 6 in connection with the VCSEL (vertical cavity surface emitting laser). The idea is based on the fact that a step in refractive index between two semiconductor layers acts to reflect a small fraction of light passing through it and, if a large number of such steps can be arranged in a regular sequence, the total reflection may add up to something approaching 100%. What is more important for our current interest, is that a regularly repeating sequence of layers can be designed to reflect only a single wavelength, the wavelength in question being determined by the periodicity of the layers (strictly, the periodicity of the refractive index steps). The idea, in this case, was for the Bragg mirror to select the laser wavelength, rather than a Fabry-Perot cavity.

All very well – but there was a difficulty. In the case of the VCSEL, it was a straightforward matter to grow the appropriate layers epitaxially because the laser emitted light normal to the surface. In the case of the communications laser, light was emitted from the edge – that is, it travelled parallel to the surface – and the question arose as to how one could form a Bragg mirror to reflect in a direction parallel to the surface. The answer was simple (even though the question wasn't!) – a layer of InP was first grown in the ordinary way, then it was etched so as to remove a sequence of narrow regions in lines running across the sample, then the missing gullies were filled in by growing a second layer of a different composition. When looked at in a direction normal to the lines and in the plane of the sample, the required index steps were plain to see. What was more, the stimulated light within the laser cavity saw these steps and the laser emitted at a wavelength appropriate to the Bragg periodicity, rather than at several wavelengths determined by a Fabry-Perot cavity. It was only necessary to form a pair of Bragg mirrors, one at each end of the laser, to replace the plane mirrors used in the conventional laser. Such an arrangement was known as a 'distributed Bragg reflector', or DBR laser. An alternative which has been widely used is referred to as a 'distributed feedback', or DFB laser, in which the Bragg reflector runs right through the active region and reflection occurs in small amounts at each index step, resulting in continuous feedback, rather than merely at the ends of the structure. An important advantage of the DFB laser is the excellent control of wavelength with only a small dependence on ambient temperature. As we shall see, this was of special merit when wavelength division multiplexing became commonplace and lasers were required to maintain a constant position within the spectrum of wavelengths employed.

Until the middle of the 1980s, long wavelength lasers appear to have been limited to the conventional DH structure and, while quantum well GaAs lasers had demonstrated numerous advantages over corresponding DH lasers, little attempt seems to have been made to incorporate such structures into the communications field. It may reasonably be assumed that the long wavelength people had quite enough to keep them occupied, without adding the initials MQW to their worksheet. However, it just had to happen sometime and it all came to pass in a few short years round about 1985. Given the flexibility of the InGaAsP material system, it was straightforward to incorporate quantum wells and, while the opportunity presented itself, the growers took advantage of a suggestion from the University of Surrey. Professor Alf Adams pointed out that strain in the well could be used to improve laser performance in respect of reduced threshold current and improved dynamic performance – it was possible to modulate the laser at higher frequency, thereby facilitating higher bit rates. The idea was deliberately to mismatch the quantum well material so it was strained in one sense, while mismatching the barrier material in the opposite sense so that the average strain was close to zero. Such subtleties merely added a few more straws to the crystal growers' burden but the resulting gain in performance was well worth the effort – threshold currents came down and modulation frequencies went up. Communication engineers, seeking ever greater bit rates and ever longer transmission distances were only too pleased at such news. It was an excellent example of the way in which basic science, materials technology and device performance are inextricably linked. The thorough understanding of the physics of laser function which led to Adams' suggestion and the crystal growers' ability to produce whatever new device structure might be required represented a perfect marriage whose progeny was innovative device design. We now take it all for granted but it was an exciting development at the time.

The next 'unfair' demand to burden the over-stressed crystal grower came from the rapid development in wavelength division multiplexing towards the end of the 1990s. The idea involved setting up a number of lasers at fixed wavelength intervals and modulating each at some prodigious bit rate – typically 10 Gb/s – then transmitting the information from each one down a single fibre. Suppose, for example, that the wavelength band near 1.55 µm is limited to the range 1.48–1.60 µm. Each individual channel would require a bandwidth of 10 GHz and the separation between channels should be at least 40 GHz, to avoid 'cross-talk' (mutual interference between adjacent channels), allowing approximately 400 possible channels. The overall bit rate available thus amounted to 4 Tb/s (1 Tb/s = 1000 Gb/s = 10^{12} b/s) but it demanded 400 separate lasers, each operating at its own wavelength which was specified to an accuracy of a few parts in 10^5! What was more, the wavelength had to remain within tight limits even though the ambient temperature might drift upwards or downwards by as much as 15 °C. To make 400 different single-mode lasers to such a specification scarcely merited serious thought – nor did the spares problem if each one had to be backed up by a ready replacement! It was clear that the only practical solution lay in designing lasers which could be tuned over the range of wavelengths required – at least, they could then be all alike. But tuning over such a bandwidth was a completely new problem. How was it to be achieved? Two quite different approaches were adopted, one based on the DBR structure that we discussed above, the other based on the VCSEL, which we have yet to discuss as a long wavelength device.

As we saw earlier, the wavelength of a DBR laser is controlled by the period of the Bragg mirror but we omitted to add that it also depends on the refractive index of the semiconductor material from which that mirror is made. This latter observation led to a very neat idea, whereby the index could be changed by the injection of free carriers into the mirror region. The laser was made in two sections, one being the active region with a standard plane mirror at one end, the second being a Bragg mirror at the other end. Two independent current leads were provided, one to drive the laser, the other to inject free carriers into the mirror region. This latter current was used to vary the refractive index and thus the operating wavelength. It was all very simple and worked remarkably well (given one or two minor modifications to ensure an acceptably smooth tuning curve). It also provided the possibility of electronic control to adjust the wavelength to that of an external standard, if so desired.

In describing the VCSEL in Chapter 6, we saw that, because the laser cavity was very short, by comparison with the 500 μm or so of the conventional laser, the longitudinal laser modes were so far apart in wavelength that only one lay within the gain spectrum and therefore only this single mode was excited. In other words, the VCSEL was automatically a single-mode laser and this feature would surely endear it to the fibre optic community. In principle, it was also much more straightforward to make because this involved simply the growth of various epitaxial layers – there was no need for the complicated etching and infilling needed to form the Bragg mirror in the conventional structure. There was, however, a drawback – it turned out that the refractive index variation available within the InGaAsP material system was too small for making effective Bragg mirrors and, for several years, no long wavelength version of the VCSEL appeared possible. Nothing daunted, the semiconductor technologist finally found a way to bypass the problem by incorporating AlGaAs/GaAs Bragg mirrors into the structure. Initially this was achieved by using wafer-bonding techniques (gluing different sections together!) but, later, the whole structure could be grown epitaxially by incorporating strain-relieving layers. It was all a trifle delicate but eventually everything worked and it provided the basis for a highly tuneable long wavelength laser. The tuning method used was quite remarkable – it involved attaching the top mirror to the device only at one point, so as to form a cantilever, allowing the spacing between it and the main structure to be varied. The variation could be achieved simply by applying a modest voltage – because the III–V compounds are piezoelectric, this voltage produced an appropriate force which bent the cantilever as required. Large tuning ranges were obtained rather readily.

While the development of tuneable lasers could solve the problem of WDM, the corresponding problem of TDM set yet another challenge to the device engineer. We have, so far, assumed that generating the necessary short pulses of light could be achieved simply by modulating laser drive current but it turned out that this was limited by two unexpected phenomena. These rejoice in the names of 'relaxation oscillations' and 'frequency chirping'. In a word, attempts to generate very short pulses resulted in the laser output oscillating up and down at a frequency of several giga-Hertz while the wavelength jumped about by a small amount, thus increasing the effective linewidth of the emission. Very roughly, these effects limited direct modulation to frequencies of 10 GHz, or bit rates of about 10 Gb/s, performance which was acceptable until sometime in the late 1980s. However, by 1995, bit rates of 100 Gb/s became the norm – something had to be done, this 'something' being the introduction of separate modulators which modulated the output from a CW laser. Not only did this have the effect of increasing bit rates, it also allowed the laser to operate at its true intrinsic linewidth, which minimised wavelength dispersion in fibres. As with the push for tuneable lasers, it increased complexity, and therefore expense but satisfied the urgent demand for greater BL product.

The first modulators used were not semiconductor devices at all but depended on a long-established phenomenon known as the electro-optic effect (discovered by Kerr in 1876). Certain crystals such as lithium niobate ($LiNbO_3$) had the ability to rotate the plane of polarisation of a light beam when a voltage was applied to them, and the amplitude of the light could be modulated by combining the crystal with a pair of 'crossed polarisers'. The first polariser produced a light beam polarised in a defined plane and the second polariser, which was set at right angles to the first allowed no light through. However, if a $LiNbO_3$ crystal was interposed between the polarisers, it could rotate the plane of the light so as to allow light through the second polariser. With the right voltage applied, so as to produce a rotation of exactly a right-angle, all the light would be transmitted – with zero voltage, none of it. The question remained; how fast could the combination switch the light on and off? Suffice it to say that a considerable effort was needed to achieve the necessary rate but, by 1995, speeds of 100 GHz were available – the only drawback being one of clumsiness. The niobate crystal was about a centimetre in length, which compared somewhat unfavourably with the half-millimetre length of the laser. It was an adequate short-term solution to an urgent difficulty but the semiconductor technologist was not disposed to leave it at that. Not only must there be a neater solution, it should also be one which could be fully integrated with the laser. It had come to be recognised that optoelectronics was at a stage of development comparable to that of transistor electronics in the 1950s, prior to the invention of the integrated circuit, and the analogy only emphasised a growing need for what came to be called 'photonic integration'. It wasn't good enough only to make something work – it had to be done cheaply and efficiently, as well!

The semiconductor answer to the $LiNbO_3$ modulator made its appearance during the 1990s in the shape of a 'quantum well electro-absorber'. It had been known since the early 1980s that optical absorption by a quantum well could be changed by applying an electric field across it, a phenomenon referred to as the 'quantum confined Stark effect'. Application of a suitable small voltage shifted the absorption edge downwards in energy. Suppose, therefore, that a beam of light with a photon energy just below the quantum well absorption edge is sent through the well. At zero voltage no absorption occurs but, under the influence of an applied voltage, the well absorbs strongly and we have the capability of making an optical modulator. The effect was first discovered in the AlGaAs/GaAs system but there was every reason to suppose that the InGaAsP system could be persuaded to function in the same way and, by the middle of the 1990s, excellent modulators had been demonstrated using strained InGaAs wells with InGaAsP barriers. This had the obvious appeal of employing the same material used to make long wavelength lasers so it should be straightforward to integrate laser and modulator in the same slice of material. Alas, it wasn't! The idea was to use a conventional edge-emitting laser with a modulator employing the same quantum well, alongside it. Notice that this meant the light beam travelled along the well, rather than through it and this provided a much longer absorption length (typically 100 μm), which represented an important advantage. However, utilising the same quantum well for laser emission and for electro-absorber meant that laser light was strongly absorbed in the absence of an applied voltage and applying a voltage only made matters worse. To make a modulator, it was necessary to arrange the absorber band edge to lie slightly above the emission energy. Was there any mechanism by which the band gap of the well could be shifted in one region of the slice, while leaving it unchanged in another? Once again, experience with AlGaAs/GaAs pointed the way. Subjecting a quantum well to suitable heat-treatment resulted in inter-diffusion of the well and barrier material so as to change the shape of the well and push the band edge to higher energies but this alone could provide no solution because both laser emission and absorber band edge were shifted equally. The key lay in the discovery that the effect could be influenced by depositing a film of silicon oxide on the sample surface – immediately underneath the oxide, the inter-diffusion occurred very much more rapidly so here, at last, was a means for differentiating between laser and absorber. Arranging for the silicon oxide to cover only the absorber, shifted its absorption edge slightly above that of the laser emission and allowed both to work as desired. Photonic integration had taken its first faltering steps.

An even tidier method of achieving high bit rates has been developed during the early years of the twenty-first century, making use of the polarisation properties of long wavelength VCSELs. The output of a typical VCSEL has a circular pattern which makes it ideal for launching light into a circular fibre but it was noticed in the early days of their development that the beam showed variable polarisation. In a perfectly circular device there is nothing to determine the plane of polarisation of the emission and in real devices there was a tendency towards random switching between different polarisation directions. It was quickly recognised that, if this could be brought under control, one would have the basis for an amplitude modulator by the simple expedient of combining the VCSEL with a polariser. It was only necessary to introduce some asymmetry into the current flow through the device by, for example, introducing two pairs of current contacts set at right angles to one another, then arrange to switch the drive current between these pairs and achieve rapid switching between mutually perpendicular polarisations. Lining up the polariser plane with one of these polarisations completed the process. This simple expedient resulted in pulse rates of over 40 Gb/s.

Our discussion, so far, has concentrated entirely on the generation of the optical signal and the perceptive reader will probably be wondering what happens to it when it reaches the end of the line. Given that telephone communication demands, an analogue electrical signal, it is clear that some more or less complicated electronic signal processing is essential before the voice waveform can be directed to the receiver. It begins with a photo-detector which converts the optical pulse train into corresponding electrical pulses, then these must be converted into their analogue equivalent by a digital-to-analogue (D–A) converter. This latter can be thought of simply as a piece of clever integrated circuitry – our concern will be to examine the properties of the photo-detector which was developed along with the optical sources already described. Its operation depends on the absorption of light by a semiconductor sample, thus converting an optical pulse into a pulse of injected electrons and holes within the sample. If left to their own devices, these electrons and holes would simply recombine (most probably with the generation of heat) so something has to be done to separate them before the inherent information is lost. The method employed by most photo-detectors is to provide a strong electric field which drives electrons in one direction and holes in the opposite direction, thus generating a 'photo-current'. This represents an electronic signal which can be amplified and used as input to the D-to-A convertor. In fact, we have already come across one form of detector in the shape of the silicon photo-diode discovered by accident at Bell Labs just prior to the Second World War. Mervin Kelly was concerned to keep the idea secret in the interest of future application to Bell's long distance communication plans. He correctly foresaw its use as a solar cell to power remote repeater stations but he certainly could have had no notion of its future use in fibre optics.

Let us briefly consider the properties required of a photo-detector. Firstly, the semiconductor must have a band gap smaller than the energy of the photons to be detected, otherwise the light would steal through it unobserved. In principle, this gap could be a great deal smaller than the photon energy but there is a good reason for it to be only slightly smaller. In the first chapter of the book we learned that free electrons and holes are generated by thermal excitation across the energy gap and it should be clear that this represents an unwanted background, interfering with the optical generation which constitutes the signal. The smaller the band gap, the greater the thermal generation rate, so it is obviously expedient to choose a semiconductor with a band gap only slightly smaller than the energy of the photons to be captured. Thus, in the early examples of fibre optic systems, based on GaAs lasers (photon energy of about 1.4 eV), it was convenient to use a silicon detector (band gap about 1.1 eV). It was also a matter of convenience in so far as silicon photo-detectors were already available off the shelf in a range of formats. Such was not the case for the long wavelength spectral region (photon energies of 0.95 eV and 0.80 eV). The need for a new material for long wavelength sources was equally pressing when it came to detectors and, unsurprisingly, the choice was also based on the InGaAsP system. In particular, the material used to absorb the incoming photons was the InGaAs alloy, matched to an InP substrate, which has a band gap of 0.75 eV.

The second requirement concerns speed of response. In order to translate faithfully between optical and electrical pulses, the detector must be capable of responding to the optical signal in a time roughly 10 times shorter than the pulse length and this is obviously intimately related to the desired bit rate. In the days of GaAs lasers and bit rates of 10 Mb/s (10^7 b/s), this implied a relatively undemanding response speed of 10^{-8} s (one-hundredth of a microsecond or 10 ns) but with modern rates pushing up to 10 Gb/s (per fibre), the speed has reached a rather frightening 10^{-11} s (one-hundredth of a nanosecond or 10 ps). What is involved in achieving such performance? The optical absorption process occurs with negligible time delay – the important criterion is the time taken for the burst of electrons and holes to be swept out of the detector and into the external circuit, where they can influence the amplifier waiting to receive them. In a typical detector, where the electric field which separates the electrons from the holes is associated with a p–n junction, the field has a value of 10^6 V/m (or 1 V/μm) and the extent of the junction is about 1 μm. Free carriers are accelerated in this field to a terminal velocity of about 10^5 m/s (10^{11} μm/s) so they take about 10^{-11} s to cross the high field region, which is gratifyingly fast. What happens when they reach the field-free region beyond the junction? Electrons end up in the n-type material where they repel some of the electrons already present and push an appropriate number out of the end of the semiconductor material and into the external circuit (rather like a dangerously dense crowd at an old fashioned football match before the introduction of seats-for-all) – ditto for holes on the other side. This mutual jostling process (which is known as 'dielectric relaxation') takes place in a time which is short, even on the scale of 10^{-11} s, so the net response time is just the time taken to separate the carriers. Clearly, speeds of the required magnitude are available, even for the most demanding cases.

However, there is yet another factor to be considered. The above calculation implies that all the carriers are generated (that is – all the photons are absorbed) within the junction region and that may not always be possible. Consider the case of a silicon photo-diode. Remembering that silicon is an indirect semiconductor, which implies rather weak optical absorption, it turns out that only a small fraction of the light is absorbed within the micron length of the junction region. In practice, it takes something like 30 μm of silicon to absorb all the light, so a simple p–n silicon photodiode would be a very inefficient detector (and, incidentally, very much slower than the above calculation suggests because free carriers would be obliged to diffuse to the junction before being swept across). The secret of designing an efficient silicon detector is to expand the high field region so as to match the absorption length of the light and this involves making a modified structure, consisting of n- and p-regions separated by an undoped (intrinsic) region of thickness comparable to the absorption length. It is known as a p–i–n diode. The downside of this modification is a slower response because it implies a longer sweep-out length. Suppose, for example, that we use a compromise thickness for the i-region of 10 μm – this implies a length 10 times larger than that appropriate to the simple p–n junction and therefore a response time 10 times longer. This would, of course, be quite acceptable for the less demanding needs of the early fibre systems and p–i–n silicon photo-diodes were widely used. As we have already made clear, silicon is not appropriate for the long wavelength region, on account of its band gap.

The choice of InGaAs for long wavelength detectors was a good one not only because its energy gap matches the appropriate photon energies but also because it has a direct gap, with an absorption length of just a few microns. However, this did present a somewhat different problem. In a conventional p–n junction in a sample of InGaAs, most of the light would be absorbed in the doped material before it could reach the junction and we should have a situation similar to the silicon p–n junction case. The solution to this problem was not, however, far to seek – it involved epitaxial growth of a double heterostructure on an InP substrate of n^+–InP – i–InGaAs – p^+–InP. The InP acted as a window through which the light passed freely, to be strongly absorbed in the undoped InGaAs. The structure behaves rather like a silicon p–i–n diode but with a much narrower high field region, thus providing the necessary high response speed of 10 ps. However, the demand for faster and faster bit rates puts even this performance under pressure and has stimulated the development of even greater response speeds, based on the use of narrower absorption regions. One example employs an InGaAs film only a tenth of a micron thick but placed inside a Fabry-Perot cavity which sends the incoming light beam backwards and forwards through the film to enhance the absorption. It depends on the use of a pair of Bragg mirrors similar to those used in VCSEL lasers. One sometimes wonders whether there will ever be an end to this particular race!

Finally, we must briefly mention recent trends which aim to rationalise the way in which communications traffic is processed. The overriding impression of the fibre business is that it all happened in too great a hurry, obliging each and every panic to be dealt with empirically, and what is now needed is a period of relative calm in which system design can take on a greater element of long-term planning. We have already referred to one essential aspect of this – a move towards photonic integration, whereby different functions are performed on a single slice of semiconductor – but even more significant are attempts to process signals optically, rather than electronically. A good example of this is the use of fibre amplifiers at repeater stations, rather than electronic regenerators. For a long time, these stations functioned by first converting the optical signal to electronic form (using a photo-diode), then regenerating the electronic signal, before converting it back into its optical consciousness (with a semiconductor laser). This was all very well prior to the introduction of wavelength division multiplexing but once this became standard procedure it placed tremendous pressure on the required circuitry – each channel had to be dealt with independently, multiplying the complexity of the electronics considerably. Looked at dispassionately, it was a veritable nightmare! Fortunately, the solution was already at hand, in the shape of a fibre amplifier, invented by David Payne at the University of Southampton, as long ago as 1987. This amplifier was made by introducing a small amount of the rare earth element erbium into the fibre core and pumping it optically in much the same way that Maiman's original ruby laser had been driven. Remarkably, it amplified at wavelengths around 1.55 μm, the very region where fibre losses were at their minimum and where WDM experiments were soon to be concentrated. Payne's original work used an inconvenient argon gas laser as pump but it was soon clear that a semiconductor laser emitting at 980 nm wavelength would represent a huge improvement. This took the form of a specially developed high power InGaAs/GaAs strained quantum well laser and, almost at once, the fibre amplifier became a hugely practical component of many fibre systems. It has the great advantage of amplifying over a wide band of wavelengths, which avoids the need for each channel to be individually processed at every repeater station – the resulting reduction in complexity is dramatic. In 1996, the TAT-12 trans-atlantic telephone cable used fibre amplifiers to considerable advantage, proof, if it were needed, that the technology had successfully met all the necessary tests.

More recent advances in optical signal processing involve the development of photonic integrated circuits based on silica waveguides (see Fig. 8.5). A film of silicon dioxide is deposited on an active semiconductor slice and patterned in narrow stripes which serve to confine light waves and direct the signals to any desired part of the sample. They can be thought of as optical equivalents of the metal lines which conduct electrons on conventional electronic integrated circuits and have the advantage that light can readily be coupled to and from optical fibres. Optical amplifiers are made in the shape of InGaAsP laser structures immediately beneath the guides (but without the mirrors which provide feedback in the conventional laser oscillator). These function in very much the same manner as the fibre amplifiers referred to above, though they are very much smaller, it being possible to make several on a single semiconductor slice. They can be used in various ways. For example, it is possible to change the wavelength of an incoming signal by using it to cause gain saturation in an amplifier which is amplifying a second CW optical wave train at a different wavelength. A positive signal pulse from the first input saturates the amplifier gain and results in a low signal level

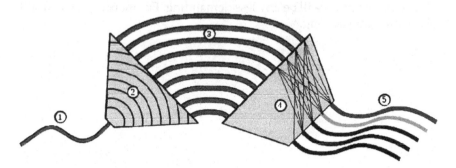

Arrayed-Waveguide-Grating.svg (SVG file, nominally 680 × 241 pixels, file size: 27 KB)

FIG. 8.5 Schematic diagram of a photonic integrated circuit used in wavelength division multiplexing (WDM). The heavy lines represent optical waveguides. Light input along waveguide 1 contains (say) five different wavelengths, each carrying its own data signal. After transmission through the circuit, these five wavelengths are separated and sent down the five separate output waveguides 5. This process is known as de-multiplexing. The circuit can also be used in reverse to send five different wavelengths down a single fibre, operating, in this case, as a multiplexer. Note that the processing is done entirely with light – there is no need to convert any signal into electronic form. Courtesy of WikiMedia.

at the second wavelength – while a zero level allows full gain and results in a high level at the second wavelength. Thus, the information on the first wave is transferred to the second wavelength – admittedly, it is upside down but that is of no serious consequence. Such a process can readily be accomplished at a bit rate of 40 Gb/s. Other functions that have been demonstrated include optical regeneration, optical switching between different output ports, optical memory (i.e. storage of optical signals) and logic operations such as are used in computers. Such photonic circuits clearly have considerable possibilities in simplifying and rationalising the design of optical systems. Regeneration, in particular, is very important for long haul communications – while the use of fibre amplifiers is extremely helpful, noise inevitably builds up and signal regeneration is still essential to really long transmission distances.

As I have hinted already, the fibre business is entering a period of consolidation and, while it is likely that the demands on the technology will continue to grow, there can be little doubt that future innovation will concentrate more on convenience and cost saving, rather than out and out acceleration of bit rate. The fibre revolution of the late twentieth century was certainly a wondrous thing, involving a rate of technological development quite breathtaking in its urgency, and arguably, changing lives as dramatically, even, as the integrated circuit had done in its day (try to imagine an internet at 1% of the speed!). Though the new millennium can expect to witness rather more sedate progress, this is not to say the required innovation will be any less demanding. Future progress will still be well worth watching.

CHAPTER 9

Seeing in the Dark: Image Tubes and Thermal Viewers

In the previous chapter, we became familiar with one particularly valuable application of infra-red (IR) radiation. A tiny band of wavelengths in the range 1.3–1.6 μm has enabled a quite remarkable global network of communications. But, important though it certainly is, this represents only one of many uses to which the IR has been turned. Night vision, seeing through smoke, thermography, intruder alarms, remote control, missile guidance, industrial process control, meteorology, climatology, astronomy, spectroscopy, pollution monitoring, art history and IR photography are a few others which spring readily to mind, nor should we overlook the best known of all its applications – keeping ourselves warm. Apart from the last, most of these depend on the use of semiconductors, either as sources or as detectors so it behoves us to explore further the realms of the IR – as we shall see, they provide a surprisingly wide spectrum of interest.

If you have ever watched with fascination those eerie greenish-grey videos of unsuspecting wild animals going about their nocturnal business, you have experienced the imaging power of IR radiation. If you have ever pressed the appropriate button on a television remote control to witness the daytime habits of unsuspecting humans (can there be anyone who hasn't?), you have harnessed its silent servitude to your wishes. If you have installed a burglar alarm against unsuspecting intruders within your home you are, once again, relying on its stealthy subservience to work to your auxiliary advantage. If you have ever placed yourself conveniently in front of a glowing electric fire you have sensed the warmth of its invisible presence. But, most of the time, most of us go about *our* business without even being aware of its existence. Indeed, right up to the beginning of the nineteenth century, no-one even *knew* of its existence.

Semiconductors and the Information Revolution: Magic Crystals that made IT happen
DOI: 10.1016/B978-0-444-53240-4.00009-X

It was in 1800 that the Hanoverian émigré, William Herschel (see Fig. 9.1) first demonstrated the existence of radiation at wavelengths beyond the red end of the visible spectrum. He had come to Bath as a struggling musician and became fascinated by the delights of astronomical observation which, eventually, led him to discover the planet Uranus and to catalogue an impressive array of nebulae. It brought him the favour of the King, George III (another Hanoverian in exile) who famously paid for the construction of Herschel's master-work, a gigantic 40-ft reflecting telescope (from which an astronomer friend, Giuseppe Piazzi fell and broke his arm!). Herschel was not only a thorough observer but a skilled maker of optical equipment and it was while studying the heating effects of the different visible wavelengths in a beam of sunlight, dispersed by transmission through a prism, that he noticed the warming presence of radiation in what we now call the IR part of the spectrum. The amount of heat increased slightly from violet to blue to green, then decreased from green to yellow to red, but, to Herschel's surprise, it continued well beyond the red. There was

FIG. 9.1 A contemporary portrait of William Herschel, the Hanoverian émigré who came to Britain in 1756 as a talented musician and developed a lifelong interest in astronomy. He is remembered mostly for his discovery of the planet Uranus in 1781, following which he was elected to Fellowship of the Royal Society and, the following year, to the post of 'King's Astronomer'. It was almost as a sideline that in 1800 he discovered the existence of radiation in the infra-red part of the electromagnetic spectrum. Courtesy of WikiMedia.

obviously more to the electromagnetic spectrum than met the eye! (though its origin as 'electromagnetic' was not appreciated until James Clerk Maxwell published his famous equations in 1864).

Herschel had detected the presence of radiation using the blackened bulb of a standard mercury-in-glass thermometer which proved adequate for his purpose, if somewhat slow to respond and possessing only modest sensitivity. The next important advance came with the use of the thermocouple detector by two Italian physicists Leopoldo Nobili and Macedonio Melloni in 1831, which provided improved sensitivity and greater convenience. (A thermocouple consists of a pair of junctions between two different metal wires, one kept cold, while the other 'sees' the radiation to be detected. This generates an electric voltage proportional to the temperature difference between the junctions.) They were able to demonstrate reflection, refraction and polarisation effects similar to those of visible light but, perhaps of even greater significance, were the experiments by two French physicists A.H.L. Fizeau and J.B.L. Foucault who discovered interference effects in 1847, thus confirming that IR radiation was a wave motion just like light. This led the American astrophysicist Samuel Pierpont Langley, in 1880, to develop a diffraction grating for dispersing IR 'light' into its constituent wavelengths and allowed these wavelengths to be accurately measured. Langley also developed yet another type of detector, based on the fact that the electrical resistance of metals (platinum in his case) varies with temperature – he called it a bolometer. Initially, IR studies had been confined to near-IR wavelengths but by Langley's time they had been extended as far as 7 μm and by the turn of the century to well beyond 100 μm. Indeed, it gradually became clear that there was no distinction to be made between long wavelength IR radiation and short wavelength radio waves – the two finally met during the 1930s, somewhere in the microwave region.

The first attempts to apply IR techniques to solving practical problems can be traced to the First World War, when serious efforts were made to develop military systems. Perhaps the most significant development in this context was the use of photoconductivity, in the semiconductor material thallium sulphide, to detect radiation in the near-IR part of the spectrum. This, so-called 'thalofide cell' was the brain child of an American, T.W. Case, of the Case Research Laboratory at Auburn, New Jersey. Working for the U.S. Navy, Case discovered that thallium sulphide could be activated by careful oxidation to produce detectors with much improved sensitivity and response speed, compared with the earlier 'thermal detectors' which were essentially detectors of heat. Semiconductor materials such as this, worked in a fundamentally different manner, depending on the generation of free electrons and holes by absorbed photons (just like the fast photo-diodes we discussed in the previous chapter). In consequence, they came to be known as 'photon detectors' and were sensitive to a specific range of wavelengths determined by their band gaps, differing again from thermal detectors which were sensitive to heat radiation at all wavelengths. As we shall see, the future of IR systems was to depend heavily on the development of a range of semiconductor materials with properties appropriate to specific wavelength ranges.

Not a great deal happened between the Wars but the onset of the Second World War stimulated considerable further development, in particular the extension of the useful spectrum out to wavelengths of about 8 μm (by the introduction of new semiconductors such as the lead chalcogenides – lead sulphide, lead selenide and lead telluride), the design of acceptable thermal imaging systems and the use of IR photo-cathodes to enable night-vision image tubes. When combined with IR headlights, these tubes provided an indispensable aid to night driving and allowed military hostilities to proceed in relative darkness, militated by the aptly named 'sniperscope' rifle sight and 'snooperscope' surveillance equipment. However, bearing in mind that the semiconductor revolution was a post-war phenomenon, we should recognise that these wartime developments were based on an essentially empirical approach and the real application of modern science to IR systems was yet to materialise. For example, the only available lead salts were in the form of naturally occurring crystals whose impurity levels were totally uncontrolled. So, once it became clear, as wartime experience had demonstrated, that both near- and far-IR techniques could offer impressive gains in fighting capability, there was little doubt that the advent of new materials would be put to immediate use and military funding for appropriate IR technology would rarely be difficult to find. Even when the West was not actually engaged in war, it was usually preparing for it! But we must not forget that the availability of sensitive night-vision tubes for nocturnal observation of wild-life was an important consequence, as was, for example, the development of thermographic equipment for breast cancer screening and searching for hot spots in integrated circuits.

Both night-vision tubes and thermal imaging systems demanded improved semiconductor materials. In the former case, this led to the introduction of multi-alkali cathodes such as caesiated sodium potassium antimonide ($Na_2KSb(Cs)$) which, again, were developed on a largely empirical basis, followed by the so-called 'negative electron affinity' cathode, an offshoot of some esoteric research into the surface physics of compound semiconductors. In the latter case, artificial crystals of the lead salts (first successfully grown in the early 1950s) were followed by the development of the narrow gap indium antimonide, then those of two highly specialised 'long wavelength' materials in the shape of mercury cadmium telluride and lead tin telluride and, latterly, by the application of low dimensional structures. All these have been applied to the problems of detecting IR radiation with ever improving sensitivity and covering an ever widening range of wavelengths but we should acknowledge that, at the same time, they have also revolutionised the manufacture of IR sources. Prior to the 1960s, the only way of generating IR energy was a by-product of heating suitable materials to high temperature, which had the disadvantage of producing a wide spectrum of wavelengths, including visible and UV radiation which was usually unwanted and had to be removed with suitable filters. This changed abruptly following the introduction of the first GaAs lasers. Almost immediately semiconductor lasers were reported in the narrow gap materials indium arsenide and indium antimonide, then, gradually a wide range of other materials was brought to bear, providing monochromatic (single wavelength) sources over a range of wavelengths matching that available to detectors. Needless to say, the evolution of reliable, continuously operating room temperature laser sources did not occur overnight (evolution doesn't work like that) but gradually the technology developed until, today, there exist lasers at wavelengths all the way from the near-IR out to hundreds of microns. It is a fascinating story of new materials allied to inspired device design. However, we are allowing our collective imaginations to roam untrammelled – first, we need to learn a little about IR methods in order to appreciate some of the problems that had to be overcome.

First, we must return to the question of imaging. We take it for granted that our eyes form images of objects around us using visible wavelengths. The fact that sunlight peaks in intensity at a wavelength of about 550 nm in the green part of the visible spectrum led to our eyes evolving with appropriate sensitivity – they are limited to detecting wavelengths between about 400 nm in the blue and 700 nm in the red, an altogether minute portion of the electromagnetic spectrum, and mankind (particularly the military arm of the species) has long longed to be able to extend the range. But, if we are to form images using any other wavelength range, we have to rely on clever opto-electronic imaging systems which only became available in the twentieth century – the IR represents an important example. However, it is well to appreciate the distinction between two very different kinds of imaging. In normal vision, our eyes make use of visible light which is *reflected* from objects ('the scene') into our eyes and we see in colour because matter, in general, reflects different colours to different degrees. More correctly, matter *absorbs* different colours to different degrees, and what is not absorbed is, of course, reflected. A red shirt appears red because the shirt material absorbs the green and blue part of the visible spectrum much more efficiently than the red part, so predominantly red light is reflected into the viewer's eye. Very much the same phenomena occur in the IR and allow the formation of images via reflected radiation, provided it can be transformed into visible wavelengths. It is this latter function that is performed by the night-vision tube, though 'colour' in this context is a somewhat arbitrary concept and the resulting images are normally seen as 'green-and-black' – 'green' simply because the visible light is produced by a green-emitting phosphor. (As an aside, we may note that a Dutch military technique exists in which artificial colour is imposed on the image, based on the nature of the scene. For example, a woodland scene may be 'naturalised' by making tree trunks blackish, grassy ground greenish, shrubs brownish, a lurking fox reddish and so on. It is remarkably successful but obviously depends heavily on our being able to recognise the nature of the scene. In no way does it imply the existence of real colour in the IR image, itself.) Clearly, the ability to form a sharp image depends on the intensity of IR 'light' available. In one scenario, this may be provided from an artificial source such as the IR headlights referred to earlier but the drawback to this approach (seen from a military viewpoint) is that one's enemy can readily detect the fact that he is under surveillance. It may be preferable to rely on ambient IR radiation such as is available from the night sky – we shall look at this in a moment.

An altogether different type of imaging system relies on the fact that a body radiates an amount of thermal energy, depending on its temperature. This forms the basis for the guidance mechanism used in heat-seeking missiles – an aero-engine must, of necessity, be at a moderately high temperature, compared with its surroundings and the missile, equipped with a suitable IR detector, is programmed to home in on this source of heat. This simple detection method makes no use of imaging but similar techniques *can* be used to form images of terrestrial objects. A tank, for example, is likely to be at a temperature of, say 500 K, some 200 K hotter than the surroundings, and it radiates energy, accordingly. An imaging system based on a detector designed to absorb this thermal radiation has the ability to 'see' the tank against its surroundings. Similarly, human beings tend to be warmer than the trees in a wood and can similarly be caught on a suitable video – particularly if they move. Unlike the reflective imaging described above, this so-called 'thermal imaging' is quite independent of background illumination and depends simply on temperature differences between different parts of a scene. Thermography similarly detects the difference in temperature between cancerous cells and normal cells in the human body. Contrast in such images is considerably enhanced because the intensity of thermal radiation increases as the fourth power of the temperature, a relation known as Stefan's Law (after the Austrian scientist who proposed it). Again, there is no meaningful colour in the scene but many thermal imaging systems apply artificial colour by correlating different colours with specific temperature ranges. The results may often be garishly unreal but the technique certainly enhances image contrast.

From the viewpoint of seeing in the dark, these two types of imaging tend to employ very different wavelength ranges and, in consequence, very different imaging techniques. Systems relying on reflection of ambient IR radiation make use of the near-IR because, on a dark night, there tends to be considerably more ambient 'light' from the night sky in the region between about 0.8 and 1.3 μm, whereas thermal systems are based on the fact that radiation from a body at room temperature (300 K) peaks at a wavelength close to 10 μm, while a 500 K body has its maximum at 6 μm. Another important factor in designing thermal image systems concerns absorption of radiation by the atmosphere. It is obviously necessary to use wavelengths to which the air is transparent and there are two suitable 'windows' available, one between 3 and 5 μm and another between 8 and 14 μm. The vast majority of thermal imagers use one or other of these windows and therefore use detectors which operate either as photoconductors or as photo-diodes with band gaps in the region of 0.25 eV (for the 3–5 μm band) or 0.12 eV (for the 8–14 μm band). On the other hand, for wavelengths in the near-IR, it is possible to use the phenomenon of electron photo-emission (electrons are emitted from a semiconductor into vacuum when photons impinge on it) and this leads to the use of the detector as the cathode in an image-intensifier tube. We clearly need to examine these very different techniques separately. Historically, the image tube came first, so we begin with it.

Photo-emission (often referred to as 'the photoelectric effect') played an important part in the development of quantum theory. Its discovery is usually attributed to Hertz in 1887 – in his use of a spark gap to generate radio waves, he noticed that, if he shone ultra-violet light onto one of the metal surfaces, the electric discharge could be significantly enhanced. For some years, it mystified scientists until Einstein, in his *annus mirabilis* of 1905, explained it on the basis of light quanta (photons). In fact, this was the first clear statement that light was quantised – Planck, who introduced the idea in explaining the spectrum of thermal radiation, believed it to be nothing more than a mathematical convenience. That apart, the effect can be seen as an important step towards the development of light detectors – if the metal (or, later, semiconductor) was made the cathode in a vacuum diode, the emitted photo-electrons could be detected as a current in the anode circuit. In effect, this meant replacing the hot cathode used in the original Flemming diode with a photo-cathode, and the material from which this was made could be chosen to suit particular wavelengths. In the case of the semiconductor cathode, the band gap was selected in order to absorb appropriate photon energies and the material was doped p-type to ensure the conduction band was empty of electrons. Absorption of light then generated free electrons in the conduction band which could be emitted from the semiconductor surface into the tube vacuum and thence be attracted to a positively biased anode. Anode current flowed only when light impinged on the cathode, thus allowing the device to function as a photon detector and it could be made considerably more sensitive by turning it into a so-called 'photomultiplier tube'. This involved the addition of extra electrodes known as 'dynodes', each being made more positive than its predecessor so that electrons were passed from one to the next in sequence. This may, at first, sound rather pointless but there *was* a serious point – when a high energy electron strikes a dynode, it loses its energy and, in doing so, stimulates the emission of a cascade of low energy electrons. These are then accelerated towards the next dynode, where the process is repeated. At the final anode, therefore, the electron current is hugely enhanced, compared with that emitted from the cathode – hence the name 'photo*multiplier*'.

As should be clear from this description, the photomultiplier is essentially a *detector* of radiation, not an imaging device but a very similar approach was adopted to turn the idea into an imager. Think of a photo-cathode made from a thin film of semiconductor deposited on a glass plate and suppose an image to be focussed onto it, by a suitable lens. Suppose, too, that the anode takes the form of another glass plate, parallel to the cathode and coated on its inner surface with a light-emitting phosphor. Electrons emitted into the tube vacuum space retain the image pattern and transfer it across the vacuum to the phosphor film which emits visible light with intensity proportional to the local electron flux, which, in turn, is proportional to the intensity of the original image on the cathode. If this image is formed from IR 'light', it is clear that the tube functions to convert an IR image into a visible image and is therefore known as an 'image convertor'. Such a simple diode arrangement lacked the power to intensify the image but during the 1970s, a clever adaptation of the 'gastroscope' principle was introduced in the shape of a glass 'channel multiplier' which could be inserted between the cathode and anode, to do for imaging what the dynodes in a photomultiplier did for detection. It consisted of a stack of fine bore glass tubes, fused together, each one forming a picture point (or pixel) of the electron image, and arranged to have a high voltage between its ends. An electron striking the input end of a channel stimulated further electron emission (as in a metal dynode) and, as these electrons bounced down the channel, the current increased by several orders of magnitude, the resulting light emission from the phosphor being intensified, accordingly. Not only did the tube convert from IR to visible, it produced a much brighter image (and was therefore known as an image-intensifier tube). Here was just the device needed to make use of the near-IR radiation from the night sky and a deal of classified research and development went into optimising its performance (see Fig. 9.2).

FIG. 9.2   The electronic recording and distribution of visible images and videos is an essential aspect of the modern world and we accept the intrusion of the television camera into most aspects of our lives. What is, perhaps, rather less familiar is the existence of infra-red cameras which are capable of seeing in the dark, thereby turning night into day. Courtesy of iStockphoto, © Volker Kreinacke, image 3278201.

From the point of view of making IR-sensitive image tubes, the choice of cathode material was crucial and, as is so often the case, was a question of compromise. The narrower the semiconductor band gap, the further out into the IR could the device work but, alas, the harder was it for the free electrons to escape from the semiconductor surface. Ideally, it was desirable to extend the response as far as 1.3 μm wavelength, so as to make use of as much IR radiation as possible but this meant that the sensitivity fell to almost negligible proportions. In the first instance, cathodes were developed on a purely empirical basis, there being little understanding of semiconductor behaviour on which to base a more scientific approach, and it was only after many years of research, with the application of much scientific understanding allied to some inspired material science that the limit could be pushed as far as 1.2 μm. It was a seriously attritional battle but the final user appeared to think it worth the struggle (i.e., was prepared to pay for it!) and it is, after all, such 'attritional battles' that keep materials scientists off the streets, so once again everyone was happy.

Interestingly, the first photo-cathode to make news, the so-called S1 cathode, *was* sensitive out to wavelengths of 1.3 μm but only with very low efficiency (of order 0.1% – one electron for every thousand photons). This was discovered in 1929 by L.R. Koller at the GE laboratory in Schenectady and was made from a magic mix of silver, oxygen and caesium. It represented a considerable breakthrough in terms of its sensitivity in the visible and near-UV-spectral regions but, even after years of study, it still remained far from clear exactly how it worked! Success depended on finding an optimum formula for depositing the film of silver, which was then oxidised and, finally activated with caesium. However, its discovery had the effect of stimulating further research into alternative materials and in 1936 Goerlich, in Germany, obtained excellent photoemission results in the visible and UV from the compound caesium antimonide (Cs_3Sb), leading to the development of the S11 cathode. However, this cathode showed no significant response in the IR so was of no interest for night-vision applications. This was followed, at RCA Princeton, in 1955 by A.H. Sommer's discovery of the S20 multi-alkali cathode – sodium, potassium, caesium antimonide (Na_2KSb-Cs) – which showed both excellent sensitivity in the visible and modest sensitivity at wavelengths out to about 0.9 μm. These two materials were later shown to be p-type semiconductors with band gaps of 1.6 and 1.0 eV, respectively; and their crystal structures were also established but, again, it was necessary to follow an empirical approach to cathode fabrication, starting with a film of antimony which was then treated with appropriate vapour pressures of the necessary alkali atoms. Optimum performance depended on careful adherence to a far-from-well-understood procedure which tended to differ somewhat between different laboratories and, bearing in mind the military importance of the S20 cathode's near-IR response, these procedures acquired a secretive nature which militated against wider scientific discussion.

Both the S1 and S20 cathodes were used enthusiastically during the Second World War but neither made optimum use of the night sky radiation and the struggle to obtain better IR performance went on long after hostilities had ceased. Nevertheless, the search for IR-sensitive photocathodes appeared to have run out of steam by the 1960s. In spite of the very many efforts to find empirical improvements in multi-alkali materials, the best performance had stagnated with the development of the S20 cathode, which had an effective cut-off at about 900 nm wavelength and maximum efficiency of 1% at 700 nm. It began to look as though the night-vision community might be obliged to make the best of this rather modest performance, when, out of the blue, in 1965, came the breakthrough that most interested parties had practically given up hope of ever achieving. It resulted from a study of fundamental surface science at the Philips Natuurkundig Laboratorium (Physics Laboratory – usually shortened to 'Philips Nat Lab') in Eindhoven by two high-flying Dutch scientists, Jacob Jan Scheer and Johannes van Laar. The 1960s was a period in the development of semiconductor science which was concerned with gaining a proper understanding of surface properties. In particular, there had been a long struggle to understand the detailed behaviour of the contact between a semiconductor and an evaporated metal film – the, by now, well documented Schottky barrier – and the contact between two different semiconductors – the heterojunction. It gradually became clear that their electrical properties depended crucially on the nature of the interface between the two constituents. In particular, Schottky barrier characteristics were influenced by the presence of impurity atoms on the semiconductor surface – atoms such as oxygen, nitrogen, carbon, sulphur, chlorine, etc which were naturally present on chemically cleaned surfaces – so it became of interest to explore the nature of contacts made on atomically clean surfaces. One technique for achieving this was to cleave (i.e., split in two) a single crystal of semiconductor in an ultra-high vacuum environment and evaporate a metal onto the cleaved surface before it could be contaminated by residual gas atoms. In this general context Scheer and van Laar were studying the effect of depositing caesium atoms on a clean surface of GaAs. Caesium having been a long-standing component of photocathode structures, it was of interest to explore its effect on a semiconductor surface which was accurately specified – that is to say one in which the nature of the surface atoms was reliably known. These were gallium atoms and arsenic atoms in well specified positions – and nothing else which might confuse the picture.

They characterised the surface in terms of a parameter known as the 'electron affinity', a measure of how easy (or more usually how difficult!) it was for electrons to escape from the conduction band into the vacuum space. In simple terms, one can think of electrons having to jump over a barrier in order to escape. Scheer and van Laar explored the way in which this barrier depended on the amount of caesium they had deposited, and they made the (at the time) remarkable discovery that, when just a single mono-layer of caesium covered the surface, this barrier had almost vanished. Electrons needed very nearly no energy at all for escape. It immediately pointed to the possibility for realising much more efficient photocathodes than any previously known, and their pioneering work was soon to yield practical benefits for the future of night-vision equipment, far removed from the original interest in fundamental surface properties of gallium arsenide. It clearly represented yet another example of the beneficial interaction between basic and applied sciences.

Ironically, perhaps, the practical breakthrough materialised not in Eindhoven but in the Philips sister laboratory in England. Andrew Turnbull and Geoff Evans were working in the Redhill laboratory which, at that time, was still known as the 'Mullard Research Laboratory', in support of a night-vision tube activity in the Mullard factory at Mitcham. They quickly realised that the tricky task of vacuum cleaving a single crystal of GaAs was hardly compatible with the manufacture of image-intensifier tubes so they explored the possibility of using a chemically cleaned surface, instead. The secret of success lay in finding an effective method of re-cleaning the GaAs surface, by heat-treating it, in the vacuum environment of the tube, then depositing the correct film of caesium. Needless to say, the results were not as good as achieved by Scheer and van Laar under their ideal conditions but they still looked promising enough to pursue further. This led them to try the effect of treating the GaAs surface with a combination of caesium and oxygen – a technique already well established from earlier work on the S1 type of cathode – and they eventually established a procedure which gave results close to those obtained originally at the Nat Lab.

This work was of considerable significance because it demonstrated the possibility of making efficient photocathodes from epitaxially grown GaAs, which ultimately proved essential to the manufacture of transmission cathodes. This latter aspect was to prove of major importance for tube manufacture. As we saw above, the concept of an image tube required the light input to reach the active part of the cathode by transmission through a suitable window, whereas all the work on GaAs cathodes had so far been done in reflection – that is, with the light impinging directly on the active surface. Attempts were made, initially, to grow epitaxial GaAs films on glass substrates but, as indicated by widespread earlier experience with other device structures, the quality of such material was totally unsatisfactory – something had to be done to obtain better quality epitaxy. The next step was to try growth on single crystals of GaP, which had a wider energy gap than GaAs and was transparent to photons with energies below about 2.2 eV but even this proved unsuccessful. Clearly, it would be necessary to grow the active material on a lattice-matched substrate (as had proved essential for the growth of laser structures) and this implied the use of AlGaAs. Bulk single crystals of this material were not available so thought turned to the use of thick epitaxial layers which could be grown by liquid phase epitaxy. Finally, the problem was brilliantly solved in 1975 (10 years after the initial breakthrough!) at the Varian laboratory in Palo Alto. George Antypas and John Edgecombe grew a multi-layer structure starting with a GaAs substrate which took the form: AlGaAs/GaAs/AlGaAs. Having first stuck it down on a quartz window, they then proceeded to etch away the substrate and the first AlGaAs layer by using selective chemical etches, leaving a GaAs layer outermost which could be activated with caesium and oxygen to produce a negative electron affinity. The quartz and the AlGaAs were transparent to photons in the near-IR, thus providing a transmission cathode, as required. Quantum efficiencies as high as 30% were obtained – another materials triumph!

As if this was not sufficiently ingenious, the Varian team was at the same time in process of addressing a second important aspect of the IR photocathode – how to extend the response to even longer wavelengths. The band gap of GaAs limits its response to wavelengths shorter than 880 nm (0.88 μm), which, as we have already seen, leaves much to be desired from the view point of using night-sky radiation. Ideally, one would like to extend the limit as far as 1.3 μm. First thoughts towards solving this problem suggested the use of the InGaAs alloy but no suitable substrate was available (to obtain the necessary lattice match) so some other material system had to be found and, once again, Antypas came to the rescue. We saw in the previous chapter how the quaternary alloy InGaAsP proved invaluable to the development of lasers for optical fibre communications but it was, of course, 'invented' in the interests of near-IR photocathodes. Grown lattice-matched to an InP substrate, which acted as a suitable window, it offered a range of band gaps from 1.35 to 0.75 eV (wavelengths of 0.92–1.65 μm), making it eminently suitable for the task of extending cathode response to 1.3 μm and the Varian team made full use of its capability. We must, however, emphasise the important compromise referred to above. In reducing the band gap to absorb longer-wavelength radiation, it became more difficult to reduce the barrier to electron emission (the electron affinity), so, as the response was pushed to longer wavelengths, the quantum efficiency was inevitably degraded. In practice, it was possible to reach a wavelength of 1.2 μm but with less than half a percent efficiency. Alternatively, a cathode responding only to 1.1 μm achieved something like 2%. It nevertheless constituted a remarkable improvement in night-vision performance and proved especially valuable for use in conjunction with a high power YAG (yttrium aluminium garnet) laser, which emits at a wavelength of 1.06 μm (though not a semiconductor device, this laser has proved of enormous value in many applications). Quantum efficiencies as high as 7% were achieved in this application.

The resulting night-vision tubes, employing transmission cathodes and glass channel plate multipliers, were extremely compact, favouring their use as goggles, a very convenient arrangement for both military and civilian applications (see Fig. 9.3). By far, the heaviest component was the battery required to power the multiplier and this could readily be carried in a back pack. Both hands were, therefore, free for other activities such as note-taking or sketching, in the case of nature watching, or holding a steering wheel or shooting a rifle in the case of the wartime activities which paid for the original tube development. We shall pursue the matter no further – our interest must now return to the matter of thermal imaging, which we left in limbo some pages back.

It will be apparent from the above account that thermal imagers, which work at wavelengths much further out into the IR, cannot be

FIG. 9.3 The development of infra-red image tubes which convert invisible radiation into visible took quite a long time. Starting with the discovery of infra-red-sensitive photocathodes in the 1930s, followed by that of the so-called negative-electron-affinity cathode in the 1960s and the invention of the channel plate multiplier at about the same time, it became possible to build small, convenient image tubes which could be fashioned into light-weight night vision goggles as shown in the photograph. Courtesy of WikiMedia.

based on the phenomenon of photoemission – they have, all along, depended on semiconductor photoconductors or photodiodes, their development having to wait on the technology of suitable narrow-gap materials. They have also been obliged to employ quite different imaging methods, the important distinction being the need for some form of scanning. In the image tube, an image of the scene is focussed onto the cathode and each point of this image produces an appropriate number of electrons in the vacuum space which may be multiplied in the channel plate but the spatial arrangement of electrons is maintained until they strike the phosphor, thus replicating, in visible photons, the IR photon distribution incident on the cathode. There is no equivalent of this process available in solid-state devices such as photoconductors or photodiodes and this demands that the image of the scene be scanned in some way so as to convert the spatial image into a time-dependent electrical signal. This signal can then be used to control the intensity of the electron beam in a cathode ray tube, which is similarly scanned over a phosphor screen so as to reproduce the original spatial image in visible form on the screen. This scanning process is then repeated continuously to produce a TV-like picture of the scene and the whole of each scan must be completed within a time short enough to fool the observer's eye into seeing it as a continuous moving picture, rather than a sequence of jittering snap-shots. The eye retains an image for a time of about one twenty-fifth of a second, so the scan should last no longer than about a fiftieth of a second, to avoid flicker. In an equivalent TV picture, there are upwards of a million picture points, so each point is scanned for about 20 ns. In the case of a thermal image, resolution may well be considerably less than this, but we are still talking of perhaps a few microseconds for each pixel, and this requires that the detector should have an appropriately rapid response. As made clear in the previous chapter, a photodiode might well respond very much faster than this so it makes no great demand on this type of detector; however, photoconductors tend to be much slower. They were used in many early types of thermal imager but have been largely phased out in recent times.

Let us just look briefly at scanning methods before returning to the question of suitable semiconductor detector materials which will be our main concern. The earliest systems made use of mechanical scanning – a pair of rotating mirrors scanned the IR scene across a single detector, one mirror scanning horizontally, while the second scanned vertically. In other words, they rotated about mutually perpendicular axes. The vertical scan mirror needed to rotate much more slowly than its horizontal compatriot, effectively moving from one line to the next in the time taken by the horizontal mirror to scan a whole line of picture elements. As each snap-shot moved across the detector, the detector output registered the intensity of radiation from that particular element, so the overall output consisted of a time-dependent signal suitable for 'programming' an associated cathode ray tube. It was far from ideal, but proved adequate to demonstrate the viability of the thermal imaging process and encourage engineers to develop more convenient electronic scanning systems. An initial improvement took the form of a single line of detectors that was scanned in the same manner, the individual signals being added electronically to give a much larger signal at the end of the line. Various other scanning techniques were used which need not concern us here but the ultimate aim was to do away with mechanical scans altogether, in the interest of achieving a much more convenient and compact imaging system. This led to the introduction of so-called 'staring arrays' (or 'focal-plane arrays') in which a two-dimensional array of detectors stares unblinkingly at the scene, each detector recording the photon flux from one particular point in the scene. In order to record a moving picture, it is obviously necessary to sample the signal from all the detectors repetitively and pass the information on to a suitable display device on a time scale of one-fiftieth of a second and this is done by feeding the output of each detector into a readout circuit (a silicon integrated circuit designed specifically for the purpose). The rather clumsy mechanical scanning is thus replaced by a much more convenient electronic scan. The fine detail need not concern us but it is clear that the integration of the detector array with the readout circuit is an important aspect of system design, bearing in mind that the detectors are inevitably made from a very different material from silicon. We should also note that building a large array of detectors (typical size being of the order 640 × 480 elements, over a quarter of a million in total) presents serious difficulties. Ideally, each and every one should perform identically, to avoid distorting the final image but this has turned out to be virtually impossible and obliged the system engineer to include correction circuitry as part of the image processing stage. Each array must be characterised by shining a perfectly uniform 'scene' onto it and using the resulting image to define an appropriate correction algorithm. Given that such schemes actually work, one can only admire both the skills of the designer and the capabilities of silicon integrated circuits.

This is all very well, but could apply to almost any solid-state imaging system. We must now concentrate our attention on the materials which have made possible the construction of high quality images from thermal radiation in the 3–5 and 8–14 μm wavelength bands. The quest for an optimum semiconductor to satisfy a demanding set of criteria makes a fascinating story which takes in something like 10 different materials, some developed specifically for the task, others 'borrowed' from earlier technologies aimed at quite separate targets. Before plunging into detail, let us first emphasise some of the essential requirements: firstly, we are concerned with semiconductors having unusually narrow band gaps – roughly 0.1 eV for the 8–14 μm band and 0.2 eV for the 3–5 μm band – and, of course, these gaps should be 'direct' in the interest of effective optical absorption; secondly, we need material for photoconductive detectors which is highly pure so that the optically generated free carriers outnumber those which make up the background level; thirdly, for photodiode detectors we need material which can be controllably doped both n- and p-type; fourthly, we need material with recombination lifetimes which can be reliably controlled; fifthly, we need material which can be prepared reproducibly to a specific recipe; sixthly, in the interest of making large area imaging systems, we need material which can be prepared with a high degree of spatial uniformity. Oh, and as with any other device requirement, it should be stable over time, even when repeatedly cycled between room temperature and the operating temperature, which might typically be some hundred or more degrees below room temperature. The fact that it took something like 60 years to satisfy most of these demands indicates just how difficult was the task – there was no shortage of urgency and some of the world's best scientists were engaged in the chase.

The story really begins with the introduction of the lead chalcogenides, PbS, PbSe and PbTe during the Second World War. Lead sulphide came first – it had, after all, been known for many years as a valuable contributor to the development of radio (in the form of natural crystals of galena) but it was only now that it was recognised as having a band gap appropriate to IR detection. Work on both sides of the North Sea demonstrated its response to wavelengths in the region of 2–4 μm – 2 μm at room temperature, increasing to about 4 μm when cooled to 20 K in liquid hydrogen – and its sensitivity was found to be about a hundred times better than those available with thermal detectors, which accounted for the surge of wartime interest. The selenide and telluride were also investigated with a view to extending the response farther into the IR, in this case with considerably greater success on the western side of the watery divide. Work at RRE Malvern showed that PbSe was sensitive at wavelengths out to 8 μm but, oddly, PbTe showed a cut-off at about 5 μm – 'oddly' because the chemistry of these materials suggested the sequence 'ought' to be sulphide, selenide, and telluride. Nature sometimes moves in mysterious ways to frustrate us – it would have been nice if PbTe could have shown a response into the 8–14 μm window but it was not to be. The end of the war came with no progress in that direction.

I should, at this point, explain that these lead chalcogenides were prepared as thin films, deposited on glass or quartz substrates and were therefore in the form of polycrystals – large agglomerates of microcrystals with so-called grain boundaries between them. As we shall see in more detail in Chapter 10, these boundaries play an important role in limiting electrical conduction through such films and therefore influence their photoconductive behaviour, too. It was not at all clear at the time exactly *how* they functioned but there was no question that they did indeed function as photon detectors and that was enough to spark more than passing interest in the use of long wavelength IR radiation for thermal imaging. There was, however, a problem here with regard to the response time of these films. It was probably due to the 'trapping' of optically generated free carriers at grain boundaries which very much lengthened their effective lifetimes, making them far too long for the type of scanning system described above. At all events, it provided strong motivation to develop a method of growing single crystals and in the early 1950s Bill Lawson, at RRE Malvern, was successful in growing crystals from the melt – the so-called Bridgman method. It had the immediate effect of stimulating measurements of fundamental properties (band gaps were in all cases found to be direct) but, alas, did little to improve IR imaging. It also led to a number of erroneous results in relation to important parameters like band gaps – the problem being one with which we are already familiar – the resulting crystals contained unacceptably high densities of impurities. It was not until the late 1970s that sufficiently pure crystals could be grown to allow accurate data to be obtained – an indication, perhaps, of just how difficult materials science can sometimes be!

High impurity concentrations also militated against the use of these crystals as photoconductors and led to the suggestion that photo-diodes might show more promise as imaging devices. This, of course, demanded the ability to control doping, both n- and p-type, not the simplest of challenges when background impurity levels are high. In any case, such discussions were overtaken by the emergence of a new contender, in the shape of the newly discovered III–V compound, indium antimonide (InSb). As we saw earlier, the III–V compounds were first explored during the 1950s, rapidly challenging even silicon in certain application areas, and InSb proved to be particularly interesting for IR imaging. It's (direct) band gap at room temperature was measured as 0.17 eV (corresponding to a wavelength of 7.3 µm), increasing to 0.23 eV (5.4 µm) at 77 K, which made it suitable for the 3–5 µm window, and it soon became available in the form of high quality single crystals which could be doped both n- and p-type. Recombination lifetimes were found to lie within the range 10^{-7}–10^{-9} s, more than short enough to satisfy scanning requirements – though actually too short when viewed from the aspect of detection sensitivity (which, in a photoconductor, improves with increasing lifetime). Lifetimes of, say, 1–10 µs would have been ideal but such complaint sounds a little like unfair carping – InSb certainly made an immediate impact on the field of IR imaging. It was used in both photoconductive and photovoltaic modes, usually operating at the temperature of liquid nitrogen (77 K).

The use of reduced operating temperature was not, by the way, a matter of adjusting the band gap – it had a far more important justification than that. The lowest light level detectable by any photon detector is limited by the presence of thermally generated free carriers, the intrinsic density n_i that we have met several times already, and we should remember that n_i tends to be rather large in semiconductors with narrow band gaps (see, for instance, Table 1.2 in Chapter 1). We should also recognise that n_i depends strongly on temperature – intrinsic carriers are generated by the thermal vibrations of lattice atoms in the semiconductor and these vibrations become more insistent as the lattice temperature increases. Conversely, by cooling the sample to 77 K, n_i is strongly reduced and this means that the detector is able to 'see' a correspondingly smaller density of optical photons, against this much reduced background. This, in turn, makes it more sensitive to small temperature differences in the thermal scene and helps it to distinguish men from trees, for example. Most IR imaging systems have therefore made use of cooled detectors, even though it makes for serious inconvenience in their mechanical construction. We shall have a little more to say about this later but, for the moment, we shall continue our quest to find the ideal detector material.

Indium antimonide clearly represented a significant advance but it could offer no hope for a detector to work in the 8–14 μm window, nor, as we noted above, was it quite perfect for application to the 3–5 μm window. It may have satisfied an immediate need but there was still a requirement for a more flexible material which could work in both windows and was able to provide the ultimate in sensitivity. After a long wait (as with London buses), not one, but two came along! In 1959, the RRE group under Bill Lawson reported the growth of bulk single crystals of a totally new ternary II–VI compound, mercury cadmium telluride, HgCdTe (MCT) which had the remarkable property of allowing band gaps to be engineered all the way from 0 to 1.6 eV by suitable choice of the cadmium/mercury ratio. Twenty percent cadmium resulted in a band gap of 0.1 eV, suitable for the 8–14 μm window, while 30% achieved the value of 0.2 eV appropriate to the 3–5 μm window. It was a brave and dramatic development, not least because of the hazardous nature of the constituent elements, all three of which are well known poisons! What was more, the vapour pressure of mercury at the melting point was high enough to threaten serious explosive tendencies (tendencies which, in several cases were transposed into realities!). Then, some few years later an American group, under Ivars Melngailis at MIT Lincoln Laboratory demonstrated somewhat similar properties (opto-electronic, rather than explosive) for the IV–VI compound lead tin telluride, PbSnTe. Again, it was possible to select appropriate band gaps by controlling the ratio of lead to tin and again single crystals could be grown from the melt. These two materials appeared, on the face of things, to offer very similar performance and for several years they fought toe-to-toe, as it were, until MCT gradually crept ahead, largely as a result of the introduction of epitaxy. At all events, by 1991 one could observe something like 40 different groups, world-wide beavering away on MCT growth, while PbSnTe had fallen some considerable distance behind (though it still had a future as a material for making IR lasers). In the interest of brevity, therefore, I shall concentrate, in the following, solely on MCT.

As with most other semiconductors (with the notable exception of silicon), the growth of bulk crystals proved extremely difficult from the point of view of obtaining adequate control over composition, uniformity and impurity concentration, while an additional complication manifested itself in that small departures from stoichiometry (too much or too little mercury) served to dope the material, thus introducing unwanted free carriers. These problems, allied with the (by now well known!) chemical hazards stimulated the introduction of epitaxial growth methods. During the 1970s there were also demands from the systems people for larger cross-sectional areas in the move towards staring arrays and the availability of relatively large cadmium telluride substrates provided additional encouragement. An important advantage of the CMT material system is the fact that the lattice mismatch between CdTe and CdHgTe is acceptably small so good quality epitaxy is possible in much the same manner as applies in the AlGaAs/GaAs system. However, this was very far from saying that epitaxy would be easy.

First attempts were made towards the end of the 1960s by M. Rodot of CNRS in France. He introduced the 'close-spaced' method whereby a polycrystalline sample of MCT was placed close to a CdTe substrate at a temperature of about 550 °C – it was really nothing more than simple evaporation and suffered from being extremely slow and from non-uniformity resulting from mercury diffusion from the growing film into the substrate. Something more practical was clearly needed so in the 1970s efforts were devoted to various versions of vapour-phase epitaxy and liquid-phase epitaxy, both of which offered some improvement, though still failing to achieve adequate uniformity or sufficiently low background doping levels. Indeed, up to about 1980, most of the device applications were still being met by bulk crystal growth. Really effective epitaxial growth demanded a process that operated at significantly lower temperature and led, in the early 1980s to the introduction of first MOCVD (by Brian Mullin and co-workers at RSRE Malvern) then MBE (by J.P. Faurie at the University of Illinois – another Frenchman but far from home). These processes could occur at temperatures as low as 300 and 200 °C, respectively, low enough to minimise mercury diffusion, and brought with them another important advantage, that of controlled impurity doping (rather than having to rely on the less controllable method of stoichiometry). This latter feature made diode fabrication much easier. At last, the technology of IR device making was beginning to look more like those earlier established for the GaAs laser or for light-emitting diodes.

Background doping levels were still something of a problem and it became clear that there was a subtle aspect to this. We have already met the idea that detector sensitivity may be limited by thermal generation of free carriers – the subtlety appears when we consider the mechanisms by which such carriers can be generated. The simplest and most direct process is that whereby lattice vibrational energy is transferred directly to a valence band electron, lifting it into the conduction band. This process occurs in absolutely pure material and is fundamental. Suppose, however, that the conduction band contains additional free electrons due to unintentional doping. These electrons are distributed in energy above the bottom of the band due to thermal energy and, in a small band gap material, a few of them will have energies as large as the band gap. There is then the possibility that one of these 'high energy' electrons might give up its energy to a valence band electron so that both of them end up at the bottom of the conduction band. It may sound pretty fanciful but it does actually happen! The process is known as 'Auger generation' (after the French scientist Pierre Victor Auger who discovered a similar Auger process in atomic spectroscopy). All such generation processes result in noise (i.e., interference) in the detector which adversely affects its sensitivity and this represents yet another reason why background doping levels should be reduced to a minimum. Materials science can be quite complex at times. On the other hand, materials scientists can also be incredibly clever.

Two excellent examples of this were demonstrated by the RSRE group under Tom Elliott during the late 1980s. Their first piece of inspiration consisted in designing the contacts to an MCT detector in such a way as literally to suck out free carriers from the active region, thus reducing the Auger process and allowing detectors to be operated at significantly higher temperatures than was normally possible. The ultimate goal was to obtain ideal performance at room temperature and that appears still to be just out of reach but the advantage of being able to rely on a simple thermoelectric cooler, rather than cooling with liquid nitrogen is well worth having. Their second contribution was a simple geometrical one, involving the use of an array of tiny lenses which focussed the incoming radiation down onto a matching array of even smaller detector elements. The reduced size of these elements meant that the generation noise was correspondingly reduced and the signal-to-noise ratio correspondingly increased, yet another advance in the quest for the holy grail of room temperature operation.

The other principal objective, has been that of making large area staring arrays which demand large area substrates for epitaxy and has led to a sustained attack on the problems of growing MCT films on alternative substrates such as silicon, germanium or gallium arsenide that are available in sizes much greater than those of cadmium telluride or the other frequently used substrate cadmium zinc telluride. There is an obvious advantage to be had from the use of silicon in so far as it can also serve as the host for the appropriate readout circuitry. Though growth of any II–VI or III–V material on silicon has always been fraught with difficulty, some worthwhile progress has certainly been made in this instance and acceptable performance achieved from quite large arrays. However, the use of a silicon base has stimulated yet another approach in which the detector elements take the form of silicon/platinum-silicide Schottky barriers (a whole range of silicides exists – PtSi, AlSi, NiSi, CoSi, etc., all of which behave as metals and can be deposited as thin films on silicon). This represents yet another 'promising' development but has yielded detectors, still giving less than ideal performance even at liquid nitrogen temperatures. Nevertheless, relative ease of manufacture has encouraged the development of quite large arrays. Overall, the subject gives the appearance of getting ever more complicated as new ideas proliferate – one of the problems associated with a serious demand for superlative device performance combined with a seriously difficult materials challenge and a near-infinite source of funding!

For our part, we shall concentrate attention on just one further approach which came to prominence in the mid-1980s. The development of low-dimensional structures in the 1970s gradually came to influence more or less every aspect of semiconductor device physics and IR detection was never likely to escape its all-embracing tentacles. The idea came originally from Esaki and Sasaki at IBM – why not use the optical transitions *within* a quantum well to absorb IR photons as a simple means of detection? It had the obvious advantage that the energy separation between adjacent confined levels (and hence the optical wavelength) could very easily be 'tuned' by simply varying the well width – much easier than controlling the composition of a 'difficult' material such as MCT! And, being based on a well developed material system, such as that of AlGaAs/GaAs, it also had the advantage that large substrates were readily available for the manufacture of staring arrays. (A further advantage lay in the availability of a handy acronym, QWIP – quantum well IR photodetector.) The idea was to 'design' an n-type quantum well with two confined energy states, the upper level being close to the top of the well, so that, when an absorbed photon excited an electron from the lower state, it could very easily escape into the continuum of states above the well and be swept out to an electrical contact and thence into the external circuit where it could be 'counted'. A later refinement did away with the upper confined state and arranged for the electron to be excited directly into the continuum but the principle remained the same. Note, too, that it was necessary to employ as many as 50 wells in order to absorb all the radiation. As with silicon/silicide devices, QWIPs had to be cooled to 77 K or below but the advantage of uniformity was certainly demonstrated in practice. In the early 1990s, a team at Bell Labs under Levine reported on a 128 × 128 pixel array showing excellent uniformity and resolution at a wavelength close to 10 μm and many other groups have followed this lead. It is now possible to buy IR cameras based on QWIPs with well over 300,000 pixel resolution (640 × 512 pixels). Operation has been extended to the 3–5 μm band by using alternative materials such as (strained) AlGaAs/GaInAs or (unstrained) AlInAs/GaInAs or GaInAs/InP which allow the formation of deeper wells. Yet another recent development is that of dual wavelength detectors, containing two different wells. These have the interesting property of providing an actual measurement of scene temperature by comparing the signals at the two wavelengths.

How can one sum up the present situation? 'Only with difficulty' is the answer. It is probably true to say that MCT still offers the ultimate performance but, largely because it suffers from relatively poor yields (particularly in the case of large arrays), it turns out to be considerably more expensive than the QWIP. On the other hand, QWIPs have to be run at lower temperatures if they are to challenge MCT in terms of detection sensitivity. Both systems can be used to form dual (or multiple) wavelength detectors but this is significantly easier with QWIPs. And if that isn't confusing enough, there is yet another contender for what is being touted as the 'third generation' detector – this being based on an exotic new material system known as a 'type-II superlattice'. It is made from a multilayer structure of indium arsenide and gallium indium antimonide. I shall make no attempt to describe its modus operandi but simply warn the reader that it might just steal the show in the not-too-near future. But by far the most likely outcome is a scenario where different material systems find distinct niches in the spectrum of applications – it being too soon to make unequivocal statements as to which. What one can say, though, with total confidence is that the performance of modern IR imaging systems is, so far, in advance of their early predecessors as to make them virtually unrecognisable. It has certainly been a long haul and progress (for good reasons) has not been particularly rapid but the achievement has, without doubt, been dramatic.

CHAPTER 10

Large Area Electronics: Flat Screen TV and Solar Electricity

In very nearly every example we have looked at so far, success in semiconductor technology has rested on the application of single crystals of the appropriate material. It all started with the use of natural crystals of lead sulphide to make cat's whisker radio detectors (though these were certainly made up of largish polycrystals) but the real breakthrough came, of course, with the development of artificial crystals of germanium and silicon for making transistors. Success here, led to the growth of enormous boules of silicon (up to 12 in. in diameter) for cost-efficient manufacture of integrated circuits and it gradually became clear that most other semiconductor devices would require similarly high-quality single crystals. When it came to the development of III–V compounds for microwave devices, light-emitting diodes and lasers, the only means for achieving adequate quality proved to be the use of epitaxial films but the principle was the same – single crystal material was essential to ultimate success (and, of course, epitaxy rested both literally and figuratively on single crystal substrates). Again, when sophistication increased to the incorporation of heterostructures, as, for example, in the semiconductor laser, it was crucial that the various layers were well matched in lattice parameter so that the overall structure could be seen as a kind of pseudo-single crystal, at least in structural terms. The startling revolution of fibre-optic communications would have been stillborn without the availability of long-wavelength lasers based on single crystal material in the indium gallium arsenide phosphide materials system and in our last chapter we recognised the importance of single crystal mercury cadmium telluride in the development of effective thermal imaging systems. Without doubt, the crystal grower was a key figure in the whole saga of both twentieth century electronics and opto-electronics. Must it always be so, one

Semiconductors and the Information Revolution: Magic Crystals that made IT happen © 2009 Elsevier B.V.
DOI: 10.1016/B978-0-444-53240-4.00010-6

wonders? Are there *any* examples where non-single crystal material can play a role? I ask the question rhetorically, of course – there are indeed just a few such examples and this final chapter is concerned to examine their (by no means negligible) contribution.

Most readers will be familiar with the revolution that has taken place in the field of television and computer display. The twentieth century was dominated by the cathode ray tube, while the twenty-first century has so far been dominated by the flat panel display (see Fig. 10.1). The search for some satisfactory flat panel technology was a long and difficult one – for years, the dream of a display panel that could be hung on the sitting room wall remained just that – a dream. Then, all of a sudden (once again like the proverbial London buses – doesn't it happen in other cities?) two came along. In the space of just a very few years, the cathode ray tube virtually disappeared, to be replaced by the plasma panel at large sizes and by the liquid crystal panel at somewhat smaller sizes (though liquid crystal television (LCTV) screens are now available with diameters as large as 50 in.). The plasma panel depends for its operation on a myriad of tiny gas discharge tubes and makes no special demands on semiconductor devices, other than the routine circuitry required to scan the TV signal

FIG. 10.1 We take for-granted the availability of laptop computers and mobile phones with sizeable display screens which tend to rule our lives. How many of us remember that the display technology which made these indispensable instruments possible came to fruition as recently as the turn of the century. Flat screen display had been a kind of holy grail of scientific research for at least fifty years when success was finally achieved. Courtesy of iStockphoto, © Jeffrey Smith, image 2095641.

over the screen. We shall welcome its undoubted contribution to man's enjoyment of his leisure hours but say no more about it here. The LCTV display, on the other hand depends crucially on semiconductor devices which control the brightness of each and every picture point (pixel) on the screen so, in the context of our exploration of the impact of semiconductors on modern life, we shall have much more to say about it. For the moment, though, suffice it to note that this vast array of switches (their number running well into seven figures) is constructed from a strange variety of semiconductor, known as 'amorphous silicon', an entirely new form of material and totally different from the single crystals we have become used to. More of that in a moment, but first we should look at another application for non-crystalline semiconductors, that of photovoltaic solar panels for converting sunlight directly into electricity.

This, by its nature, represents a very large area application. While display panels may be characterised by diameters of say 40 in. (TV) or up to 20 in. (computer), one must envisage a solar panel covering much of the roof of a house or office building, or possibly spread out over a sizeable area of the Sahara desert. The requirement, therefore, is for a large semiconductor diode which absorbs sunlight and generates an electric current in an external circuit, and there are two rather different approaches to achieving this. Either one can assemble a large number of modestly sized single crystal cells or one can plump for an alternative technology, based on either amorphous or polycrystalline material, which lends itself to deposition over large areas. In practice, both approaches have been tried and both show promise. Briefly, single crystal cells are more efficient but more expensive, while non-crystalline cells are cheaper but less efficient. After many years of trying, it still remains to be seen what the long-term solution to this particular challenge will turn out to be. I shall try to outline some of the issues in more detail in what follows. Firstly, we should take a closer look at the nature of electrical conduction in amorphous and polycrystalline materials.

While discussing the nature of infra-red detector materials, we came across the example of the lead chalcogenides which seemed to work best in the form of polycrystalline thin films deposited on glass substrates and we looked very briefly at the physics of their electrical conduction behaviour. The time has now come to examine more carefully the electrical properties of non-single crystal materials so that we may better understand their role in the two important areas of endeavour referred to above. We start with the structure of these materials which, in turn, determines their electrical properties. Single crystals are characterised by a regular arrangement of atoms – indeed, if we know the position of a single atom in a perfect crystal, together with the appropriate lattice parameters, we can predict the positions of all the other atoms in the sample. That such idealised crystals do not exist in practice should surprise no one with experience of this imperfect world but real crystals can show amazingly close approximation to the ideal and such a model serves to explain the electrical properties remarkably well. In the absence of an electric field, electrons buzz about with 'thermal velocities' of order 10^5 m/s in completely random directions and when an electric field is applied, the 'electron cloud' moves with a 'drift velocity' of typically a few meters per second in the direction of the field. It is this drift which characterises an electric current. The distance an electron moves freely before it is scattered (i.e., deflected) by an imperfection in the crystal is typically about 100 nm (0.1 μm) which represents several hundred lattice spacings, so the crystal behaves rather like a 'sea' through which electrons can move easily – they appear to be more or less oblivious to the presence of individual atoms. This is in marked contrast to the situation in an 'amorphous' semiconductor which is characterised by an almost total lack of order in its atomic arrangement. Knowing the position of any one atom in this case no longer allows one to predict the positions of other atoms – the concept of a lattice no longer applies and the single crystal 'sea' no longer exists. Electrons are scattered after only very small distances – in other words, they are conscious of more or less every atom and their progress is correspondingly very much slower. We express this by saying that their 'mobility' is much smaller. What is more, amorphous materials show major imperfections in chemical bonding. In a single crystal, each atom is bonded to a well defined number of neighbours and all its chemical bonds are satisfied. In an amorphous semiconductor this can no longer be so – some atoms are left with unsatisfied bonds (known in the jargon as 'dangling bonds') and these bonds have a tendency to grab passing electrons and trap them (i.e., hold on to them for a significant length of time), thus further slowing their progress through the sample. Mobilities in amorphous semiconductors may therefore be several thousand times smaller than in single crystal materials.

Perfect single crystals and amorphous materials are at opposite extremes of the conductivity spectrum. Between them lies a range of polycrystalline materials, made up of agglomerations of small single crystals bonded tightly together but misoriented with respect to their individual crystal lattices. The electrical behaviour of such polycrystalline samples depends strongly on the physical size of the crystallites – in general, this may vary from only a few atomic spacings (closely similar to the amorphous state) all the way to millimetre sizes which result in behaviour not unlike that of single crystals. In fact, the physics of polycrystalline samples is intriguing because their electrical conduction depends on the size of the crystal grains in a particularly interesting way and the key to understanding this requires that we recognise the part played by the boundaries between crystallites (usually known as 'grain boundaries'). These regions contain atoms which are not properly chemically bonded (rather like the amorphous materials we discussed earlier) and this implies that they tend to trap electrons and thereby contain excess electric charge. This trapped (i.e., immobile) charge presents an electrostatic barrier to the flow of free electrons between grains, rather like the Schottky barrier existing at the interface between a semiconductor and a metal. Electrons must acquire a certain amount of energy to surmount such barriers and this implies a limitation to the effective conductivity. What is more, these barriers also have a finite width which is related to the doping level of the crystal grains – the more heavily they are doped, the narrower are the barriers – and the resulting electrical behaviour depends on the size of the grain, compared to the width of the barriers. Typically, barrier widths lie in the range of sub-micron to micron dimensions so the really interesting properties are seen in samples with grain sizes in this same range. Remember, too, that electron scattering lengths in the single crystal grains are also of order one-tenth of a micron. Could the semiconductor physicist possibly wish for a more exciting set of circumstances to challenge his ingenuity in unravelling Nature's complexities? Readers will hardly be surprised when I say that many an esoteric research paper has been published in the struggle to understand the electrical conductivity of polycrystalline semiconductor samples. Nor is this the whole story. In considering the behaviour of polycrystalline solar cells, we must also take account of the fact that the absorption of light generates minority carriers which may also be trapped at grain boundaries, thus changing the amount of fixed charge and, in consequence, the height and width of the inter-grain barriers. While there may be considerable commercial advantage in preferring a polycrystal technology to one based on high quality single crystals, it can hardly be said that it eases the scientific understanding of the resulting device performance. The reader will doubtless appreciate that when solar cell efficiency turns out to fall short of that hoped for, the poor bewildered technologist, given the task of sorting it out, has rather too many avenues down which to get lost!

So much, then, for this brief introduction to the vagaries of non-single crystal semiconductor science. As must be apparent, a great many high class physicists have worked for many years to build an adequate working knowledge of the properties of amorphous and polycrystalline semiconductors and a very wide range of materials has been investigated. Nor should we overlook the fact that an equally wide range of chemical processes has been explored in the preparation of suitable samples. We have seen that grain size is a vital parameter in determining electrical behaviour and, unsurprisingly, it is also a function of the deposition process employed, the latter being vital, indeed, to determining whether one obtains polycrystalline or amorphous films. I shall say no more of all this in general terms but concentrate in what follows on the specific problems associated with the two applications I have chosen to explore – those of LCTV display and the photovoltaic solar panel.

Commercially viable television display was based on the cathode ray tube, first invented by our old friend Carl Ferdinand Braun in 1897. Its operation depended on a beam of electrons emitted from a heated cathode (the electron gun) being scanned over a screen on which was deposited a suitable phosphor. Energy from the beam was thereby converted into light whose intensity could be controlled by varying the intensity of the electron beam, itself. In the original black-and-white TV sets, only one gun was used and the phosphor emitted a mixture of colours approximating to white light. The much more exciting (and demanding!) goal of colour TV was made possible by the invention in 1949 at the RCA laboratory of the so-called 'shadow mask tube' which used three electron guns (one for each of three colours) together with a steel plate full of tiny holes (the shadow mask). On the screen were deposited triads of red, green and blue-emitting phosphors and the holes in the plate (amazingly!) arranged so as to guide only the 'red' electrons to the red phosphor spots, the 'green electrons' to the green spots and the 'blue' to the blue spots. It was a triumph of mechanical design and dominated domestic television for over 50 years, producing, as it did, bright, satisfying colour images on screens of up to about 30 in. diagonal. However, good though it was, it was not without drawbacks – it was bulky, operated with seriously high voltages, generated significant amounts of X-radiation which had to be screened from its viewers by means of a heavy lead glass screen and struggled for years to achieve an acceptably flat picture. Herein lay more than adequate motivation for display technologists to seek an alternative – it should be truly flat, thin, light in weight and operate at low voltages compatible with the integrated circuits which would be used to 'drive' it – and it must, of course, provide a picture quality comparable with that of its predecessor, at a competitive price. As we have already noted, success was a long time in coming. The LCTV display emerged only gradually from an inspired coming together of a range of initially unconnected developments. It first saw the light of day in the shape of small personal displays such as were introduced in long-haul aircraft to help while away the tedious hours separating far-distant airport concourses. Nor should we forget the stimulus provided by the mobile phone which, in many instances, helps to minimise the tedious minutes separating one unnecessary communication from its predecessor. In the 1980s, doubts were expressed concerning the likelihood of screens reaching dimensions large enough even to satisfy the (then) modest needs of personal computer displays but the huge potential market proved tempting enough to stimulate the necessary investment and in the early years of the present century my old company, Philips announced its first LCTV screen with a 50 in. diagonal. They were not alone, of course – even a casual glance round any well stocked electronic store, today, will reveal a multitude of choice at sizes from 15 to 50 in.. How did it all happen?

The story began in the post-transistor rush to develop 'digital' displays for scientific instruments, pocket calculators, transistor radios, wrist-watches, etc. First came the still familiar red, seven segment alpha-numeric displays which announced the arrival of GaAsP and GaP LEDs (see Chapter 7). They were just bright enough (as long as the sun wasn't shining directly on them!) but their low efficiencies constituted a serious problem for battery-operated equipment. Something better was certainly needed for use in digital watches where battery power was at an obvious premium, and the propitious arrival of the liquid crystal in the 1960s soon swept that particular board. The concept of a *liquid* crystal is, perhaps, a strange one. Can there really exist the kind of regular molecular order that we associate with crystals in a liquid which is essentially characterised by free flow? The answer is essentially a compromise – given the right molecules, it is possible for a certain degree of order to exist within a liquid under special circumstances. There are, in fact, many different types of liquid crystal and many examples of each. Research into their properties began as long ago as 1888 when the Austrian bio-chemist Friedrich Reinitzer discovered this new form of matter while studying the properties of cholesterol, and in 1991 the French theoretical physicist Pierre-Gilles de Gennes was awarded the Nobel Prize for his work in elucidating their unusual behaviour. Work at RCA, the University of Hull and at RSRE Malvern led to the development of materials designed specifically for display applications.

Out of the whole gamut of possible liquid crystal behaviours, we shall refer to only one particular type, known as 'twisted nematic' (from the Greek word 'nema', meaning 'thread') which is characterised by long, thin molecules that can be persuaded to take up specific orientations with respect to the cell containing them. The method of using such twisted nematic liquid crystals to make a display employs two clever techniques, one involving the interaction between the thread-like molecules and the internal surfaces of the container and a second concerning the effect of an electric field on their orientation. It also depends on the fact that the molecules can interact with a light beam shining through the cell. Let us examine each in turn. Imagine a situation where the liquid crystal material is sealed between a pair of optically flat glass plates separated by a very small distance (something like 10 μm). Firstly, the molecules can be oriented at the wall of the cell by making fine scratches along the glass surface (a process known as 'rubbing') and the inspirational aspect consists of rubbing top and bottom plates in mutually perpendicular directions. This has the effect of causing the molecules to change their alignment gradually through the thickness of the cell and, remarkably, this has the property of rotating (by 90°) the plane of polarisation of a polarised light beam passing vertically through the cell. Suppose now that a pair of optical polarisers with parallel orientation is placed immediately above and below the cell. The lower polariser produces plane polarised light whose plane is rotated by the liquid crystal molecules so that it is now at right angles to the plane of the upper polariser. The result is that no light is transmitted through the cell–polariser combination. However, if an electric field is applied between the plates of the cell, the molecules tend to reorient along the direction of the field, so they no longer interact with the light, which is therefore transmitted without attenuation. Thus, application of an electric field has the effect of switching the transmitted light between zero and maximum intensity – the structure acts effectively as a 'light switch'. So, to complete our description of the switch we merely need to add a pair of transparent electrodes at top and bottom of the cell – these usually take the form of evaporated films of an unusual oxide, indium tin oxide (ITO), which conducts like a metal but transmits light like an insulator. A voltage applied between these two electrodes serves to operate the switch. (This description applies to the case of a 'back-lit' display – i.e., with a light source behind it – which is the mode of operation appropriate to a TV display, but a suitably modified version can be designed to work with reflected light, as used in a digital watch.)

In order to make a seven segment display (see Fig. 10.2), for example, the space between the plates must be divided off into appropriately shaped sections and the top electrode patterned to match. A common electric connection is then made to the back electrode, while individual connections are made to the various top electrodes in order to switch the individual segments on and off as required. In the case of a colour television display, the sections take the form of individual pixels, each pixel consisting of three 'pixelettes' combined with appropriate colour filters. There are, of course very many of them and they are very much smaller than in the seven segment format but the operating principle is the same. As we shall see in a moment, the essential, and extremely important, distinction is one of 'addressing' – that is, the method of imparting the necessary picture information to the individual cells.

In the seven segment display, each segment is either 'on' or 'off' – there is no question of any intermediate state – and this demands a liquid crystal with a so-called 'switching characteristic'. In other words, it requires only a small change in voltage to change it from dark to light.

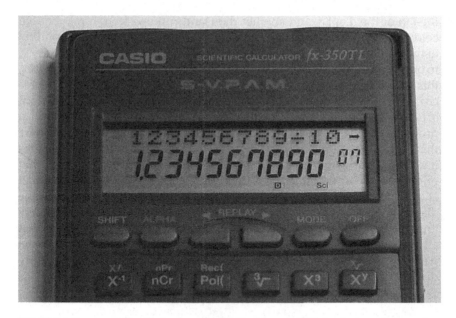

FIG. 10.2   One example of a seven-segment liquid crystal display in a modern pocket calculator. Being a passive form of display, liquid crystals took over from LEDs as the preferred technology during the 1960s, largely because they made far less demand on battery power. The simple on/off nature of the display elements made for relatively easy addressing circuitry. Courtesy of the author.

In the case of the TV display, the situation is very different because the actual brightness of the transmitted light is important. It must be possible to adjust the voltage over a wide range to obtain the necessary range of light intensities, and each pixelette has to be provided with its own individual brightness instruction – that is, its own specific signal voltage. This seemed to imply that each one must have its own electrical connection in the same manner as applied to the seven segment display but herein lay a major problem. As we have already remarked, there are several million pixels in a TV display, each one being roughly a tenth of a millimetre in linear dimension. Individual wires to each one could only be fashioned from thin lines of evaporated metal which would somehow have to be squeezed through the even tinier spaces between pixels, and a simple calculation shows that they would need to be impossibly thin, resulting in their having impossibly high electrical resistance. Clearly, some alternative means of addressing the pixels had to be invented – it is known as 'matrix addressing' and it requires only one wire per row (or column) of pixels, rather than the thousand or more required by the direct connection method. Suppose we have to address a million pixels in a square array. Imagine a thousand wires running from left to right, across the top of the display and another thousand running from top to bottom below the display. There will be a million crossing points, corresponding to the million pixels in the display and, if we apply a positive voltage, say, to one of the upper wires and a negative voltage to one of the lower wires, this will effectively address the pixel at the point where these two wires cross. The two voltages can be adjusted to produce just the right brightness from the pixel in question. Wonderful! But there is still a problem. All the pixels along that particular row and all the pixels along that particular column will also be addressed with the individual voltages involved. If it were simply a matter of switching a pixel on or off, that might not matter (because the individual voltages could be chosen small enough to be below the switching voltage) but, as we noted above, the display pixels have a characteristic which produces an approximately linear variation of brightness with applied voltage and this implies that all the pixels along the row and all those along the column would be partially switched on, a totally unacceptable situation in a high quality display. Impasse!

It was here that semiconductors were to come to the rescue. Suppose that at each pixel point there is a switch which can be operated by the application of an electric voltage. If we apply a suitable voltage along one row, this will switch on all the switches in that row and if those switches are correctly designed they may make connection between the relevant pixel and the column wire which is to address it. All we then have to do is arrange for the appropriate signal voltages (brightness information) to be applied along the column wires to light up each pixel correctly during the time for which the switch is closed. The switching voltage is then moved down one row and the next line of switches activated, while the signal voltages are reset to those values appropriate to this next row. Repeating the sequence for all the rows allows us to scan the necessary brightness information to all pixels in the display. Then we can simply repeat the whole sequence at regular intervals to maintain the picture, the essential criterion being that this whole 'frame scan' is completed in less than one twenty-fifth of a second (the integration time of the eye). Suppose we complete the scan in one-hundredth of a second. This implies that each row is addressed for a time of $[1/100 \times 1/1000]$ s (assuming a thousand rows), or 10^{-5} s which implies that the signal circuitry (or 'drive circuitry') should operate at a speed corresponding to a frequency of about 1 MHz. Standard integrated circuits have no difficulty with this. So, all we have to do is design a suitable switch and replicate it several million times over the area of the display – another way of saying this being that we make a giant integrated circuit with several million transistors (no big deal in itself) but over an area of about half a square metre! In integrated circuit terms, the resolution problem might be trivial but the area problem appeared horrendous. That it has been successfully overcome reflects creditably on a whole body of technologists, the world over.

This exciting story now lacks only two minor features – how was the switch to be designed and what was it to be made from? Clearly, it was not to be made from single crystal silicon – even 12 in. boules (which, in any case, didn't exist in the 1980s when all this was happening) would be inadequate for the size envisaged. (It would not be possible in this application to stack several rectangular slices together because the joints would be certain to show – the eye is remarkably clever at picking out such detail.) Some novel technology was called for and all the signs pointed to either a polycrystalline or an amorphous semiconductor, though it was far from clear which. It was even less clear what the material might turn out to be.

To some extent, the decision as to what kind of switch should be used was determined by some earlier work at RCA. In 1962, Paul Weimer demonstrated a thin film transistor (TFT) made from polycrystalline cadmium sulphide (CdS). It was a field-effect transistor made by evaporating a thin film of silicon oxide (the gate oxide) onto a glass substrate, followed by a second film of CdS to serve as the channel and he was able to achieve remarkably promising performance from it. There were even claims that it might challenge the silicon MOSFET (which was then still in its infancy) as the workhorse of the semiconductor industry – remember that, in the early 1960s, RCA were also very deeply involved in silicon MOSFET development. In the event, such claims proved to be over-optimistic but, nevertheless, showed the new technology to be surprisingly potent. It was clear that such a TFT could be used to switch the rows of pixels in a liquid crystal display by applying the switching voltage to the TFT gate electrodes, while the source-drain channels were connected between the signal lines and the pixels. Indeed, in 1973, a team at Westinghouse laboratories demonstrated a working monochrome display with cadmium selenide (CdSe) TFTs at each pixel point. It was modest in size and in resolution but more than adequate to prove the feasibility of this approach.

The second question to be answered was what material would be most suitable for the switches in a practical TV display. Research had already been done on several other polycrystalline semiconductors such as silicon, gallium arsenide, indium arsenide, indium antimonide, the lead chalcogenides and tellurium so there was plenty of choice. However, in the end, the decision was made to use none of these but rather the altogether unlikely material amorphous silicon. Why? Like so much else in this saga, it was a complicated story.

The best known example of an amorphous material is that of glass and its transparency indicates that amorphous materials, like their crystalline counterparts, may have wide energy gaps. Furthermore, these wide gaps result in insulating behaviour – glass being an excellent electrical insulator. However, many amorphous materials show semiconducting properties, behaviour that was explored during the 1950s in the shape of glassy forms of arsenic, selenium, tellurium, arsenic selenide and a bewildering range of more complex compounds. The subject built up a seriously dedicated following, much high powered effort being devoted to developing a theory of electrical conductivity, during the 1950s and afterwards. Certain rather mysterious switching behaviour in a number of these materials led to some tentative commercial interest but the subject remained a rather cosy, esoteric branch of semiconductor science until the 1980s when amorphous silicon came to the notice of those whose future careers depended on commercial profit, rather than on improved scientific understanding. The fame of amorphous silicon resulted in a severe distortion in the make-up of amorphous semiconductor conference programmes and even more so in the make-up of attendance lists!

Just how and why it came to pass makes a fascinating story in itself. Early work on amorphous silicon films deposited by evaporation or by sputtering showed that this material, like the majority of other amorphous semiconductors, was characterised by high resistivity and high densities of electron trapping centres (due to the dangling bonds we mentioned earlier). In short, it scarcely looked like a material appropriate to any useful electronic application. The situation changed abruptly, however, in 1969 when a group at Standard Telephone Laboratories in Harlow, Essex produced a new material which came to be known as 'hydrogenated amorphous silicon' (or a-Si:H) by deposition from a plasma of silane (SiH_4). This involved a container of silane gas excited by a radio-frequency discharge which broke down the gas into silicon and hydrogen ions, the silicon being deposited onto a glass substrate. The key to the new material's properties, though, was the fact that some of the hydrogen was incorporated in the silicon, effectively mopping up the infamous dangling bonds. No longer were electrons trapped – it was now possible to dope the material in much the same manner as appropriate to crystalline silicon and control the conductivity, accordingly. Of even greater significance, it also turned out to be possible to make a field-effect transistor. Here was an alternative material for making the switches required in the LCTV display. But, in saying all this, I am running ahead of myself – the virtues of a-Si:H were not immediately appreciated. Indeed, its virtues were not appreciated at all by the STL management, who promptly shut down the activity, as unlikely to lead to anything of commercial value. (Their decision could be perfectly well understood – in 1969, there appeared nothing of interest in such material when seen from the viewpoint of telecommunications.) This might well have been the end of the story but, fortunately, the running was taken up by two physicists at the University of Dundee in Scotland, Walter Spear and Peter LeComber, who successfully explored the electronic properties of a-Si:H for several years before passing the baton back again to industry. It was at their suggestion that the idea of using the material in LCTV display was seen to be worth pursuing and, by the time their proposal came to be widely accepted during the 1980s, numerous new research activities had mushroomed, not only in Europe but also in the USA and in Japan.

The fact that hydrogen incorporation resulted in a significant reduction in dangling bond density was clearly demonstrated by Spear and LeComber's work on field-effect measurements. By applying an electric field, they were able to modulate the conductivity of an a-Si:H film in just the same manner as occurs in a TFT and they were able to show that the density of deep trapping states was very much lower than in evaporated or sputtered amorphous silicon. Later work in several laboratories then showed that the addition of hydrogen to sputtered films could reduce the deep state density by orders of magnitude, thus confirming the idea. The corollary of these experiments was, of course, that a-Si:H was an excellent material for making TFTs and, from the point of view of making LCTV displays, it turned out to be preferable to any of the polycrystalline semiconductors. In Spear and LeComber's original argument, they placed emphasis on the idea that an elemental material like a-Si:H should be better with regard to uniformity, reproducibility and reliability, all features of importance in any commercial application, but the reality was that a-Si:H provided the most suitable switching ratio between its resistance in the 'off' state and the 'on' state. We now need to look at this point in a little greater detail.

When the TFT switch is 'on', the signal voltage in effect charges a parallel-plate capacitor which is made from the top and bottom ITO electrodes, with a dielectric (the liquid crystal) sandwiched between them. This charging process must occur in a time much shorter than the 'line time' (the time for which each line of pixels is 'on') and this puts an upper limit on the 'on' resistance of the TFT channel. On the other hand, when the TFT is 'off' the charge must remain on the capacitor for a time much greater than the 'frame time' (the time for a whole picture scan). This puts a lower limit on the 'off' resistance and allows us to calculate a minimum value for the 'switching ratio' (the ratio of 'off' to 'on' resistances). This turns out to be about a factor of 10^4, a ratio which could be achieved with either amorphous or polycrystalline TFTs. However, when one looks at the actual values of the two resistances, it becomes clear that the very much lower mobility of amorphous silicon gives a-Si:H a considerable advantage over its polycrystalline rivals – it just happens that this low mobility matches well with the geometry of a TFT for this particular application. It was this that clinched the decision, rather than any other consideration, though uniformity, reproducibility and reliability were certainly appreciated by the engineers responsible for turning such esoteric principles into a working consumer product.

The TFTs were made by depositing a film of silicon nitride over the evaporated gate electrode, to act as gate insulator, followed by a film of a-Si:H to provide the channel and n^+ a-Si:H to serve as source and drain, the drain making contact with the top ITO layer and the source with the signal line. The deposition equipment (plasma source) had to have dimensions large enough to contain the whole screen, of course and the individual TFTs had to be defined by photolithography on the grand scale. It all started with screen sizes of a few inches and gradually crept up to the present giants which, not so very long ago, seemed quite beyond the realms of reason (see Fig. 10.3).

FIG. 10.3 The new millennium saw liquid crystal, flat panel display take over remarkably rapidly from the long-standing cathode ray tube in television and computer display. The picture shows a typical LCDTV set such as has come to dominate most of our living rooms in the twenty-first century. Courtesy of the author.

So much, then, for the display, itself – there is one further aspect which demands our attention, that of the circuits used to 'drive' the panel. As we saw above, the speed with which the TFTs must be switched on and off is relatively modest and this process can be undertaken by a line of a-Si:H transistors along the edge of the display. However, the brightness signals change much too rapidly for such treatment and must be handled by some considerably more sophisticated devices. In the early days, this simply took the form of standard integrated circuits mounted in a hybrid fashion (i.e., a combination of two separate technologies) at the top edge of the display. It worked perfectly well but was expensive and aesthetically unsatisfactory (let us not forget that science and technology have aesthetic feelings, every bit as much as literature and the arts!), leaving a lingering desire to do something better. The simplest approach would obviously be to make use of the amorphous silicon, already available, but to convert it into polycrystalline material so as to achieve the somewhat greater electron mobility required. Two processes were tried. Firstly, it was found possible to effect the conversion by using a heat treatment at temperatures of about 600 °C but it was a slow process and was barely compatible with the use of glass plates for confining the liquid crystal. A much better solution made use of a process referred to as 'laser annealing' whereby the amorphous silicon was melted very briefly with pulses of light from a high power laser, scanned over the surface. The silicon recrystallised in polycrystal form and the resulting mobility turned out to be high enough to make drive circuits adequate for the purpose. It sounds very simple when tossed out like this but it certainly wasn't – it took something like 15 years to achieve success! But, at last, all was well – even the more sensitive engineers were happy that they had a product they could be proud of and time has made clear that customers have a product they are happy to show off in their living rooms. Further refinements are inevitably creeping into the design at regular intervals but we shall mention only one such. The display screen we have described relies, of course, on a suitable back-light to produce an adequately bright picture and, in the first instance; this took the form of a custom-designed fluorescent tube. More recently, following the development of white light LED sources, there has been a move to replace it with LEDs. They operate at just a few volts, compatible with the voltages used in other parts of the display and their extremely high efficiencies are an obvious bonus in these energy conscious days. We shall leave it there – it seems that the fundamentals will not change very much and our story is historically acccurate, no matter what may happen in the future – it is now time to take up our second major topic, that of photovoltaic solar cells.

In the case of the LCTV display, it is clear that amorphous and polycrystalline semiconductors play an essential role – single crystals are totally inappropriate. On the other hand, in the case of the solar cell, single crystals have so far played a major role, while amorphous and polycrystalline semiconductors have struggled somewhat to compete. Hydrogenated amorphous silicon has found a niche and several forms of polycrystalline cell have threatened to break into commercial contention but, at the present time, single crystal silicon still dominates the market. However, in the long-term, it seems likely that the market will be split between thin film cells, on the one hand, and sophisticated (and expensive!) single crystal cells made from compound semiconductors, on the other. The vital parameters in this complicated balancing act are those of conversion efficiency – how many watt-hours of electrical energy can be generated from any given amount of solar energy – and cost. These two may then be rolled into a single quality factor – the cost of electricity in units of dollars per watt. The cost is very largely one of initial capital outlay, while the power generated depends on the effective area over which the solar radiation is collected, together with the efficiency of conversion. As we remarked earlier, it is generally true that single crystal cells, while more expensive, are also more efficient. It is also true that, the larger the collection area, the greater the capital cost, though within such generalisations there exist many subtle qualifications which tend to render *any* generalisations unreliable! Consider, for example, the fact that many collection systems make use of some method of concentrating solar energy by focussing it onto a relatively small area. Consider also the fact that solar energy makes itself available over a wide spectrum of wavelengths (or photon energies), while solar cells tend to be selective in their responses – in particular, long wavelength radiation may pass right through a cell without being absorbed, while short wavelength radiation is likely to be converted with only low efficiency. Cells tend to show maximum conversion efficiency for photon energies just greater than the semiconductor band gap. If all this has left readers totally confused, I make no apology – the subject is indeed rather complicated. Hopefully, some degree of clarification will emerge from the following discussion, though it would be arrogant of me to suppose that I can do more than provide some elementary guidance. Perhaps our best hope is to begin by examining the history of photovoltaic solar energy conversion, then look in greater detail at the technicalities.

As we saw in Chapter 3, the solar cell dates back to a fortuitous observation made by Russell Ohl at Bell Labs in 1939. He accidentally produced a piece of silicon containing a p–n junction which showed remarkable sensitivity to visible light, producing an electrical output whenever light was shone onto it. The Bell management immediately saw an important future for such a phenomenon, in providing electric power to remote telephone switching stations, and in 1954 Calvin Fuller and Gerald Pearson succeeded in making the first 'Solar Battery' based on a diffused p–n junction. It was the beginning of a lengthy and important development which is still far from reaching its ultimate conclusion (though many instances of similar remote power supplies exist in today's high tech world – telephones along French motorways, for example – and even more in the low tech Third World where solar power provides remote villages with the fundamentals of civilised existence). The modus operandi of a solar cell is easy to understand – absorbed light of photon energy greater than the semiconductor band gap generates free electrons and holes which are separated by the electric field in the p–n junction and swept out to metal contacts on either side of the junction, thus generating an electric current in an external circuit. All this was understood very early on by the Bell workers and the theory of the cell was also well understood. In 1961, Shockley and Queisser published a paper which analysed the 'single junction' cell and showed that its maximum efficiency for converting light energy into electrical energy was about 31%. Practical values for the early Bell devices were little more than 5% so, leaving aside Bell's special interest in telephone switching, there was only mild interest in the rest of the world. It was far too expensive and far from good enough to excite widespread interest in power generation.

However, this degree of apathy was soon modified by one important development – the coming of the satellite age. Vanguard I was launched in 1958 and the first satellite communication experiment by way of Telstar took place in 1963, both powered by silicon solar cells. It was soon followed by others but the real boost to solar cell development came with the cold war rivalry which led to the man-on-the-moon race, culminating in Armstrong and Aldrin's moonwalk in June 1969. The race was on to develop ever more efficient power sources. Not only did the efficiency of single crystal silicon cells improve significantly, but other materials were investigated too. First came gallium arsenide heterostructures which seriously challenged silicon with efficiencies of over 20%, then came even more sophisticated III–V devices with two or even three different materials (each containing a p–n junction) which gradually achieved efficiencies of over 30%. The point to appreciate here is that this was an application where cost was of secondary importance – efficiency and reliability were paramount, acting as admirable stimulants to research. In fact, it was a small-scale replica of the corresponding race to develop integrated circuits, which proved so timely for the American electronics industry. Let it not be thought, however, that silicon could not fight back – the efficiency of single crystal cells appears now to have levelled off at about 24%, closely similar to the best GaAs cells (and approaching the theoretical limiting value). Only multi-junction cells are able to do significantly better than that – the best results reported to date being close to 40%. There is a modest industry dedicated to the development and manufacture of solar cells for satellite applications but it will never set the world alight in commercial terms. The long-term future clearly lies with terrestrial power generation, in both large and small scale installations – and that is a different ball game altogether. Cost is even more important here than efficiency and ultimate solutions will surely depend on quite different materials.

Having said that, it is interesting to note that, initially, single crystal silicon succeeded in cornering the lion's share of the terrestrial market. Not only was it widely available, it was also reliable, moderately efficient and able to rely on the background support of silicon integrated circuit development. Larger and larger silicon boules were being produced in the interest of reducing integrated circuit costs, and solar cell manufacturers were happy to swim with the tide. What was more, they were often able to make use of material not quite up to specification for IC manufacture, which helped their material costs considerably. Nevertheless, costs were (and still are) a serious problem – assuming an operating life of 20 years, a practical, commercially advertised domestic system in the United Kingdom (see Fig. 10.4) (including a DC-to-AC inverter and power meter)

FIG. 10.4 A typical photovoltaic solar panel installation on a domestic roof. This form of renewable energy would appear to be ideal in so far as it can make a serious contribution to the energy requirement of the average household, while having no discernible disadvantage in the shape of spoiling the environment or causing distress to wildlife. Courtesy of iStockphoto, image 6947649.

will generate power at a cost approaching 40 pence/ kWhr, some three to four times more than I currently pay for my 'conventional' electricity (even after recent increases in 2008). Clearly, research effort had to concentrate on bringing down the cost of photovoltaic panels if they were to gain widespread acceptance in the domestic market. Indeed, similar comments applied to the development of systems for large scale power generation.

Two quite different approaches to this problem were pursued. In the case of large systems, emphasis was placed on the use of optical concentration – lens or mirror arrays designed to focus sunlight down onto much smaller solar cells, where the cost of the total system is dominated by the concentrators, rather than the cells. Concentration factors of up to a 1000 times are typical. This means that it is acceptable to employ highly efficient cells, even though these might cost very much more than standard silicon cells. In fact, the trend has been towards using similar, multi-junction cells to those developed for satellite applications, providing efficiencies as high as 40%. This had the effect of transferring the development challenge to the design of the concentrator – it was necessary to design a tracking system capable of pointing the array directly at the sun and holding it there to an accuracy of about 1°. It was also important to provide sufficient mechanical stability to meet the aggressive intent of sudden surges in wind strength. On the other hand, attempts to tackle the 'house roof' problem were based largely on the use of less expensive approaches to the manufacture of the cells themselves. A range of techniques was applied first to improving the efficiency of single crystal cells but it soon became clear that something more drastic was needed, leading to the development of several forms of thin film cell. Chief among these were polycrystalline silicon, amorphous silicon, copper indium di-selenide ($CuInSe_2$) and cadmium telluride (CdTe), the latter pair being heterostructures, with cadmium sulphide windows. In most of these, the materials were deposited by simple evaporation or chemical means on glass or steel substrates, processes that were very much cheaper than the growth of single crystals of silicon. In round figures, all these so-called 'second generation' cells cost something like a factor of three times less than their predecessors, while offering conversion efficiencies of 10–15%, comparable with the earlier crystalline silicon cells. Nevertheless, the overall performance of these cells appears still to fall short of that necessary for their widespread adoption. To meet the criteria laid down by the United States Department of Energy, for example, it seems necessary for a further improvement in the efficiency-cost ratio by about a factor of four. What is more, these cells have yet to prove themselves capable of working reliably over a period of 20 or more years. It is important to recognise that the cost of electricity is essentially determined by the capital cost of the system, divided by the operating lifetime, and if the latter falls short, the generation cost is correspondingly increased.

At the time of writing, this is more or less where things stand. There is obviously a need for considerably more research into new materials and new fabrication methods, which, in turn, implies considerably greater investment. That photovoltaic solar panels should and, indeed, must play a much greater role in solving the world's energy problems is transparently obvious. It is a simple fact that more solar energy falls on the earth each day than the whole of mankind uses in a year. It is also true that the amount of energy falling on the roof of each and every building is nearly 10 times that used within. Nor should it be thought that the sun only shines in warm southern climes. The difference in *the yearly average* solar insolation between the equator and the north of Scotland is only a factor of three, which means there is more than sufficient radiation in the United Kingdom to make photovoltaic electricity a viable possibility. It is already clear from data collected over the past two and a half decades by Martin Green of the University of South Wales, Australia (who has been responsible for several important solar cell developments) that a major factor in bringing down the cost of photovoltaic solar electricity is one of investment in generating capacity. As the total global solar generating capacity has increased, the cost of solar electricity has fallen by slightly more than a factor of two for each order of magnitude increase in installed capacity, and, as the present capacity represents no more than one-ten thousandth part of the world's energy usage, there is still plenty of scope for further investment! At a rough estimate, assuming present trends continue, the factor of four reduction in solar energy cost required to meet the cost of conventionally generated electricity should be achieved by about the year 2020, even with no further improvement in cell performance. However, it could happen much more quickly if sufficient effort were to be invested in appropriate research. This, of course, demands political will – the solar resource is far greater than that of windpower, yet the United Kingdom has chosen to concentrate all its investment in this latter, rather than speed up the rate of solar cell development. Some countries do better – Germany, for instance has evolved a scheme whereby private owners of solar generators are allowed to sell their surplus power back to the national grid at a price determined by what it costs them to generate it (even though that may be almost four times the going rate for conventional power). In consequence, Germany is far ahead of the United Kingdom in its investment in solar generating capacity and, incidentally, in its industrial capacity to exploit the solar power business. California is leading the way in the USA in establishing incentive schemes to stimulate the installation of small scale solar generating plant, while a significant amount of research is funded by the European Union into high efficiency concentrator systems. These are merely a few typical examples and it is surely no more than a matter of time before photovoltaics supply an important fraction of the world's energy needs. However, it would be nice to feel that this time might be shortened considerably by even greater investment in appropriate research and development.

So much for politics – let us now examine the ins and outs of solar cell function and fabrication in a little more detail. Firstly, we should be clear about the wavelength distribution of the radiation reaching the earth, because it is necessary that the majority of this spectrum be utilised in converting light into electricity. In terms of photon energies (which can be more easily related to semiconductor band gaps), the spread covers the range from about 0.5 eV in the infra-red (2.5 μm) to 4 eV in the ultra-violet (0.3 μm). There is a fairly sharp peak at 2.3 eV in the green part of the visible spectrum and a steady fall-off into the infra-red. In order to absorb most of this radiation we might think of choosing a semiconductor like germanium, with a band gap of 0.67 eV but such a choice would be far from optimum because the high energy photons would be utilised very inefficiently. I should explain. The output power from the cell is measured by the product of the current generated times the cell voltage, so we must be clear how these two quantities are determined. To a good approxima-tion, the current is determined by the number of photons 'collected' (i.e., converted into electron-hole pairs and swept out into the external circuit) per second. However, each photon absorbed, *no matter what its energy*, results in only a single electron-hole pair, the problem being that photons with energies larger than the band gap produce electrons well above the conduction band edge (and holes well below the valence band edge) and the excess energy is lost in the form of heat as these 'hot' electrons drop down to the bottom of the conduction band (holes rise up to the top of the valence band). This means that, in a germanium cell, a flux of 4 eV photons will yield the same current as the same flux of 1 eV photons.

On the other hand, the cell voltage is determined by the band gap of the semiconductor employed (the precise relationship is rather compli-cated but we can say that the maximum possible cell voltage is equal to the bandgap voltage, E_g/e). Thus, the output voltage of a germanium cell will be typically about 0.5 v, whereas, if we had chosen a semiconductor with a band gap of 4 eV (which could equally well collect 4 eV photons), we might expect an output voltage of perhaps 3 v. Clearly, the power output from the 4 eV photons would be roughly six times greater for our 4 eV bandgap material, a tremendous improvement. This would be fine, except that a 4 eV bandgap would allow most of the solar radiation to pass straight through, unabsorbed. The overall efficiency would be minimal! There is obviously a need for compromise. If we decide to use a single semiconductor material, its bandgap should be something like 1.3 eV – assuming everything goes well within the cell, this can be shown to give a maximum overall efficiency close to 30%. Silicon (1.12 eV) falls a little below the optimum, GaAs (1.43 eV) falls a little above, so the best we can hope for from either of them in simple, single cells is about 25%, efficiencies which have now been demonstrated in practice.

However, the above discussion suggests an obvious strategy to utilise more of the solar spectrum and thereby do significantly better. We could use a wide gap material to deal with the high energy photons, a middle gap material for the mid-energy photons and a small gap material for the low energy photons, thus optimising the powers from each. If we were to stack them one above the other, with the wide gap material on top and the small gap material at the bottom, the upper cell would absorb the high energy fraction, allowing the rest to pass through, to be absorbed selectively by the two lower cells. All we then have to do is arrange that the three outputs are added together and we have the best of all worlds! Actually, this is not quite true – to achieve *that*, we need to use an infinite number of cells in tandem (which can be shown to give about 60% efficiency), rather than just three, but 'three' represents a practical compromise (yielding theoretical efficiencies of about 50%). As we have already mentioned, the best results obtained so far, in practical structures, are close to 40%. What, then, of these practical structures?

In order to achieve a convenient stack, it seems sensible to grow the various semiconductor layers epitaxially, one on top of the other and this has been the approach so far adopted. A typical sequence uses InGaP (band gap 1.85 eV) on top, InGaAs (band gap 1.35 eV) below and germanium (0.67 eV) at the bottom. The whole structure can therefore be grown on a single crystal germanium substrate, to which the top layer is precisely lattice-matched, while the InGaAs is only slightly mismatched. The amount of indium is kept fairly small to minimise this mismatch. There are two significant difficulties with this kind of tandem cell. The first relates to the fact that each layer contains a p–n junction to separate the photo-generated electrons and holes (to prevent their recombining). If the p-layers are on top, we have a top-to-bottom sequence: p–n, p–n, p–n, which makes clear that between each different material there is an n–p junction, which happens (inevitably!) to be reverse-biased, and some care is necessary to minimise the associated losses. In practice, this means using heavy doping in both layers, while making the individual collector junctions in the form $(p^+ - i - n^+)$. The second problem concerns the fact that the current flowing through the structure must be the same in all of the cells so it is necessary (for optimum performance) to design each cell to produce this current. To do so means matching each cell to that portion of the solar spectrum that it is designed to use, a fairly tricky procedure, at best, but made much more difficult by the fact that the wavelength distribution tends to change during the course of a day, leading to inevitable variation in overall conversion efficiency (though this is not a problem in satellite applications). That overall efficiencies as high as 40% have been achieved can only be seen as reflecting positively on those burdened with such design problems!

Incidentally, it is nice to think that germanium might now be on the point of making a grand come-back. Having been seriously involved in the original invention of the point-contact transistor, it has been obliged to take a back seat on account of its rather small band gap, with the consequent tendency to thermal runaway. However, in the case of high efficiency solar cell development, this band gap seems well matched to tandem cell requirements, while germanium's good lattice match with gallium arsenide must be seen as an added bonus. As the element was discovered in 1886 by a German chemist, Clemens Winkler, who named it after his fatherland, it is appropriate that it should return to prominence in a context where Germany is making a major contribution.

Another design problem which we have ignored so far is that of light absorption. As noted previously on several occasions, direct gap semi-conductors are much better at this than indirect materials such as silicon and germanium, so it is perhaps surprising that silicon should have cornered the terrestrial market so effectively (the explanation, of course, concerns price). It did, however, lead to significant technical problems in reaching acceptable efficiencies. Initially, the technology of making silicon cells followed closely that of integrated circuits – 300 μm slices were sawn from a Czochralski-grown boule, diffused to form the p–n junction, then cut into squares suitable for mounting together to cover the necessary large area. It used a lot of silicon – much more than was required to absorb the light (30 μm would have been sufficient) – and, though silicon was a relatively cheap material, crystal growth was far from a cheap process. Thinner slices necessitated the development of a shallower diffusion process, aided by the introduction of a 'back surface electric field' to reflect minority carriers back towards the junction, rather than allow them to recombine at the back surface. Other important innovations included the use of a textured upper surface which served to reduce reflection of incident light and to contain the light within the cell. Combined with reflection of light from the back surface, the absorption process became highly efficient even in a very much thinner cell. Further improvements were effected by arranging for the p–n junction to lie in a vertical plane – a more effective method of collecting minority carriers. Conversion efficiencies rose gradually in response to these concerted efforts but the cost of single crystal material grown by either Czochralski or floating zone methods was still a worry for panel makers, so various methods of producing large grain polysilicon were tried. One such, which attracted wide interest was the EFG (edge-defined film-fed growth) method in which molten silicon was pulled through a slot in a graphite block to produce a long ribbon. Several metres of material could be produced from a single pull and the use of a single crystal seed at the start of the process ensured a modest approximation to single crystal growth. Surprisingly, conversion efficiencies were little different from those obtained with true single crystal samples. Large area commercial panels such as those shown in Fig. 10.5 showed efficiencies of typically 15% but with considerable cost advantage over earlier single crystal cells.

So much for single and pseudo-single crystal silicon, which still employed bulk material. Attempts to reduce the volume of semiconductor by using thin films deposited on glass or metal substrates had been under

FIG. 10.5 An array of photovoltaic solar panels applied to the large scale generation of electricity, as an adjunct to the national grid. In those parts of the world which receive plenty of sunlight solar electricity is close to meeting the economic challenge of providing power at a cost equal to that generated by burning fossil fuels. Courtesy of iStockphoto, © Otmar Smit, image 3762588.

investigation for some considerable time. One of the first such techniques, developed during the 1960s, used a CdS/Cu$_2$S heterojunction. Cadmium sulphide, with a band gap of 2.49 eV, absorbed high energy photons, while cuprous sulphide (about 1.2 eV) took care of the longer wavelengths. However, this was not a tandem cell – the output voltage was determined by the cuprous sulphide alone. It was originally made by evaporating a film of cadmium sulphide onto a metal substrate, then treating it with a hot solution of copper chloride to form the desired cuprous sulphide layer. Later, a spray process was developed which allowed continuous, production line deposition and led to some over-confident predictions of cheap electricity generation. Sadly, there were problems with stability and this particular structure seems to have disappeared from the scene altogether. However, it certainly stimulated efforts to develop other thin film structures, including CdS/CdTe, CdS/CuInSe$_2$ and our old friend a-Si:H, all of which came into prominence during the 1970s.

Hydrogenated amorphous silicon has established itself as a photovoltaic solar cell material, with a niche market in supplying power to pocket calculators and wrist watches so I will consider it first. Interestingly, its 'pseudo-band gap' (about 1.6 eV) is considerably greater than that of crystalline silicon but still well within the range of values appropriate to making efficient solar cells. It is not possible to speak of either 'direct' or 'indirect' gaps because these concepts apply only to crystalline semiconductors. However, light is strongly absorbed by a-Si:H, making it suitable as a thin film photovoltaic material and the fact that it can be deposited over extremely large areas is another favourable property. The first observation of the photovoltaic effect in a-Si:H was made at RCA by Dave Carlson, in 1974 and RCA pursued research into possible solar cell applications for a number of years. Initially, conversion efficiency was minimal because of problems with metal contacts and with material quality. In particular, the efficiency with which minority carriers can be collected depends on the lifetimes and mobilities of minority carriers, both of which are degraded by the presence of electron states within the energy gap. However, both difficulties were overcome and efficiencies gradually improved. By 1980, both RCA and various rivals had achieved values of 7%; then further improvements were made following the use of tandem cell structures. By alloying silicon with a modest amount of hydrogenated germanium it was possible to make a narrower gap material and by alloying with carbon a wider gap material. However, the quality of the amorphous material was paramount and careful studies of the deposition process eventually resulted in efficiencies of 12% for single layer cells and, more recently, 14% for tandem cells. Oddly enough, though the early developments were made in the USA, it was a bevy of Japanese firms that took up the running and commercialised amorphous silicon cells for powering pocket calculators and watches. Indeed, it was a Japanese inspiration that spotted the suitability of amorphous silicon for converting not only sunlight but also domestic lighting sources into battery power for domestic devices, applications where efficiency was of only marginal importance but where cost was vital. As a consequence, amorphous silicon now has a healthy 6% of the world market for solar cells.

Competition in thin film technology has come principally from two direct gap semiconductors CdTe and $CuInSe_2$, allied with CdS window layers. As CdS can only be doped n-type, it has been necessary to dope the other component p-type, fortunately not difficult in either case. However, much materials development effort had to be put into realising adequate quality in the thin films before acceptable efficiencies could be realised. Eventually, both types of cell achieved 16%, a few percent better than amorphous silicon – but is that enough to bring commercial success? Probably not – low cost and efficiency in excess of 20% is the minimum requirement to bring electricity cost down to the level of conventional generation costs. Perhaps it will be possible to make tandem cells in these systems and push performance that vital bit higher. At the moment such possibilities are nothing more than wishful thinking. The world waits impatiently for signs of practical realisation.

An alternative approach to obtaining higher efficiency in home-based solar panels might be to use concentration, but in a very different manner from that adopted in large scale generation plant. The idea of installing clumsy optical systems and accurate steering systems on one's house roof sounds totally impracticable but an interesting alternative exists in the shape of what is now quite an old idea. Proposed during the mid-1970s, the 'luminescent (or fluorescent) concentrator' makes use of a thin film of a suitable light absorber, which then emits photons at a specific (lower) energy, which are, in turn, concentrated on a matching photovoltaic cell. Because the cell is designed with a band gap slightly smaller than the emitted photons, it operates at maximum efficiency (nominally close to 90%) and, because these photons are concentrated on the cell, its area can be much smaller than in normal systems, making it correspondingly cheaper. The key to success lies in the concentrator, which is nothing more than a plane sheet of glass, about 1 mm thick which functions rather like an optical waveguide with the cell mounted along one edge. Sunlight is absorbed over the whole area of the glass, light from the luminescent material being emitted in all directions and most of it totally internally reflected so as to stay within the glass. It therefore travels towards the edges of the sheet, and if we arrange mirrors on these edges, everywhere apart from the small area covered by the solar cell, a large fraction of the light will reach the cell. It is clear that, if the process were 100% efficient, the concentration ratio would be the ratio of the area of the glass sheet divided by the area of the cell – factors of a hundred or more being easily possible. Such ratios make it possible to consider the use of relatively expensive single crystal cell material, such as InGaP, GaAs, Si or Ge. In fact, the likely practical arrangement would make use of three wave-guide sections, one on top of the other, operating at three different wavelengths, so it would be necessary to use three different cell materials. The principal is similar to that of the tandem cell. In the upper sheet, the luminescent material would absorb high energy photons and 'red-shift' them by an amount sufficient to prevent reabsorption of the emitted light. Photons of lower energy would be transmitted through the upper sheet and be incident on the sheet below. Here, the process would be repeated at lower photon energies, while the bottom sheet would operate at lower energies still. Such an arrangement would clearly make optimum use of the available solar spectrum.

By presenting the argument in subjunctive mood, I have probably made it clear that it is still more in the realm of the imagination than one of hard practical fact but this should not be interpreted as being dismissive – practical demonstrations do exist and theoretical models do encourage further development. The original proposal was concerned with using organic dye molecules as 'down-convertors' (or red-shifters) and various combinations have been explored with promising results. In this case, the luminescent material has been deposited in thin film form, on top of the glass sheet and, ideally, the glass should have the same refractive index as the thin film to ensure that all the light is transmitted into the glass 'waveguide'. An interesting alternative might be to use layers of semiconductor quantum dots (see Chapter 6), as proposed by Keith Barnham and collaborators at Imperial College, London. Several inexpensive methods of making quantum dot structures are available (not involving epitaxial growth) and it has been demonstrated that dots can act as very efficient down-convertors. They also have a big advantage in the context of solar concentration because their absorption and emission wavelengths can be controlled simply by varying the physical size of the dots. For example, InP dots can cover the visible spectrum and InAs dots the infra-red region by varying dot diameter between 2 and 6 nm, which can readily be done by adjusting growth conditions. This is a real boon because it facilitates match-making between the down-convertor and well established solar cell materials. Another key parameter is the degree of reabsorption of emitted photons by the dye (or quantum dots) because the desired large concentration ratios require that emitted photons have to travel relatively long distances in the waveguide and are therefore more likely to be absorbed again before reaching the cell. Minimising reabsorption demands a large red-shift so that the emitted photon energy is well removed from the tail of the absorption spectrum. Quantum dots have a significant advantage in this respect because the red-shift can be controlled by adjusting growth conditions, while in the case of organic dyes the red shift depends on the choice of material and is far more difficult to control. Time will tell how these various factors can be successfully optimised but it is clear that a great deal more work is necessary before practical systems can emerge. But let it not be forgotten – the prize of cheap, efficient photovoltaic conversion is well worth the effort.

This brings me to the point of trying to sum up the present position. It is generally agreed within the scientific community that present day solar panels are either too expensive or too inefficient (or both!). For many years, it has been the accepted wisdom that, in the interest of cost reduction, single crystal silicon cells must eventually be replaced by some type of thin film cell, based on non-single crystal material. However, it hasn't happened yet, the efficiencies of such cells having so far proved inadequate – best results to date hover somewhere below 20%, while the requirement for photovoltaic systems to generate electricity at, or below the cost of conventional methods probably demands efficiencies closer to 40%. A whole gamut of ideas has been put forward to overcome current deficiencies, some plausible, others rather less so, but one thing is clear – most of the problems revolve around materials. As we have seen all along in this book, materials are at the very heart of solid state electronics, be it transistors, lasers, LEDs, infra-red detectors or flat panel displays – and solar cells are no different. However, it seems likely that the next decade may see considerable progress towards the holy grail of reduced electricity prices, based both on large concentrator systems and on small roof-top installations. We should not overlook the fact that the present trend is already steadily downwards, based simply on the well established fact that economies of scale nearly always result in cost reduction. This implies that political will to stimulate the necessary expansion in photovoltaic generating capacity can play almost as important a role as high powered research to improve performance. One can only hope that the two will succeed in working together. It has been remarked before but I make no apology for repeating the sentiment that the field of solar conversion is at a truly intriguing stage of development. Perhaps, 10 years hence, commentators will no longer feel the need to continue saying so.

A particularly important development has recently appeared in the literature, in the shape of a large scale thin film solar power plant installed in Nevada by the California-based company Sempra Generation. The plant uses 167,000 cadmium telluride modules to generate 10MW of power at a price predicted to be close to that of 'conventional' electricity. The secret appears to be the use of a cheap and efficient method of depositing thin films, developed by their supplier, First Solar (based in Arizona). Such a promising development only emphasises the comments made above that photovoltaic solar panels have the potential to generate power at competitive prices. A similar technology could surely be used on every house and factory roof. All that is required is political will and a modicum of research funding.

BIBLIOGRAPHY

Bauer, H (Ed) "Automotive Electrics and Electronics" Society of Automotive Engineers, Warrendale, PA, 1999.

Bell, R L "Negative Electron Affinity Devices" Oxford University Press, Oxford, 1973.

Benda, V, Gowar, J and Grant, D A "Power Semiconductor Devices: Theory and Applications" Wiley, Chichester, 1999.

Blaabjerg, F and Chen, Z "Power Electronics for Modern Wind Turbines" Morgan and Claypool, San Rafael, CA, 2006.

Bleaney, B, Ryde, J W and Kinman, T H "Crystal Valves" Journal of the Institution of Electrical Engineers IIIA 93, 847 (1946).

Braun, E and Macdonald, S "Revolution in Miniature" Cambridge University Press, Cambridge, 1982.

Busch, G "Early History of the Physics of Semiconductors – from Doubts to Fact in a Hundred Years" European Journal of Physics 10, 254 (1989).

Cavendish, H "The Electrical Researches of The Honorable Henry Cavendish: Written Between 1771 and 1781", Edited by James Clerk Maxwell, Cambridge University Press, Cambridge, 1879 (Reprinted by Kessinger Publishing).

Dupuis, R D "The Diode Laser – the First Thirty Days Forty Years Ago" http:// www.ieee.org/organisations/pubs/newsletters/leos/feb03/diode.html

Goldstein, A and Asprey, W (Eds) "Facets: New Perspectives on the History of Semiconductors" IEEE Center for the History of Engineering, New Brunswick, NJ, 1999.

Green, M A "Power to the People: Sunlight to Electricity Using Solar Cells" University of New South Wales Press, Sydney, 2000.

Green, M A "Third Generation Photovoltaics" Springer, Berlin, 2006.

Greene, M "The Nearly Men – A Chronicle of Scientific Failure" Tempus, Stroud, 2007.

Hecht, J "City of Light" Oxford University Press, Oxford 1999 – Revised 2004.

Hilsum, C "The Use and Abuse of III–V Compounds" in Advances in Imaging and Electron Physics Vol. 91, p.171 Academic Press, New York, 1995.

Hilsum, C and Rose-Innes, C A "Semiconducting III–V Compounds" Pergamon Press, Oxford, 1961.

Hoddeson, L, Braun, E, Teichmann, A and Weart, S (Eds) "Out of the Crystal Maze: Chapters from the History of Solid State Physics" Oxford University Press, Oxford, 1992.

Hodges, H "Technology in the Ancient World" Penguin Books, Harmondsworth, Middlesex, 1971.

Holonyak, N Interview, IEEE History Center, New Brunswick, NJ, 1993.

Holonyak, N IEEE Transactions on Power Electronics 16, 8 (2001).

Hughes, A "Electric Motors and Drives: Fundamentals, Types and Applications" Newnes, Oxford, 1998.

Jenkins, T "A Brief History of Semiconductors" Physics Education 40, 430 Institute of Physics, 2005.

Johnstone, R "Brilliant: Shuji Nakamura and the Revolution in Lighting Technology" Prometheus Books, Amherst, NY, 2007.

Kilby, J S "Invention of the Integrated Circuit", IEEE Transactions on Electron Devices 23, 648 (1976).

Kurylo, F and Susskind, C "Ferdinand Braun" MIT Press, Cambridge, MA, 1981.

Madelung, O (Ed) "Semiconductors: Basic Data" Springer, Berlin, 1996.

Manchester, H "Light of Hope or Terror?" Readers Digest, February, p. 97 (1963).

Mulligan, J F "Who Were Fabry and Perot?" American Journal of Physics 66, 797 (1998).

Novikov, M A "Oleg Vladimirovich Losev: Pioneer of Semiconductor Electronics" Physics of the Solid State 46, 1 (2004).

Orton, J W "Material for the Gunn Effect" Mills and Boon, London, 1971.

Orton, J W "The Story of Semiconductors" Oxford University Press, Oxford, 2004.

Pearson, G L and Brattain, W H "History of Semiconductor Research" Proceedings of the Institute of Radio Engineers 43, 1794 (1955).

Queisser, H "The Conquest of the Microchip" Harvard University Press, Cambridge, MA, 1988.

Reid, T R "The Chip: How Two Americans Invented the Microchip and Launched A Revolution" Random House, New York, 1985/2001.

Riordan, M and Hoddeson, L "Crystal Fire: The Birth of the Information Age" Norton and Company, New York, 1997.

Schubert, E F "Light-Emitting Diodes" Second Edition Cambridge University Press, Cambridge, 2006.

Seitz, F and Einspruch, N G "Electronic Genie: The Tangled History of Silicon" University of Illinois Press, Urbana, 1998.

Shelley, T "Nanotechnology: New Promises, New Dangers" Zed Books, London and New York, 2006.

Shockley, W "Electrons and Holes in Semiconductors" Van Nostrand, New York, 1950.

Shurkin, J N "Broken Genius: The Rise and Fall of William Shockley, Creator of the Electronic Age" Macmillan, London and New York, 2006.

Sommer, A H "Photoemissive Materials" Wiley, New York, 1968.

Street, R A "Hydrogenated Amorphous Silicon" Cambridge University Press, Cambridge, 1991.

Sze, S M "Semiconductor Devices: Physics and Technology" Wiley, New York, 1985.

Welker, H J, IEEE Transactions on Electron Devices, ED-23, 664 (1976).

Zheludev, N "The Life and Times of the LED – a 100-Year History" Nature photonics 1, 189 (2007).

Absolute temperature temperature measured on the Kelvin scale. Absolute zero is at approximately −273 °C, the temperature of melting ice being 273 K.

Acceptor an impurity atom introduced into a semiconductor crystal which has the effect of generating free holes in the valence band.

Aharonov–Bohm effect an electron interference effect which has been demonstrated using one-dimensional semiconducting wires.

Analogue signal a voltage or current waveform in which the amplitude and frequency represent appropriate information – for example, a telephone voice signal.

Anode the positive electrode in a thermionic diode or triode.

Avalanche breakdown the rapid increase of reverse current in a diode when the reverse voltage exceeds the breakdown value. It results from the ionisation of atoms in the semiconductor by impact from energetic electrons.

Band gap the energy difference between the bottom of the conduction band and the top of the valence band in any semiconductor.

Bandwidth the range of frequencies employed in any communications system.

Binary system the use of arithmetic based on the power of two (rather than the more usual power of ten) to encode information. In practice, the two states used are represented by the presence or absence of a current or voltage pulse.

Bipolar transistor a semiconductor device doped so as to contain either a p–n–p or n–p–n structure, enabling it to amplify an analogue signal or serve as a switch in a digital circuit.

Bit   the basic unit of information in a digital system, usually in the form of a current or voltage pulse.

Bit rate   the number of bits per second.

Black body   a theoretical concept of a body which absorbs all the radiation falling upon it.

Bolometer   a type of infra-red detector based on the fact that various physical properties are temperature-sensitive. For example, the resistance of a metal wire increases as its temperature increases.

Bragg mirror   a mirror made up of a large number of alternating thin films having different refractive indices and separated by a quarter of the wavelength of the light for which the mirror is designed. The step in index at each interface reflects a small amount of radiation and all the reflected waves add in phase.

Breakdown voltage   the voltage at which a reverse-biased p–n junction or Schottky barrier diode breaks down – that is, at which the reverse current suddenly begins to increase rapidly.

Capacitor   a passive device consisting of a pair of parallel metal plates separated by a small distance which has the property of storing electric charge.

Carrier   the radio-frequency, microwave or optical wave which is modulated with information signals in a communication system.

Carrier confining layer   a film of semiconductor material with a wide band gap designed to confine free carriers in a thin film or quantum well by means of a step in energy.

Cathode   the negative electrode in a thermionic valve or semiconductor power device.

Cat's whisker   a fine metal wire in the form of a spring which makes contact with a semiconductor surface.

CFL (compact fluorescent lamp)   a modern form of fluorescent tube designed to serve as a direct replacement for the conventional tungsten bulb.

Chalcogenide   a generic term for sulphide, selenide or telluride compounds, such as lead sulphide.

Channel multiplier   an electron multiplier used in vacuum image tubes. It consists of a stack of fine glass tubes with a high voltage applied along the length of each tube.

CMOS (complementary metal oxide silicon)   a pair of MOS transistors, one with an n-type and one with a p-type channel which serve to store digital information.

Co-axial cable   transmission line in the form of a central conducting wire encased in an insulating material and sheathed in a copper outer.

Compound semiconductor   a semiconductor material containing two or more types of atom – such as GaAs or CdS or HgCdTe.

Conduction band   a band of allowed electron states in a pure semiconductor which, at absolute zero of temperature, would be completely empty of electrons but which, under normal conditions contains free electrons, giving rise to n-type electrical conduction.

Covalent bond   a method of holding atoms together in a crystal which involves the exchange of electrons between adjacent pairs.

Crossed polarisers   a pair of optical polarisers set at right-angles so as to pass no light.

Cross talk   mutual interference between two or more signals.

CRT (cathode ray tube)   a picture tube consisting of an electron gun and a phosphor screen, together with some electrostatic or magnetic means for scanning the electron beam over the surface of the screen.

Crystal lattice   an imaginary set of points in space which represent the nominal positions of atoms in a crystal.

Czochralski   a method of growing single crystals which involves slowly pulling a small seed crystal from a melt of the material.

Dangling bond   an unsatisfied chemical bond in a solid.

DBR laser (distributed Bragg reflector laser) a semiconductor laser which makes use of a pair of Bragg reflecting mirrors, rather than the more customary plane mirrors.

Decibel (dB) a logarithmic unit describing the ratio of two quantities. 3dB implies a ratio of two, 10 dB a ratio of ten, 20dB a ratio of a hundred, 30dB a ratio of one thousand, etc.

Delta doping a technique for doping a semiconductor in a single plane. See 'planar doping'.

Density in semiconductor parlance, 'density' is used to signify the number of free carriers or dopant atoms per unit volume. It has units of $(metres)^{-3}$ not $(kilograms-metres^{-3})$.

DFB laser (distributed feedback laser) a semiconductor laser which makes use of distributed feedback throughout its length, rather than by reflection at the ends of the cavity.

DH laser (double heterojunction laser) a semiconductor laser which uses a pair of heterojunctions to confine injected free carriers in the active layer.

Dielectric relaxation the process whereby the motion of electrons in a wire proceeds as a result of their electrostatic repulsion, movement of electrons at one end pushing electrons out of the opposite end.

Diffraction a process in which light quanta or electrons (for example) are scattered on passing through a structure with dimensions of the order of the relevant wavelength.

Diffusion gas molecules 'diffuse' down a concentration gradient – that is, they spread out from regions of high density to regions of low density. The process is used to introduce impurity atoms into semiconductors by heating in a high concentration of the atoms. Similarly, electron and hole 'gasses' diffuse away from high concentration regions.

Digital signal an electronic signal in digital form, that is, a stream of 'ones' and 'zeros', following rapidly after one another, rather than the old fashioned wavering sine waves which characterised an analogue signal.

Direct gap the energy gap in a semiconductor which favours strong absorption and emission of photons, without the involvement of phonons.

Dislocation   a defect in a crystal, in the form of a jog in the linear arrangement of atoms along a crystal direction.

Dispersion   different wavelengths travel at different velocities in an optical material. This is known as wavelength dispersion. A similar phenomenon occurs in waveguides, whether optical or microwave. Different waveguide modes travel at different velocities.

Diode   strictly, an electronic device which has just two electrodes. It frequently takes the form of a rectifier.

Donor   an impurity atom introduced into a semiconductor in order to dope it n-type – that is, to introduce free electrons.

Doping level   the density of donors or acceptors in a semiconductor.

Double heterostructure   a type of semiconductor laser structure consisting of a narrow gap semiconductor layer sandwiched between two wider gap materials.

Drift velocity   the velocity with which electrons or holes move under the influence of an applied electric field.

DX centre   a troublesome deep level formed in silicon-doped AlGaAs when the aluminium content is increased beyond about 20%.

Effective mass   the apparent mass of an electron in a semiconductor conduction band (or hole in the valence band), measured by its acceleration in an applied electric field.

EFG (edge defined film-fed growth)   a growth method used to produce long, thin strips of polycrystalline silicon for solar cell applications.

Electric field   literally, the rate of change of potential. Thus, if a voltage V is applied between the plates of a parallel plate capacitor, separated by a distance d, the electric field within the capacitor is V/d. (This is a nice simple example because it is 'linear' – not all examples are!)

Electroluminescence   the conversion of electrical energy (in the form of an electric current) into radiation.

Electromagnetic wave   in 1864, Clerk Maxwell predicted the existence of electromagnetic waves over a wide spectrum of wavelengths. Hertz discovered radio waves in the 1880s. They represent transmission of energy through space in the form of oscillating electric and magnetic waves. Light is another well known example.

Electron   a fundamental particle carrying the basic unit of electric charge (roughly 1.602×10^{-19} C) and a constituent of all matter.

Electron affinity   the amount of energy required to remove an electron from the bottom of the conduction band in a semiconductor into vacuum.

Electron gun   a source of free electrons in a vacuum tube, including electrodes designed to produce a controlled narrow beam of electrons.

Electron volt (eV)   an electron falling through a potential difference of 1 V gains an amount of energy known as one electron volt.

Electro-optic effect   a property of a crystal in which an applied electric field rotates the plane of polarisation of light propagating through it.

ELOG (epitaxial lateral overgrowth)   a technique developed by Shuji Nakamura to grow high quality nitride films for application in making blue laser diodes.

EMF (electromotive force)   the voltage supplied by a battery or other electrical generator.

Energy band   solid materials are characterised by energy bands of finite width (width in energy terms). They are somewhat similar to energy levels in isolated atoms but differ because of their width, which might be several electron volts.

Energy level   the quantum theory of atomic structure was based on the idea that only certain specific energy values were allowed in atoms (differing from classical theory which allowed all possible energies). Semiconductor science also talks about donor and acceptor 'levels' – these are the energies of donor or acceptor impurity atoms with respect to their appropriate bands.

Epitaxy   the growth of a thin crystalline film of one material on a crystal of another, usually, though not always, the same material.

Extraction efficiency   the efficiency with which radiation is extracted from a light-emitting device. It is the ratio of the external efficiency to the internal efficiency.

Extrinsic conduction   electrical conduction resulting from the introduction of donor or acceptor impurities.

Fabry–Perot mirrors   a pair of parallel mirrors, spaced by a relatively small distance, as in the case of a semiconductor laser. Light reflected backwards and forwards between the mirrors builds up in intensity and dominates the emission from the laser.

FET (field effect transistor)   a semiconductor amplifier in which a gate electrode controls the electrical conduction through a channel. In an electron device, a negative potential on the gate tends to pinch off the channel and reduce conduction.

Fibre amplifier   an optical amplifier built into a glass fibre by means of doping it with the rare earth atom erbium, pumped by a GaInAs/GaAs QW laser.

Floating zone   a molten zone in a semiconductor sample which is held in place by surface tension.

Forward bias   a p–n junction or Schottky diode is characterised by a 'forward' direction in which conduction is easy and a 'reverse' direction in which it is very much more difficult. A 'forward bias' is a voltage applied to the diode in the forward direction.

Frame time   the time taken to scan a complete set of picture points in a television display.

Free carrier   conduction in semiconductors is characterised by electrons in the conduction band or holes in the valence band, where these bands are only partially full of carriers. These carriers are referred to as 'free electrons', 'free holes' or, often simply as 'free carriers' i.e. carriers of electricity.

Frequency   an oscillating signal swings alternatively positive and negative, covering a complete cycle in a time known as the 'period'. The reciprocal of the period or the rate of oscillation is known as the 'frequency'.

Frequency chirping   an undesirable feature of semiconductor lasers. When directly modulated at high frequencies, the laser output frequency changes in sympathy.

Gastroscope   a device which can be swallowed by a patient to allow a doctor to view the patient's stomach. The first really successful version employed a bundle of glass fibres.

Glass fibre waveguide   an optical waveguide consisting of a central glass core clad with an outer glass sheath having a smaller refractive index.

Graded index fibre   a glass fibre waveguide in which the refractive index of the core glass varies gradually from the centre outwards.

Grain boundary   the interface between crystallites in a polycrystalline sample.

Grid   the control electrode in a thermionic triode valve.

Gunn diode   a microwave generator making use of the transferred electron effect, named after its discoverer J.B. Gunn.

Hall effect   if a bar of semiconductor is carrying a current along its length, while a magnetic field is applied at right-angles to the bar (let us say from top to bottom), electrons or holes are deflected sideways so as to set up a 'Hall voltage' across the bar. Measurement of the Hall voltage allows the investigator to determine the sign of the current-carrying charges and the density of free carriers involved.

Hall resistance   the ratio of Hall voltage to sample current in a Hall effect measurement.

HBT (heterojunction bipolar transistor)   an idea promulgated by William Shockley in 1948, it only came to prominence towards the end of the twentieth century. It aims to use a heterojunction as emitter, the emitter band gap being larger than that of the base.

HEMT (high electron mobility transistor)   a field-effect transistor making use of a two-dimensional electron gas as channel, in the interest of obtaining better electron mobilities.

Heterodyne   a type of radio receiver which uses a local oscillator to down-change the signal frequency to a pre-selected intermediate frequency.

Heterojunction   a junction between two different semiconductor materials, usually having different band gaps.

HFET (heterojunction field effect transistor)   an alternative name for the HEMT.

Hole   a hole is a missing electron from the (full) valence band. It acts as a charge carrier in much the same way as do electrons in the conduction band.

Indirect gap   a semiconductor energy gap in which absorption or emission of photons requires the simultaneous participation of phonons.

Integrated circuit (IC)   a semiconductor electrical circuit in which gain, resistance, capacitance and inductance are all provided on a semiconductor chip, with no need to wire in additional components.

Intrinsic conduction   in a pure semiconductor at low temperatures, the valence band is full and the conduction band empty of electrons, resulting in zero conductivity. At elevated temperatures, electrons are thermally excited into the conduction band from the valence band and conduction can take place both by free electrons and by free holes in the valence band.

Inversion layer   given the right circumstances, a surface layer of free carriers in a semiconductor may be of opposite type to that in the bulk. This inversion layer may be formed either by appropriate chemical treatment of the surface or by applying a strong electric field.

Inverter   a semiconductor circuit used to convert DC to AC.

Ion   an atom which has lost or gained one or more electrons so as to be electrically charged.

Ion implantation   a method of doping semiconductors in which dopant atoms are ionised, then accelerated to high energies in an accelerator before impinging on the semiconductor surface. Their energy causes them to burrow some distance below the surface, thus doping the material. It is a very precise doping technique.

Ionic bond   a method of holding atoms together in a crystal which depends on electrostatic attraction between a pair of oppositely charged ions.

Ionisation energy   the energy required to remove an electron from a donor atom and place it in the conduction band. (Alternatively, the energy required to take an electron from the valence band and place it on an acceptor atom.)

IR (infra-red)   usually refers to electromagnetic radiation at wavelengths longer than about 700 nm (at the red end of the visible spectrum).

Isoelectronic impurity   an impurity coming from the same column of the periodic table as a constituent atom of the semiconductor. An example is that of nitrogen as an impurity in GaP.

ITO (indium tin oxide)   a compound with high electrical conductivity, while possessing a band gap wide enough to transmit the complete spectrum of visible light.

JFET (junction field effect transistor)   a field effect transistor in which the gate takes the form of a p–n junction.

Junction (p–n)   the junction between an n-type and a p-type semiconductor is a key element in semiconductor circuitry. It acts as a rectifier, as a photodetector and as an isolating region. It also has important properties as a voltage-dependent capacitor.

Laser (light amplification by the stimulated emission of radiation)   it depends on the stimulated emission process to amplify radiation travelling between a pair of parallel mirrors, which reflect it backwards and forwards between them until the intensity builds up to laser pitch. A fraction of the energy is transmitted through one of the mirrors into the outside world.

Lattice parameter   essentially, the distance between atoms in a periodic crystal lattice, along a 'crystal direction'. It helps to define a crystalline material. It is a serious matter to consider if wishing to grow a monocrystal of one material on another.

LCTV (liquid crystal television)   a type of television display based on the use of a liquid crystal light switch.

LDS (low dimensional structure)   a semiconductor structure involving two or more layers in which one or more of the layers has a thickness in the nanometre range.

LED (light-emitting diode)   a semiconductor diode which converts electrical energy into radiative energy (usually a p–n diode under forward bias).

LEEBI (low energy electron beam irradiation)   the use of electron beam irradiation by Akasaki to activate the magnesium acceptor in GaN.

Lifetime (minority carrier) when a minority of free carriers of the 'wrong' type are injected into a semiconductor, these minorities will recombine with some of the dominant majority carriers, the rate of recombination being characterised by a 'minority carrier lifetime'.

Line time the time taken to scan a single line of picture elements in a TV display.

Liquid crystal a material which, though liquid, can occur in an ordered state.

Liquid encapsulation (LEC) a technique used in Czochralski crystal growth whereby the molten material is covered by a film of boric oxide or other suitable sealant which prevents evaporation from the surface of the melt.

LPE (liquid phase epitaxy) a method of growing thin crystalline films of semiconductors in which the constituent atoms are deposited from a melt.

Luminous efficiency the efficiency with which electrical energy is converted into visible light, measured in lumens (of light) per watt (of electrical energy).

MASER (microwave amplification by stimulated emission of radiation) a device which amplifies microwave signals by the use of stimulated emission.

Mass spectrometer an analysis tool used in UHV equipment. Atoms are sputtered from a surface in the form of ions whose mass is measured by studying their interaction with a magnetic field.

Material dispersion dispersion associated with the fact that different optical wavelengths travel at different velocities in transparent solids.

Matrix addressing a method of supplying appropriate brightness information to the picture points in a TV display, which involves a cross-bar system of control wires.

MBE (molecular beam epitaxy) a sophisticated method of growing thin semiconductor films by evaporation of atomic species from heated cells.

MESFET (metal semiconductor field effect transistor) the MESFET first came to notice as a microwave transistor in gallium arsenide. It consists of a conducting channel whose conductivity is controlled by applying a voltage to a Schottky barrier gate.

Mesoscopic   (from Greek for 'middle-sized'), refers to a range of quantum devices which are one stage up in size from the very smallest.

Micron 10^{-6} metres.

Microprocessor   a computer on a chip.

Microwave   a region of the electromagnetic spectrum characterised by wavelengths in the range of centimetres to millimetres.

Minority carrier   in an extrinsic semiconductor, the carrier type which is present in minority, for example, holes in an n-type material.

Minority carrier lifetime   the characteristic time in which minority carriers recombine with majority carriers in a semiconductor.

Mobility   a measure of how mobile a free carrier is, when subjected to an electric field. Mathematically, it is the ratio of its velocity in metres per second to the applied electric field in volts per metre.

Modal dispersion   dispersion associated with the existence of different propagation modes in a waveguide.

MODFET (modulation doped field effect transistor)   an alternative name for the HEMT.

Moore's law   Gordon Moore's observation that the number of components in an integrated circuit increases exponentially with time. It increases by about 1.6 times each year.

MOSFET (metal oxide silicon field effect transistor)   this was the first really successful field-effect transistor – a silicon channel, a silicon oxide insulator and a metal gate electrode.

MOVPE (metal organic vapour phase epitaxy)   a refined version of VPE in which the metallic species is provided as a metal-organic compound while the non-metal is provided as a hydride.

MQW   multiple quantum well.

Multiplexing   any method of sending multiple signals along a transmission line. Examples are 'time-division' and 'wavelength division' multiplexing.

Nanostructure   a multi-layer semiconductor structure in which layer thicknesses are in the nanometre range.

Negative resistance   the phenomenon in which a device shows a *decrease* in current when the applied voltage is *increased*.

Neutron   a neutrally charged fundamental particle with mass very close to that of the proton.

NMOS   a metal-oxide silicon transistor in which the conducting channel contains free electrons.

Noise   the term used in electronics or opto-electronics to describe unwanted signals (usually of a random nature) which interfere with the information signal.

Ohms law   the statement that the current flowing in an electric circuit is proportional to the voltage drop across the circuit. (Note that there are many exceptions, such as provided by semiconductor diodes.)

Optical modulator   a device which allows the amplitude of a light signal to be varied at high frequency by application of an electric field.

PCM (pulse code modulation)   the use of short pulses of current or light to represent information in a communication system.

Periodic table   the arrangement of the chemical elements in a table according to their atomic weights and chemical properties, as proposed by Mendeleev in 1869.

PHEMT (pseudomorphic high electron mobility transistor)   a version of the HEMT in which the channel consists of a pseudomorphic layer; particularly the example of an InGaAs channel, combined with an AlGaAs supply layer.

Phonon   the quantum of heat energy.

Phosphor   a material providing visible light when irradiated by a beam of electrons or ultra-violet light.

Photocathode   the negative electrode in an image tube, which emits electrons into vacuum when illuminated with light.

Photoconductivity   electrical conductivity induced in a solid by the absorption of light.

Photo-current   an electric current induced by the absorption of light in a solid.

Photo-detector   a device which converts a beam of light into an electric current or voltage, thus allowing the light to be detected.

Photoelectric effect   the removal of electrons from a solid into vacuum by the absorption of photons.

Photolithography   the process of writing an integrated circuit based on the use of a photoresist and a photo-mask.

Photon   the quantum of light energy.

Photonic integration   the use of photons, rather than electrons, in signal processing, thus allowing optical signals to be manipulated without first converting them into electronic form.

Photoresist   a chemical compound which reacts to incident light in such a way as to become more or less soluble in certain solvents. Used in defining integrated circuits.

Photo-voltaic effect   the effect whereby the absorption of light in a solid gives rise to an electric voltage, proportional to the light intensity.

Piezo-electric effect   a property of certain semiconductor materials in which mechanical strain gives rise to an induced voltage across a single crystal (and vice versa).

P–i–n diode   a type of semiconductor photodetector with an extended region of undoped material between the p-doped and n-doped contact regions, thus facilitating the absorption of photons in the high field region.

Planar doping   a technique for doping semiconductors in which the dopant atoms are confined to a single atomic plane in the crystal.

Planar technology the process of making a semiconductor circuit in which all electrical connections are made on the surface of the sample.

PMOS a metal-oxide silicon transistor in which the conducting channel contains free holes.

Proton a fundamental particle with a positive charge numerically equal to that of the electron and a mass approximately 1836 times greater.

Pseudomorphic growth an example of epitaxial crystal growth where one material grows on another, having a different lattice constant, but where the film grows with the in-plane lattice spacing appropriate to the substrate.

Quantum a fundamental unit of energy. For example, a photon is the smallest amount of light energy that can exist.

Quantum confined Stark effect a property of quantum wells in which their optical absorption can be varied by the application of an electric field.

Quantum confinement the trapping of electrons (or holes) in a quantum structure such as a quantum well, with consequent modification to the electron (or hole) energy.

Quantum dot a small volume of matter all of whose dimensions are such that confined electrons suffer quantum effects.

Quantum efficiency the ratio of the number of photons produced divided by the number of electrons which recombine in a light emitting device.

Quantum Hall effect the occurrence of a sequence of plateaux in the plot of Hall resistance against magnetic field.

Quantum well a sheet of matter with thickness small enough to introduce quantum effects in a direction normal to the plane of the sheet.

Quantum wire a length of matter whose normal dimensions are small enough to introduce quantum effects.

QWIP quantum well infra-red photo-detector.

Radiative recombination   the recombination of electrons and holes in a semiconductor, giving rise to the emission of photons (rather than phonons).

Rectifier   an electronic device which passes current much more readily in the forward direction than in the reverse.

Refractive index   the ratio of the velocity of light in a vacuum to that in a transparent material.

Regenerator   an opto-electronic circuit which detects a noisy incoming digital data stream and sends out a 'clean' version.

Relaxation oscillation   a feature of semiconductor lasers in which the output oscillates when the device is switched on. It limits the maximum possible modulation frequency under direct modulation.

Resistor   an electronic component which impedes the flow of electric current.

Reverse bias   a voltage applied across a rectifier in the reverse direction – that is, that direction which results in minimum current flow.

RHEED (reflection high energy electron diffraction)   a technique for measuring surface structure of a crystalline solid by reflecting a beam of electrons from its surface.

Rheostat   a variable (i.e., adjustable) resistor.

Scattering   electrons or holes moving under the influence of an applied electric field are said to be 'scattered' by collision with imperfections in a solid – that is, their directions and velocities are changed by the collisions. It is this process which limits electron or hole mobilities.

Schottky barrier   a contact between a semiconductor and a metal such that an energy barrier occurs at the interface. One consequence is that the contact behaves as a rectifier.

SCR (semiconductor controlled rectifier)   an alternative name for the thyristor.

Self-aligned gate   a field effect transistor structure in which the gate electrode is automatically aligned between the source and drain electrodes.

SI (semi-insulating)   semiconductor material, such as GaAs or InP, which is deliberately doped with an impurity to make it highly insulating.

Solar cell   a semiconductor diode which absorbs radiative energy and converts it directly into electrical energy.

Spontaneous emission   a process in which an excited system emits photons without external stimulation.

Sputtering   a method of depositing a semiconductor film in which the required atoms are 'chipped' from a suitable solid (or solids) by dislodging them with a high energy beam of ions (frequently argon ions).

Staring array   an array of infra-red detectors used to form an image in which each detector provides one picture point in the scene. All the detectors 'stare' continuously, there being no optical scanning of the scene.

Stimulated emission   a process in which an excited system is stimulated into emitting photons by the interaction with existing photons of the same wavelength.

Stoichiometry   in a binary compound semiconductor such as GaAs, it is possible for the material to exist in a condition where the number of gallium atoms differs slightly from the number of arsenic atoms. This is known as non-stoichiometry. Stoichiometry is a measure of this effect.

Superconductivity   the property of certain metals or compounds whereby their electrical resistivities become zero below a certain critical temperature.

Superlattice   an artificially grown crystalline structure having a regular periodicity different from that of the natural crystal structure.

Surface state   an electronic state at the surface of a semiconductor capable of trapping an electron, thereby immobilising it.

TDM (time division multiplexing)   a technique for increasing the amount of data that can be sent along a transmission line by interleaving data pulses in the time domain.

TEGFET (two-dimensional electron gas field effect transistor)   an alternative name for the HEMT.

Telegraph   the system of electrical communication based on transmission of coded electrical signals, such as the Morse code.

Telephone   communication by way of analogue voice signals, covering the frequency range 100 Hz–3 kHz.

TEM (transmission electron microscopy)   a method of analysing thin layers of material in which a beam of electrons is passed through the structure and a diffraction pattern recorded on a photographic film.

TFT (thin film transistor)   a transistor made in a suitable thin film of semiconducting material (usually in the form of a polycrystalline or amorphous material).

Thermal energy   energy possessed by a material as a consequence of its being at a finite temperature. In a solid, this takes the form of the energy of vibration of the individual atoms.

Thermal runaway   the property of a semiconductor in which its resistivity becomes smaller and smaller due to its temperature increasing as a result of the current flowing in it. The current therefore increases without limit.

Thermal velocity   the velocity of an electron or hole consequent upon its having thermal energy.

Thermionic emission   certain metals, when heated to high temperature have the ability to emit electrons into a vacuum. These are used as the source of electrons in a thermionic vacuum tube.

Thermionic valve   a vacuum device containing a heated cathode to supply electrons and one or more additional electrodes. Used as rectifier or amplifier.

Thyratron   a gas-filled thermionic valve originally used for switching large currents.

Thyristor   a four-layer semiconductor device of (p–n–p–n) or (n–p–n–p) configuration which has a switching characteristic.

Total internal reflection   light incident on an interface characterised by a change in refractive index will always be partially reflected but at a transition from a higher to lower index it is possible for this reflection to be total (i.e. there is no loss of energy). Total reflection occurs for all angles of incidence greater than a certain critical angle.

Transferred electron effect   the basis of the Gunn diode in which a negative resistance is produced by the transfer of electrons between different conduction band states.

Transistor   a semiconductor electronic device (usually with three electrodes) which acts as an amplifier or switch.

Transmission line   any line along which signals may be transmitted. It might take the form of a twisted pair of wires or a copper waveguide.

Triode   an electronic valve with three electrodes which can amplify an electronic signal.

Tunnel diode   a p–n junction diode whose current-voltage characteristic includes a negative resistance region.

Tunnelling   a quantum mechanical effect whereby electrons may pass through a thin barrier.

Twisted nematic liquid crystal   a particular type of liquid crystal possessing long string-like molecules which can be aligned by suitable surface treatment of the glass plates used to confine it and which can be realigned by application of an electric field.

Two-dimensional electron gas (2DEG)   a collection of free electrons which are constrained to move only in two dimensions, that is, in a plane, rather than in three dimensions.

UHV (ultra-high vacuum)   pressures of 10^{-9} torr or less, used in molecular beam epitaxy or for high quality surface studies.

UV (ultra-violet)   that part of the electromagnetic spectrum covering wavelengths less than about 400 nm – photon energies greater than those associated with violet light.

VCSEL (vertical cavity surface emitting laser)   a semiconductor laser which emits light normal to the sample surface. It usually employs Bragg mirrors, rather than ordinary plane mirrors and is based on a very short cavity.

VPE (vapour phase epitaxy)   a method of growing thin films of semi-conductor material on suitable substrates in which the appropriate atoms are supplied in the form of gasses.

Waveguide   a structure made from glass, semiconductor or metal which guides electromagnetic radiation along a predetermined path.

Wavelength   the distance between adjacent peaks or troughs in a sinusoidally varying waveform.

WDM (wavelength division multiplexing)   a technique for increasing the amount of data which can be sent along a transmission line by using a set of distinct wavelengths, each carrying a separate data stream.

Zone refining   the passage of a molten zone through a semiconductor sample, having the effect of sweeping out impurities.

1909 – Carl Ferdinand Braun – Wireless telegraphy, cat's whisker rectifier.

1956 – John Bardeen, Walter Houser Brattain, William Shockley – Transistor.

1973 – Leo Esaki – Tunnelling effect.

1977 – Sir Nevill Francis Mott, Philip Warren Anderson – Amorphous semiconductors.

1985 – Klaus von Klitzing – Quantum Hall effect.

1998 – Robert Laughlin, Horst Stormer, Daniel Tsui – Fractional quantum Hall effect.

2000 – Zhores Alferov, Herbert Kroemer – Heterostructures, Jack Kilby – Integrated circuits.

Printed in the United States
By Bookmasters